Applications of Hydrogen Peroxide and Derivatives

RSC Clean Technology Monographs

Series Editor: J.H. Clark, *University of York, UK*

The chemical process industries are under increasing pressure to develop environmentally friendly products and processes, with the key being a reduction in waste. This timely new series will introduce different clean technology concepts to academics and industrialists, presenting current research and addressing problem-solving issues.

Feedstock Recycling of Plastic Wastes
by J. Aguado, *Rey Juan Carlos University, Móstoles, Spain*; D.P. Serrano, *Complutense University of Madrid, Spain*

Applications of Hydrogen Peroxide and Derivatives
by C.W. Jones, *formerly of Solvay Interox R & D, Widnes, UK*

How to obtain future titles on publication

A standing order plan is available for this series. A standing order will bring delivery of each new volume upon publication. For further information please contact:

Sales and Customer Care
Royal Society of Chemistry
Thomas Graham House
Science Park
Milton Road
Cambridge
CB4 0WF

Telephone: +44(0) 1223 420066

RSC
CLEAN TECHNOLOGY
MONOGRAPHS

Applications of Hydrogen Peroxide and Derivatives

Craig W. Jones
Formerly of Solvay Interox R&D, Widnes, UK

ROYAL SOCIETY OF CHEMISTRY

ISBN 0-85404-536-8

A catalogue record for this book is available from the British Library

Published by The Royal Society of Chemistry,
Thomas Graham House, Science Park, Milton Road
Cambridge CB4 0WF, UK

For further information see our web site at www.rsc.org

Typeset by Paston PrePress Ltd, Beccles, Suffolk
Printed by MPG Books Ltd, Bodmin, Cornwall, UK

Preface

Hydrogen peroxide, as well as being an incredibly simple inorganic compound, is also a beautifully versatile one. Over the last decade it has had somewhat of a rebirth in both industrial and academic circles. The rather glib explanation for such a renaissance is due to regulatory forces causing the chemical industry to reduce, and in some instances eliminate, environmental pollution. However, such a reason does a great disservice to hydrogen peroxide. Whilst it is true that environmental agencies and legislation have caused a major shift in emphasis during the latter half of the century and polarised our efforts on so called 'green chemistry', by far the most overriding reason why hydrogen peroxide is now more popular is due to the fact that the chemical industry has learnt to employ the chemical in a safer, more efficient, and innovative manner. In addition, hydrogen peroxide and its derivatives can not only be employed for their traditional bleaching applications or for the manufacture of pharmaceutical and fine chemicals, but also have uses in a diverse array of industries. Precious metal extraction from the associated ores, treatment of effluent, delicing of farmed salmon, and pulp and paper bleaching are but a few areas where hydrogen peroxide has had a profound effect on the quality of all our lives.

The aim of this book is to allow those unfamiliar with the versatility of hydrogen peroxide and its derivatives to walk into their laboratories and to look for possible applications in their own areas of expertise where hydrogen peroxide can perhaps help increase a yield, purify a compound, or afford a more environmentally benign route to be devised. The author would also like to encourage educationalists to attempt to introduce courses on hydrogen peroxide on an academic and practical level to not only undergraduates but to those of school age studying the sciences. The introduction of topics like this coupled with an understanding of catalytic routes to industrially important chemicals will hopefully encourage future scientists to think in terms of relatively benign synthetic methodologies rather than being constrained by the chemistries of the 19th century synthetic chemist.

This book has been organised such that each chapter can be read as a stand-alone monograph in its own right. However, the author would encourage those readers unfamiliar with the use of hydrogen peroxide to read Chapter 1, which includes an important section on its safe use. In this book I have aimed to present a description of the preparation, properties and applications of hydrogen peroxide, and its derivatives. The number of different peroxygen systems, and their structural diversity, makes it difficult to gain a thorough under-

v

standing of the subject by studying individual peroxygen systems. I have, therefore, tried to emphasize general features of the properties of the peroxide bond by reference to the activation of hydrogen peroxide throughout the book.

Chapter 1 puts hydrogen peroxide in its historical context with particular emphasis on the preparation of hydrogen peroxide from the acidification of barium peroxide to the integrated generation of hydrogen peroxide. The chapter concludes with a practical approach to employing hydrogen peroxide and its derivatives in a safe manner. The activation of hydrogen peroxide is discussed in Chapter 2, and this is intended to provide a firm basis for the understanding of the chemistry of hydrogen peroxide. Chapter 3 is intended to illustrate the application of activated hydrogen peroxide towards the oxidation of important organic functions such as olefinic compounds to epoxides, diols or diol cleavage to aldehydes, ketones or carboxylic acids. Other functional group oxidation includes organonitrogen, organosulfur, ketones, alcohols, and alkyl side chains of arenes. Chapter 4 briefly describes to the reader the application of heterogeneous systems for the activation of hydrogen peroxide. It is this area of hydrogen peroxide chemistry which is likely to become of pivotal importance in relation to 'integrated pollution control' programmes. Chapter 5 summarizes the use of hydrogen peroxide for the clean up of environmental pollutants. Fenton's chemistry is discussed in this respect together with other advanced oxidation processes for the generation of hydroxyl radical. The final chapter of the book looks at the impact hydrogen peroxide has had on several industries, from the preparation of chemical pulp to the purification of industrially important chemicals.

I hope everyone who turns the pages of this book finds something which helps them in their deliverance for the sake of humankind, or discovers the rich tapestry of chemistries, and industries, that have been founded on the simple peroxygen bond.

In writing this book I have been fortunate to have had the expert guidance, and encouragement from my colleagues at the Solvay Interox R&D department based in Widnes in the UK. It is also with deep sadness that when this book is finally published the department at Widnes will no longer be in existence. It is to all those people that I say a special thank you to and dedicate this book to them, especially Bill Sanderson, Phil Wyborne, Sharon Wilson, Colin McDonagh and Gwenda McIntyre, because without their learning, understanding and good humour, this book could never have come to fruition. I would thank all those workers in the field of peroxygen technology, some of whom I have had the privilege to meet professionally, and many I have not met. It is their work which is referenced and discussed within these pages. It is their selfless dedication to the ongoing understanding of materials containing peroxygen bonds that has breathed new life into a wonderfully diverse chapter of science. My wife Helen deserves a special mention as she has typed a large proportion of this manuscript, and was a constant source of advice, encouragement, and practical assistance during its preparation.

Contents

To Helen, and the memory of Solvay Interox R&D, Widnes

CHAPTER 1

Introduction to the Preparation and Properties of Hydrogen Peroxide

1 Introduction

The following chapter will discuss the preparation of hydrogen peroxide, historically, the present day and future vistas for its *in situ* preparation. A brief introduction to the physical properties of hydrogen peroxide will also be made for the sake of completeness. Finally, the chapter will conclude with a practical approach to the safe handling of peroxygen species, destruction of residual peroxygens, and the toxicological and occupational health considerations required when handling hydrogen peroxide.

2 Industrial Manufacture of Hydrogen Peroxide

The industrial manufacture of hydrogen peroxide can be traced back to its isolation in 1818 by L. J. Thenard.[1] Thenard reacted barium peroxide with nitric acid to produce a low concentration of aqueous hydrogen peroxide; the process can, however, be significantly improved by the use of hydrochloric acid. The hydrogen peroxide is formed in conjunction with barium chloride, both of which are soluble in water. The barium chloride is subsequently removed by precipitation with sulfuric acid (Figure 1.1).

Hence, Thenard gave birth to the first commercial manufacture of aqueous hydrogen peroxide, although it took over sixty years before Thenard's wet chemical process was employed in a commercial capacity.[2] The industrial production of hydrogen peroxide using the above route was still operating until the middle of the 20th century. At the turn of the 19th century, approximately 10 000 metric tonnes per annum of barium peroxide were converted to about 2000 metric tonnes of hydrogen peroxide. Thenard's process has, however, some major drawbacks which quenched the expectant explosion of its use in an aqueous form. Firstly, only three percent m/m aqueous hydrogen peroxide solutions were manufactured using the barium peroxide

$$BaO_2 + 2HCl \longrightarrow BaCl_2 + H_2O_2$$

$$BaCl_2 + H_2SO_4 \longrightarrow BaSO_4 + 2HCl$$

$$BaO_2 + H_2SO_4 \longrightarrow BaSO_4 + H_2O_2$$

Figure 1.1 *Thenard's route to aqueous hydrogen peroxide.*

process, and hence only a limited market was afforded because production costs were prohibitively high. Further, due to the high levels of impurities present in the isolated hydrogen peroxide, subsequent stability was poor.

The disadvantages of the process discovered by Thenard were largely alleviated by the discovery in 1853 by Meidinger that hydrogen peroxide could be formed electrolytically from aqueous sulfuric acid.[3] Berthelot later showed that peroxodisulfuric acid was the intermediate formed,[4] which was subsequently hydrolysed to hydrogen peroxide, and sulfuric acid (Figure 1.2).

The first hydrogen peroxide plant to go on-stream based on the electro-chemical process was in 1908 at the Österreichische Chemische Werke in Weissenstein. The Weissenstein process was adapted in 1910 to afford the Müncher process developed by Pietzsch and Adolph at the Elecktrochemische Werke, Munich. In 1924, Reidel and Lowenstein used ammonium sulfate under the conditions of electrolysis instead of sulfuric acid, and the resulting ammonium peroxodisulfate (Reidel–Lowenstein process) or potassium peroxo-disulfate (Pietzsch–Adolph process) was hydrolysed to hydrogen peroxide. As a result of this process, production of hydrogen peroxide as 100% *m/m* rose to approximately 35 000 metric tonnes per annum.[5]

In 1901, Manchot made a decisive breakthrough in the industrial preparation of hydrogen peroxide. Manchot observed that autoxidizable compounds like hydroquinones or hydrazobenzenes react quantitatively under alkaline condi-tions to form peroxides.[6] In 1932, Walton and Filson proposed to produce hydrogen peroxide via alternating oxidation and reduction of hydrazo-ben-zenes.[7] Subsequently, Pfleiderer developed a process for the alkaline autoxidation of hydrazobenzenes in which sodium peroxide was obtained, and sodium amalgam was used to reduce the azobenzene.[8] A commercial plant based on this technology was operated by Kymmene AB in Kuisankoski, Finland.

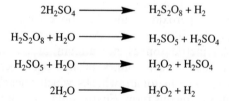

$$2H_2SO_4 \longrightarrow H_2S_2O_8 + H_2$$

$$H_2S_2O_8 + H_2O \longrightarrow H_2SO_5 + H_2SO_4$$

$$H_2SO_5 + H_2O \longrightarrow H_2O_2 + H_2SO_4$$

$$2H_2O \longrightarrow H_2O_2 + H_2$$

Figure 1.2 *Electrochemical manufacture of aqueous hydrogen peroxide.*

The major drawbacks associated with the azobenzene process, i.e. hydrogenation of azobenzene with sodium amalgam, and oxidation of hydrazobenzene in alkaline solution, were ultimately resolved by Riedl. Riedl employed polynuclear hydroquinones. Based on Reidl and Pfleiderer's work, BASF developed, between 1935 and 1945, the anthroquinone process (often referred to as the AO process) in a pilot plant with a monthly production of 30 metric tonnes. Two large plants were then constructed at Heidebreck and Waldenberg, each having a capacity of 2000 metric tonnes per annum. Both plants were partially complete when construction was halted at the end of World War Two. In 1953, E.I. Dupont de Nemours commissioned the first hydrogen peroxide plant using the AO process, and consequently the production capacity of hydrogen peroxide was greatly increased. In 1996, world capacity stood at 1.3×10^6 metric tonnes as 100% m/m hydrogen peroxide.[9]

The underlying chemistry of the AO process is outlined in Figure 1.3 and a typical autoxidation plant schematic is summarized in Figure 1.4.

The features of all AO processes remain basically the same, and can be described as follows. A 2-alkylanthraquinone is dissolved in a suitable solvent or solvent mixture which is catalytically hydrogenated to the corresponding 2-alkylanthrahydroquinone. The 2-alkylanthraquinone solution is commonly referred to as the reaction carrier, hydrogen carrier or working material. The 2-alkylanthraquinone–solvent mixture is called the working solution. Carriers employed industrially include 2-*tert*-amylanthraquinone, 2-iso-*sec*-amylanthraquinone and 2-ethylanthraquinone. The working solution containing the carrier product alkylanthrahydroquinone is separated from the hydrogenation catalyst, and aerated with an oxygen-containing gas, nominally compressed air, to reform the alkylanthraquinone, and simultaneously forming hydrogen peroxide. The hydrogen peroxide is then extracted from the oxidized working solution using demineralized water, and the aqueous extract is then purified

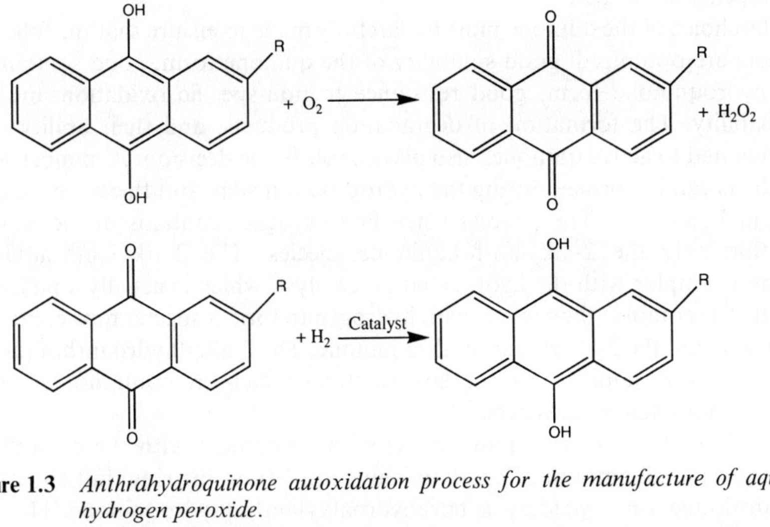

Figure 1.3   *Anthrahydroquinone autoxidation process for the manufacture of aqueous hydrogen peroxide.*

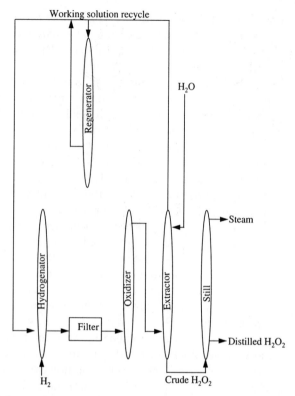

Figure 1.4 *Schematic diagram of the AO process.*

and concentrated by fractionation to the desired strength. The AO process, therefore, leads to the net formation of hydrogen peroxide from gaseous hydrogen and oxygen.

The choice of the quinone must be carefully made to ensure that the following criteria are optimized: good solubility of the quinone form, good solubility of the hydroquinone form, good resistance to non-specific oxidation and easy availability. The formation of degradation products, and their ability to be regenerated to active quinones also plays a rôle in the decision. A number of by-products can be formed during the hydrogenation step, and these are summarized in Figure 1.5. The process when first engaged, contains in the working solution only the 2-alkylanthraquinone species. The 2-alkylanthraquinone forms a complex with the hydrogenation catalyst, which is usually a palladium metal. The complex then reacts with hydrogen to form a species now containing the metal and the 2-alkylhydroanthraquinone. The 2-alkylhydroanthraquinone is subject to a number of secondary reactions which are continuously taking place during each process cycle.

The 2-alkylhydroanthraquinone (A) when in contact with the catalyst will undergo a small amount of catalytic reduction (B) on the ring, initially on the unsubstituted ring, yielding a tetrahydroalkylanthrahydroquinone. Unfortu-

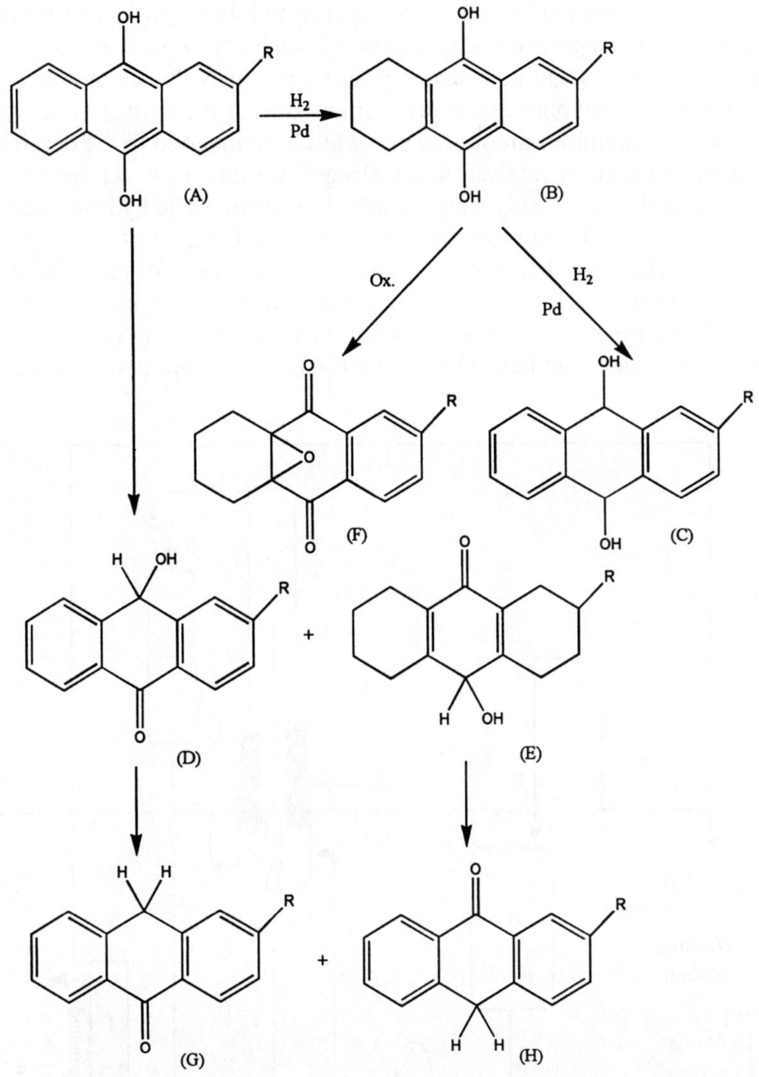

Figure 1.5 *Secondary reactions taking place in the presence of 2-alkylanthrahydroqui-nones.*

nately, once the octa-product (C) is formed, it remains until purged owing to its very low rate of oxidation. Tautomerism of the 2-alkylhydroanthraquinone yields hydroxyanthrones (D, E) which can be further reduced to the anthrones (G, H). The epoxide (F) formed from the alkylhydroanthraquinone does not participate in the formation of hydrogen peroxide, and leads to a loss of active quinone. Measures have, therefore, been suggested for regenerating the tetra-hydro compound from the epoxide.[10]

A number of additional processes are also required to maintain the AO process. For example, in order for the hydrogenation phase to run efficiently, part of the catalyst load is removed, regenerated and returned to the hydrogenator. The hydrogenation step is possibly the most important feature of the modern AO process. Quinone decomposition products that cannot be regenerated into active quinones are always formed during the hydrogenation phase. Therefore a tremendous amount of effort has been invested in the development of new hydrogenation catalysts and hydrogenator designs which have, in some cases, deviated dramatically from the BASF principle. The hydrogenation step in the BASF plant (Figure 1.6) employs a Raney nickel catalyst at a slight excess of pressure. However, because Raney nickel is sensitive to oxygen, the working solution from the extraction, drying and purification steps cannot be fed directly into the hydrogenator. The working solution at this stage still contains residual hydrogen peroxide, and has to be decomposed over a supported Ni–Ag catalyst

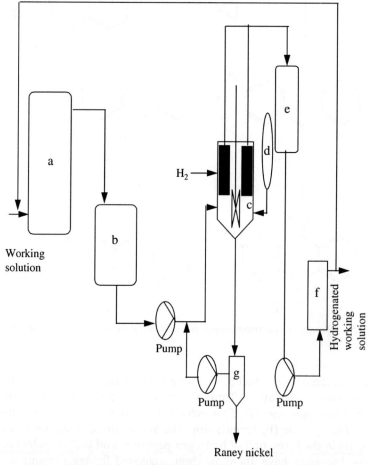

a = pre-contact column; b = feed tank to hydrogenator; c = reactor; d = catalyst feed tank; e = oxidizer feed tank; f = safety filter; g = catalyst removal tank.

Figure,1.6   *BASF hydrogenator.*

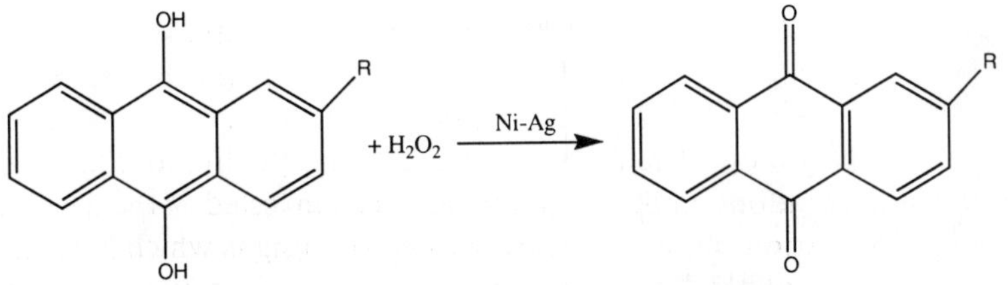

Figure 1.7   *Destruction of residual hydrogen peroxide in the BASF process.*

(Figure 1.7), together with a small amount of hydrogenated working solution (which also contains 2-alkylhydroanthraquinone). Such a step removes the hydrogen peroxide completely, thus extending the life of the Raney nickel catalyst.

The problem with Raney nickel as the hydrogenation catalyst is that it has a limited selectivity, i.e. the ratio of hydroquinone formation to the tetrahydro compound is low. BASF have largely alleviated this problem via pre-treatment of the catalyst with ammonium formate.[11] The pyrophoric properties of Raney nickel also require more stringent safety procedures when handling the material. Despite the drawbacks of Raney nickel, the catalyst is still used in some AO plants. The majority of AO plants worldwide prefer, however, to employ palladium hydrogenation catalysts because of their higher selectivity, their greater stability towards hydrogen peroxide residues and the simplified handling procedures in comparison to the Raney nickel systems. Degussa have employed palladium black as the hydrogenation catalyst in the majority of their plants.[12] The main advantages of the Degussa hydrogenation stage are: near-quantitative conversion of hydrogen, easy exchange of palladium black, the catalyst is non-pyrophoric and the palladium black is easily re-activated. Laporte chemicals made a significant breakthrough in the operation of the hydrogenation phase by employing supported palladium, which has a particle size diameter of 0.06–0.15 mm.[13] The supported palladium catalyst allows for easier filtration, and recirculation of the catalyst back to the hydrogenator. Laporte, at the same time, also employed a new design for running the hydrogenation phase.[14] Figure 1.8 illustrates the Laporte design.

The Laporte hydrogenator contains a series of tubes which dip just below the surface of the liquid. Hydrogen is then fed into the bottom of each tube, and small gas bubbles are formed. A counter current flow is set up due to the density difference between the solutions in the tube and the reactor. The palladium catalyst suspension is drawn into the tubes by a continuous movement of the working solution.

The problem with all three methods thus far discussed is the fact that the hydrogenator catalyst has to be removed prior to the formation of hydrogen peroxide. If the catalyst is not removed, then catastrophic dismutation of the hydrogen peroxide can occur. In response to the problem, FMC developed a mixed-bed hydrogenation process. The bed is impregnated with palladium, and hence the problem associated with catalyst removal is alleviated.[15]

Hydrogenator off-gas

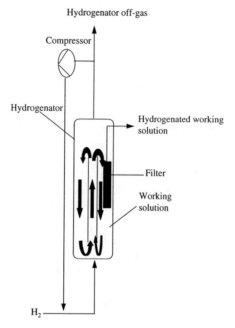

Figure 1.8 *Laporte hydrogenator.*

On an industrial scale, the catalyst-free hydrogenated working solution is generally oxidized with slight pressures of air (up to 0.5 MPa). The oxidation phase must satisfy several criteria, mainly economically driven, which include: small reactor volume to lower investment costs for equipment; efficient utilization of oxygen to reduce the volume of off-gas; and low compressor pressure to decrease energy costs. Like the hydrogenation phase, several companies have developed and used their own oxidation regimes. For example, BASF flow hydrogenated working solution through four oxidation columns arranged in series (Figure 1.9) as a cascade. The oxidized working solution then flows into an extractor tank. The nitrogen–oxygen mixture is compressed and fed into each of the four reactors.

Solvay Interox's plant based at Warrington in the UK operates a co-current oxidation in a column.[16] The whole volume of the reactor is used for air gassing (Figure 1.10). The air and hydrogenated working solution leave the top of the column and are fed into a separator. The air then reaches the two-stage activated carbon filters, which remove residual working solution and impurities. The working solution then passes to the extraction phase.

Finally, it is worth mentioning that Allied Colloids have employed a counter-flow oxidation reactor,[17] which has a residence time of hydrogenated working solution of less than 2.5 min at a partial oxygen pressure of 70–100 kPa.

Inevitably, due to the constant circulation of working solution, by-products are formed from the working solution and the solvents. The by-products have to be purged from the system to prevent destabilization of the crude hydrogen

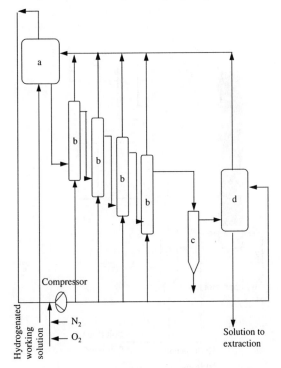

a = separator chamber; b = oxidation reactor column;
c = filtration unit; d = extractor feed tank.

Figure 1.9   *Illustration of the BASF oxidizer.*

peroxide, and an increase in density and viscosity of the working solution. Further, the impurities in the working solution cause a decrease in the surface tension, and encourage the formation of an emulsion, which can be difficult to destabilize. By-product formation can also cause deactivation of the hydrogenation catalyst, hence the working solution can be purified by a range of techniques which include treatment with alkaline solution,[18] treatment with active aluminium oxide or magnesium oxide at about 150 °C,[19] use of alkaline hydroxide such as calcium hydroxide, ammonia or amines in the presence of oxygen or hydrogen peroxide[20] and treatment with sulfuric acid.[21]

The crude hydrogen peroxide exiting the extraction phase requires purification. A number of methods have been devised for the treatment of crude hydrogen peroxide including the use of polyethylene,[22] ion-exchangers[23] and the use of hydrocarbon solvents.[24] The purified hydrogen peroxide is then fed to a distillation column where it is concentrated to the usual commercial concentration range of 35–70% *m/m*. Solvay Interox produce 85% *m/m* hydrogen peroxide, but only use it captively for the preparation of 38% *m/m* peracetic acid used for the oxidation of cyclohexanone to ε-caprolactone. Higher strengths can be achieved as hydrogen peroxide does not form an azeotrope with water, but a number of technical safety requirements must be observed.

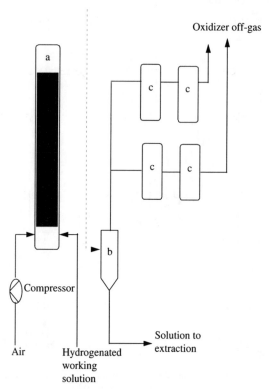

a = reactor; b = separation column; c = activated carbon
adsorption unit.

Figure 1.10   *Solvay Interox oxidation method.*

Before we leave the discussion of industrial processes, it is worth mentioning one other autoxidation process, based on the oxidation of propan-2-ol, developed by Shell Chemicals. The process was employed by Shell in its 15 000 metric tonnes per annum facility at Norco between 1957 and 1980. The process was discovered in 1954 by Harris,[25] who showed that the oxidation of primary and secondary alcohols formed hydrogen peroxide, and the corresponding aldehyde or ketone (Figure 1.11).

Only propan-2-ol has had any industrial use since the aldehydes formed in the reaction with primary alcohols are easily oxidized. The oxidation of propan-2-ol in the liquid phase with oxygen does not require a special catalyst, because it is catalysed by a small amount of hydrogen peroxide, which is added to the feed-stream of the propan-2-ol in order to shorten the induction phase (Figure 1.12).

Reduction of by-products can be achieved by only partially oxidizing the propan-2-ol, and by carrying out the oxidation in several consecutive steps, at decreasing temperatures.[26] The hydrogen peroxide yield is typically 90–94% with respect to the propan-2-ol, and the acetone yield is 92–94%.

Over the years, there have been many other methods proposed for the

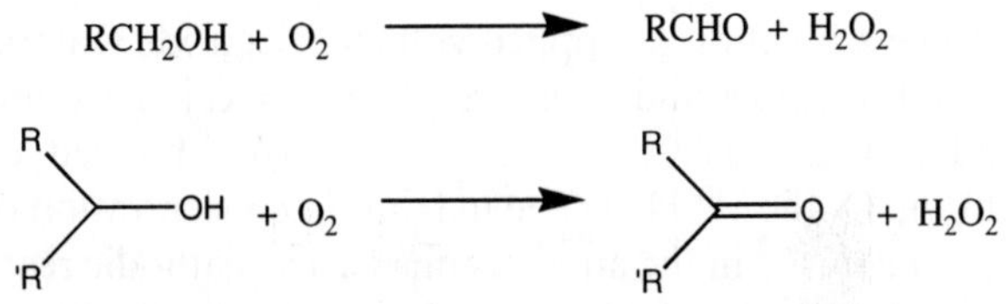

$$RCH_2OH + O_2 \longrightarrow RCHO + H_2O_2$$

Figure 1.11 *Shell process for the production of aqueous hydrogen peroxide.*

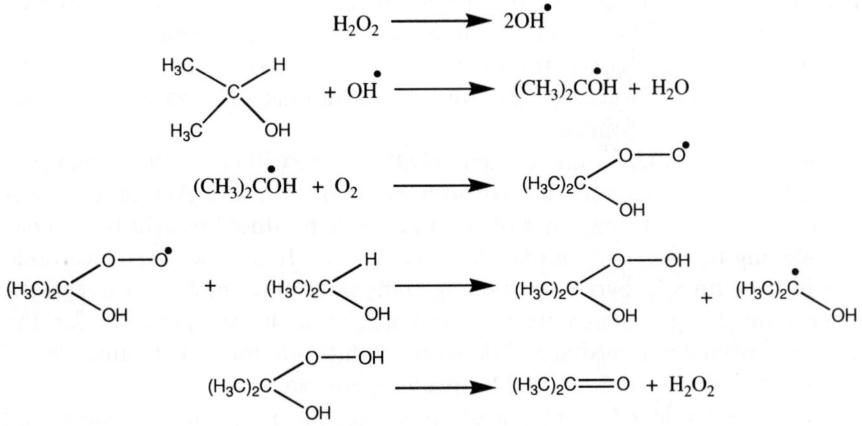

Figure 1.12 *Mechanism of Shell process for the preparation of aqueous hydrogen peroxide.*

preparation and subsequent purification of hydrogen peroxide. However, to date no industrial plants have been designed and commissioned based on such technologies. For example, Arco have devised a method for the preparation of hydrogen peroxide based on the autoxidation of methyl benzyl alcohol isomers with molecular oxygen.[28,29] The process employs ethylbenzene and water to extract the hydrogen peroxide from a mixture of methyl benzyl alcohol and other oxidation by-products. For safety reasons, the water is supplied as a downward-flowing stream in the reactor, together with an upward flow of ethylbenzene. The process also contains one further feature worthy of note, which is that the crude aqueous hydrogen peroxide is passed through a cross-linked polystyrene resin which has a macro-reticular structure. This resin purification step has the advantage that subsequent concentration stages are inherently safer due to the lower organic contents. A number of novel electrochemical processes for hydrogen insertion reactions into molecules have also been applied to the preparation of hydrogen peroxide.[30–32] One process worth describing involves the electrochemical production of hydrogen peroxide together with the simultaneous production of ozone.[32] The preparation of ozone is from the anode and of hydrogen peroxide from the cathode. The oxidants are generated from water and oxygen in a proton-exchange membrane (PEM) reactor. The optimum conditions for generating the oxidants were found by the workers to be a function of applied voltage, electrode materials, catalyst loadings, reactant flow-rates and pressure. The ozone is generated at room temperature and pressure using lead dioxide powder bonded to a proton exchange membrane (Nafion® 117). The maximum concentration of the ozone formed is about 3 mg dm^{-3} in the aqueous phase. The cathodic reaction during the preparation of the ozone is hydrogen, which is oxidized with oxygen at 15 psi and a flow-rate of 100 ml min^{-1}. The electrocatalysts investigated were various loadings of gold, carbon and graphite powders which are bonded to the membrane or to a carbon fibre paper pressed against the membrane. Hydrogen peroxide was evolved from all the catalysts studied, with the graphite powders yielding the highest concentration (25 mg dm^{-3}). This process may have potential for the destruction of low concentrations of hazardous organic compounds in water courses.

For the conceivable future it is unlikely that there will be a radical change in the industrial production of hydrogen peroxide, i.e. the AO process will continue to dominate and the hydrogen peroxide produced bought by companies wishing to effect certain oxidation chemistries. It is, however, conceivable that in the future, progressive-thinking companies may employ an integrated process involving the manufacture and use of hydrogen peroxide for the oxidation of key intermediates. Therefore, with this in mind, the remainder of this section will be dedicated to this area of operation.

Arco have developed an integrated process for the production of industrially important epoxides via an adapted AO process (Figure 1.13).[33,34] A sulfonic acid substituted alkylhydroanthraquinone alkylammonium salt is reacted with molecular oxygen to form the alkylanthraquinone and hydrogen peroxide. The hydrogen peroxide is then reacted with an alkene in the presence of a titanium zeolite catalyst (TS-1; see Chapter 4). The epoxide product is then separated, and the anthraquinone salt recycled to a hydrogenator for reaction with

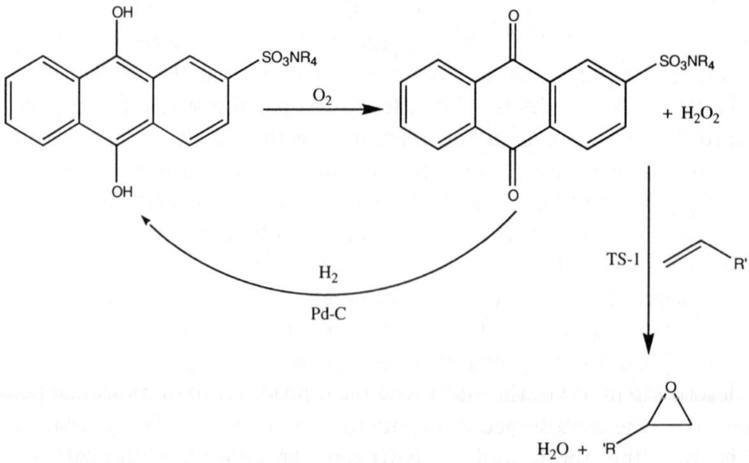

Figure 1.13 *Integrated production of epoxides via the* in situ *generation of hydrogen peroxide.*

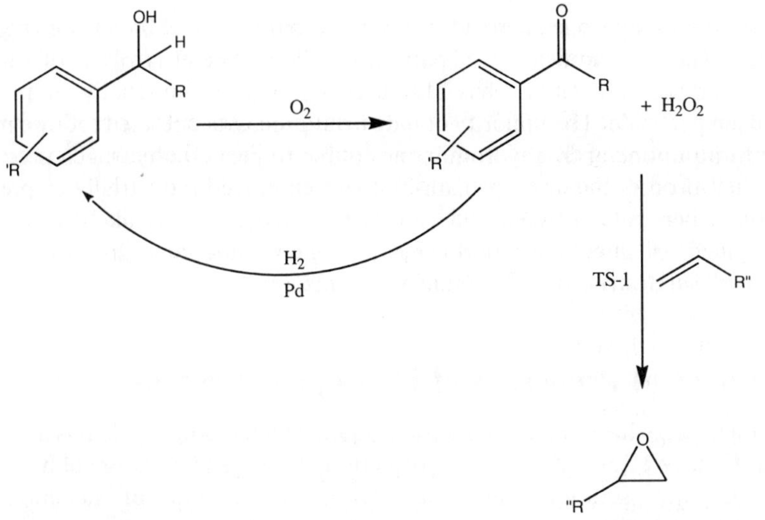

Figure 1.14 *Integrated production of epoxides via the* in situ *generation of hydrogen peroxide.*

hydrogen in the presence of a transition metal. The advantage of this system is the high solubility of the alkylammonium salts employed, thus allowing reactor volumes to be minimized, and higher concentrations of hydrogen peroxide to be produced. Further, no prior treatment or fractionation of the oxidation product is necessary before its use in the catalysed reaction.

Epoxides have also been prepared in a similar fashion to that described above, except an aryl-substituted alcohol is used as one-half of the redox couple (Figure 1.14).

The advantage of the above two methods are high yields of epoxides, and the titanium silicalite catalyst is not deactivated or poisoned by the contaminants in the crude oxidation mixture. Hence, the processes are commercially attractive. The *in situ* hydrogen peroxide generation based on the AO process from either the anthraquinone/anthrahydroquinone or ketone/alcohol redox couples has also been used for the following synthetic reactions:

- ammonia to hydrazine hydrate;[35]
- ammonia and a nitrile to ketazines;[36]
- alkanes to alcohols, aldehydes and ketones;[37]
- phenol to hydroquinone and catechol;[38]
- benzyl alcohols to hydroxybenzoic acids.[38]

A number of electrochemical processes have been employed in an integrated approach for the production of hydrogen peroxide which is subsequently used to oxidize organic functional groups. The electrochemical processes have not only been employed for the preparation of fine chemical intermediates,[39] but also for the destruction of organic pollutants in water courses.[40]

In summary, hydrogen peroxide was first prepared over 180 years ago by L. J. Thenard via the acidification of barium peroxide. The electrolysis of sulfuric acid or ammonium sulfate has also been employed industrially to prepare hydrogen peroxide. The majority of industrial processes operated today employ an anthraquinone/anthrahydroquinone couple to generate hydrogen peroxide. The Shell process based on propan-2-ol was employed industrially to prepare hydrogen peroxide between 1957 and 1980. The future is likely to see the employment of integrated approaches to organic functional group oxidation and low-level destruction of organic pollutants.

3   Physical Properties of Hydrogen Peroxide

Hydrogen peroxide is a clear, colourless liquid which is completely miscible with water. Figures 1.15–1.20 contain information on the general nature of hydrogen peroxide–water solutions, and Table 1.1 compares some of the important properties of hydrogen peroxide–water mixtures. Hydrogen peroxide and its highly concentrated aqueous solutions ($>65\%$ *m/m*) are soluble in a range of organic solvents, such as carboxylic esters.

Hydrogen peroxide and water do not form azeotropic mixtures and can be completely separated by distillation. Most workers, however, obtain 100% *m/m* hydrogen peroxide by fractional crystallization of highly concentrated solutions. Pure 100% *m/m* hydrogen peroxide is usually only of academic interest, and is not produced on an industrial scale, although some niche uses may

Table 1.1 *Physical properties of hydrogen peroxide and water*

Property	Hydrogen peroxide	Water
Melting point (°C)	−0.43	0.0
Boiling point (°C)	150.2	100
Heat of melting (J/g)	368	334
Heat of vaporization (J g^{-1} K^{-1})		
25 °C	1519	2443
b.p.	1387	2258
Specific heat (J g^{-1} K^{-1})		
liquid (25 °C)	2.629	4.182
gas (25 °C)	1.352	1.865
Relative density (g cm^{-3})		
0 °C	1.4700	0.9998
20 °C	1.4500	0.9980
25 °C	1.4425	0.9971
Viscosity (mPa s)		
0 °C	1.819	1.792
20 °C	1.249	1.002
Critical temperature (°C)	457	374.2
Critical pressure (MPa)	20.99	21.44
Refractive index (η_D^{20})	1.4084	1.3330

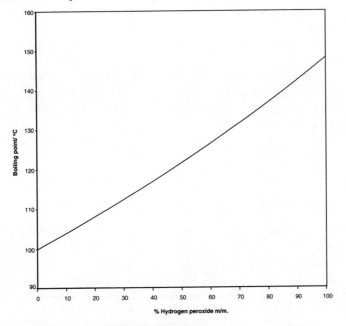

Figure 1.15 *Boiling point range of hydrogen peroxide–water mixtures.*

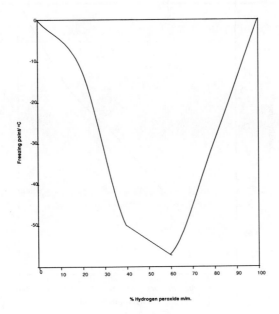

Figure 1.16 *Freezing point range of hydrogen peroxide–water mixtures.*

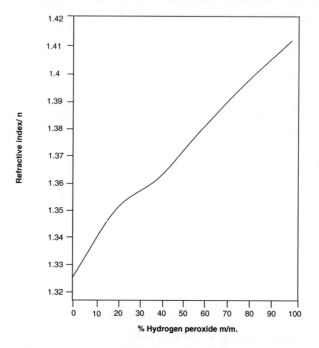

Figure 1.17 *Refractive index range of hydrogen peroxide–water mixtures.*

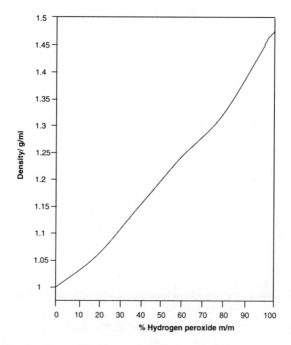

Figure 1.18 *Density range of hydrogen peroxide–water mixtures.*

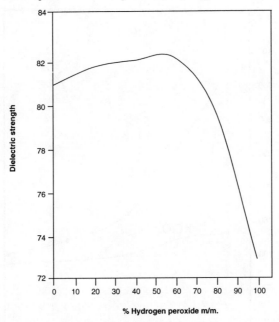

Figure 1.19 *Dielectric strength range of hydrogen peroxide–water mixtures.*

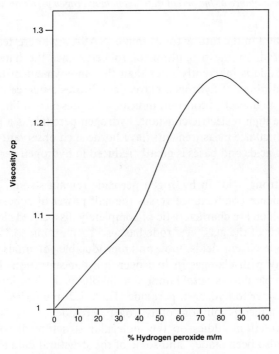

Figure 1.20 *Viscosity range of hydrogen peroxide–water mixtures.*

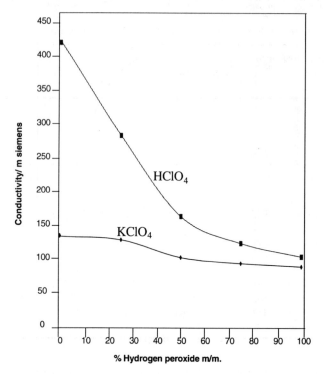

Figure 1.21   *Conductance of the perchlorate anion in hydrogen peroxide–water mixtures.*

become important in the future; for example, NASA are interested in the use of the pure material for the propulsion of rockets, since the handling of pure hydrogen peroxide is inherently safer than the employment of liquid oxygen. The long liquid range of hydrogen peroxide indicates a degree of association, and the very high dielectric constant indicates the presence of linear chains. On the basis of the high dielectric constant, hydrogen peroxide is a good ionizing medium. Conductance measurements have borne such observations out.[41] The conductance of acids and bases is greatly reduced in hydrogen peroxide (Figure 1.21).

Generally, strong acids in hydrogen peroxide remain strong. For example, plots of equivalence conductance versus the half-power of concentration yield straight lines which are characteristic of completely dissociated electrolytes.

The behaviour of the glass electrode has also been examined.[42–43] The glass-calomel electrode system yields stable and reproducible potentials which vary in the normal way with changes in hydrogen ion concentration. However, the EMF of the couple shifts several hundred millivolts as the solution composition changes from water to hydrogen peroxide. Table 1.2 summarizes the apparent and true pH of aqueous solutions of hydrogen peroxide.

Neutron diffraction studies on the molecular structure of solid hydrogen peroxide have also been made[44] and some of the structural data are outlined in Table 1.3.

Table 1.2 *Apparent and true pH of aqueous hydrogen peroxide*

Concentration of hydrogen peroxide solution (% m/m)	Equivalence point[a]	True pH	Correction factor
35	3.9	4.6	+0.7
50	2.8	4.3	+1.5
70	1.6	4.4	+2.8
90	0.2	5.1	+4.9

[a] Measured using a calomel–glass electrode.

Table 1.3 *Molecular dimensions of hydrogen peroxide in the gas phase*

Characteristic	Measurement
Bond length O–O	0.1453 ± 0.0007 nm
Bond length O–H	0.0998 ± 0.0005 nm
Bond angle O–O–H	$102.7 \pm 0.3°$
Azimuthal angle	$90.2 \pm 0.6°$

The vapour pressure and partial pressure of aqueous hydrogen peroxide are illustrated as a function of temperature in Figures 1.22 and 1.23 respectively.

Figure 1.24 shows the vapour–liquid equilibrium curve for aqueous hydrogen peroxide.[45] The solid–liquid phase diagram shown in Figure 1.16 shows eutectic points for the mixtures ice–$H_2O_2 \cdot 2H_2O$ at 45.2% *m/m* hydrogen peroxide, and for solid H_2O_2–$H_2O_2 \cdot 2H_2O$ at 61.2% *m/m* hydrogen peroxide with a congruent meeting point for the compound $H_2O_2 \cdot 2H_2O$ between them. Numerous other

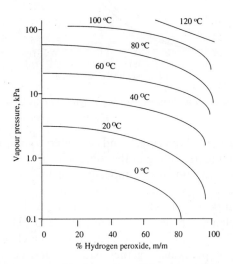

Figure 1.22 *Vapour pressure of hydrogen peroxide–water mixtures.*

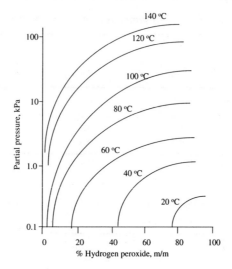

Figure 1.23   *Partial pressure of hydrogen peroxide–water mixtures.*

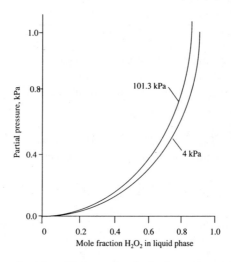

Figure 1.24   *Vapour–liquid equilibrium curve for hydrogen peroxide–water mixtures.*

physical data appear in the literature,[46] and such literature should be consulted for a more thorough understanding of the subject.

The heat of formation and of decomposition of hydrogen peroxide are as illustrated in Figure 1.25. The decomposition equations and heat-generated data are extremely important to know when working with hydrogen peroxide, because safety problems can occur. Decomposition is pH, temperature and impurity sensitive. The remainder of the chapter will, therefore, discuss the safe handling of hydrogen peroxide together with its destruction from process liquors.

$$H_2(g) + O_2(g) \longrightarrow H_2O_2(g) \qquad\qquad -136.2 \text{ KJ/Mol}$$

$$H_2(g) + O_2(g) \longrightarrow H_2O_2(l) \qquad\qquad -187.9 \text{ KJ/Mol}$$

$$H_2O_2(g) \longrightarrow H_2O(g) + 0.5O_2(g) \qquad -105.8 \text{ KJ/Mol}$$

$$H_2O_2(l) \longrightarrow H_2O(l) + 0.5O_2(g) \qquad -98.3 \text{ KJ/Mol}$$

Figure 1.25 *Heat of formation and decomposition of hydrogen peroxide.*

4 Considerations for the Safe Use of Hydrogen Peroxide

The basic hazardous properties and causes of incidents when working with hydrogen peroxide can be attributed to the following:

- Decomposition to oxygen and water with the evolution of heat. The decomposition rate increases with temperature at about 2.3 times per 10 °C rise.
- Pressurization due to oxygen evolution. Hydrogen peroxide, in all forms, is thermodynamically unstable, and continuously dismutates to water and oxygen. Typically, commercial material loses less than 1% *m/m* of its active oxygen per year, however, 20 metric tonnes of 70% *m/m* hydrogen peroxide losing only 0.3% *m/m* of its active oxygen per year will evolve 13 dm^3 of oxygen per day, enough to pressurize sealed equipment or give oxygen enrichment in the headspace of the container.
- Decomposition due to contamination or contact with active surfaces. The rate of decomposition can be increased by the presence of soluble impurities and/or contact with active surfaces. High and low pH will also destabilize hydrogen peroxide. pH affects the activity of the catalytic impurities and the stabilizers which are present.[47] Self-heating can rapidly accelerate the decomposition rate of destabilized hydrogen peroxide. Large amounts of oxygen and steam can be formed quickly (Table 1.4).
- Formation of explosive hydrogen peroxide/organic mixtures. Hydrogen peroxide is a very reactive chemical, and an extremely powerful oxidizer under certain circumstances. Hydrogen peroxide of strength higher than

Table 1.4 *Decomposition data for hydrogen peroxide*[a]

Concentration of hydrogen peroxide solution (% m/m)	Isothermal volumes of hydrogen peroxide	Adiabatic volumes of oxygen and steam	Adiabatic decomposition temperature (°C)
100	512	6643	996
60	263.3	1672	100
40	163.5	932	100
10	36.6	45	89

[a] Volume of oxygen evolved isothermally and volumes of total gas evolved adiabatically from unit volumes of hydrogen peroxide solutions at atmospheric pressure.

Table 1.5 *Explosive power and sensitivity of various substances*

Substance	Explosive power	Sensitivity (kg cm)
Nitroglycerine	52	2–5
85% *m/m* hydrogen peroxide/glycerol	46	10–15
70% *m/m* hydrogen peroxide/polyethylene	30	22
Picric acid	32	75
Trinitrotoluene	30	150
97% *m/m* hydrogen peroxide	17	Insensitive

about 40% *m/m* can also form explosive mixtures with organic compounds. Such mixtures can equate to conventional high explosives in power, but may be much more sensitive in terms of detonation (Table 1.5).

- Spontaneous reaction of hydrogen peroxide/sulfuric acid/water/organic mixtures. These reactions can accelerate rapidly and terminate violently, and can be outside the predicted explosive area.
- Vapour phase hydrogen peroxide explosions. Hydrogen peroxide vapour of concentration above 39% *m/m* at atmospheric pressure is explosive (Figure 1.26).[48]

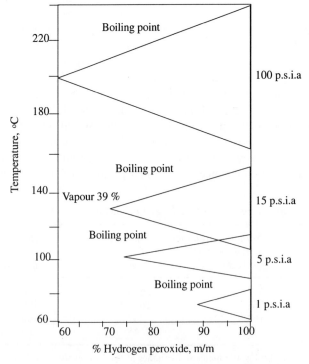

Figure 1.26 *Compositions of liquid hydrogen peroxide which form ignitable vapour.*

Table 1.6 *Minimum ignition energies (mJ) in air and oxygen*

Substance	In air	In oxygen
Methane	0.3	0.003
Acetone	1.15	0.0024
Diethyl ether	0.20	0.0013

- Oxygen enrichment. Oxygen evolved by decomposition may give rise to atmospheres with a high oxygen content. Oxygen-rich flammable atmospheres have low ignition energies (Table 1.6). The rate of burning is also increased. Flame arrestors and other flame-proofing provisions may be ineffective. Explosive limits are widened, but there are no significant changes to the lower explosive limit and flash point. Some vapours and gases which are not flammable in air are flammable in oxygen, for example, some chlorinated solvents.[49]

The above discussion is not meant to deter any interested parties from employing hydrogen peroxide during their work, but it is intended to allow one to embark on studies using peroxygen compounds via an understanding of the main hazards and how to avoid them. The remainder of the section will discuss the practicalities of employing hydrogen peroxide, and peroxycarboxylic acids (RCO_3H).

Hydrogen peroxide and peroxycarboxylic acids (RCO_3H) are normally relatively stable, but as mentioned above they can be rendered unstable by a wide variety of contaminants, particularly at excessively high temperatures. Cleanliness, good housekeeping and proper storage are therefore essential. The major contaminants that cause decomposition are combustible organic materials (*e.g.* cotton, wool, paper) or metals, particularly transition metals and their salts (Table 1.7).[50] Alcohol thermometers or stainless steel thermocouples

Table 1.7 *Effect of added metal ions on the decomposition rate of hydrogen peroxide*

Metal ion	Amount added (ppm)	Active oxygen lost in 24 h at 100 °C[a] (%)
None	—	2
Al(III)	10	2
Sn(IV)	10	2
Zn(II)	10	10
Fe(III)	1.0	15
Cu(II)	0.01	24
Cr(II)	0.1	96

[a] Hydrogen peroxide not stabilized.

should be used for determining temperature and stainless steel ball hydrometers used for density measurements.

Copious supplies of water should always be at hand for washing spillages including contact with skin. The latter should be avoided by wearing adequate protection, particularly for the hands and eyes. Before commencing experiments, glass equipment should be thoroughly washed with water (and a little detergent) and rinsed. Washing with dilute nitric acid (< 5% m/m) followed by distilled water will remove acid-soluble impurities (*e.g.* metals). Hydrogen peroxide and peroxycarboxylic acids should be stored in vented polyethylene bottles and kept cool. Peroxycarboxylic acids should not be kept for longer than six months.

Prior to using active oxygen compounds, the selection of solvents for experiments and for cleaning purposes should be considered carefully. A number of incidents have occurred in the past due to the use of acetone, a result of the formation of acetone peroxides, some of which are highly explosive and crystallize readily. For this reason, acetone or other low molecular weight ketones should never be employed as solvents for extraction or as cleaning agents. Chlorinated solvents, esters and alcohols can, however, be safely employed.

The importance of planning reactions of peroxygen compounds with a pre-knowledge of potential hazards and their control is self evident. In a new field or with a new reaction, the user should conduct initial experiments on a small scale and pay adequate regard to the reaction conditions employed. Standard practical techniques should be used to observe and understand the reaction as it is occurring.

To prevent incidents caused by the rapid decomposition or explosion of active oxygen compounds in reactions with organic substances, a set of safety rules should be followed. These rules have been derived logically and quantitatively. The approach may be illustrated by reference to a three-component mixture of hydrogen peroxide, organic substance and water. The three-component mixture for typical organic compounds is represented in the triangular diagram in Figure 1.27. The diagram was obtained by the deliberate detonation of different mixtures using the blasting cap test. Outside the heaviest shaded area of detonable composition, mixtures could not be exploded. The results above refer to tests with glycerol,[51] however, an extensive range of other organic compounds, for example, acetic acid, ethanol, aniline and quinoline, have been shown to behave similarly.

When working with active oxygen compounds, steps should be taken to ensure that mixtures do not occur in the detonable area during the reaction or processing phases. It should be noted that when using 35% m/m or less hydrogen peroxide then it is unlikely that detonable compositions will be formed. Therefore, use of 35% m/m or less hydrogen peroxide should be employed wherever possible and higher strengths discouraged. Peroxycarboxylic acids can be broken down to their organic acid and hydrogen peroxide components for comparison with the diagram. It is recommended that reactions are carried out in such a way as to prevent the hydrogen peroxide content (or

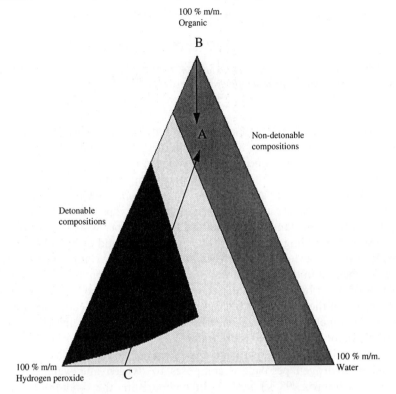

Figure 1.27 *Preferred operating region for peroxygen processes.*

equivalent) exceeding 20% *m/m*. If the reaction has two or more phases, this recommendation should be applied to each phase. Proper attention should be paid to ensure adequate mixing of all phases takes place.

There are, however, systems where the general triangular diagram represented in Figure 1.27 does not hold. Replacement of the inert diluent, water, with significant quantities of sulfuric acid will alter the position and area of the explosive region, bringing it closer to the 20% *m/m* hydrogen peroxide zone. This and other similar situations occur during certain procedures for preparing peroxycarboxylic acids, and not normally when using percompounds as oxidizing agents in reactions. Further information can be obtained from Solvay Interox.[52]

The importance of the order of addition is also illustrated in the triangular diagram. For example, assume an experiment is being conducted with 70% *m/m* hydrogen peroxide and an organic compound. The final composition of the final mixture is represented by point A if no reaction occurs. Addition of organic compound to hydrogen peroxide would result in a reaction mixture with an initial composition at point C. As the organic mixture is added, the reaction mixture composition will pass through the detonable region before reaching A. By adding hydrogen peroxide to the organic compound, the composition would

$$(RCO)_2O + H_2O_2 \longrightarrow RCO_3H + RCO_2H$$

$$RCO_3H + (RCO)_2O \longrightarrow (RCO_2)_2 + RCO_2H$$

Figure 1.28   *Formation of diacyl peroxides in systems containing an excess of acid anhydride compared to hydrogen peroxide.*

change from point B to A, without the composition passing through the detonable area. It is therefore always advisable to carry out reactions in this way even when using low concentration hydrogen peroxide. There is, however, one specific circumstance where the rule of adding hydrogen peroxide last is changed. This is where acid anhydrides are present, when the hydrogen peroxide must be in molar excess to avoid the formation of diacyl peroxides (Figure 1.28).

It is equally important to prevent the hydrogen peroxide (or equivalent) concentration exceeding 20% *m/m*. Diacyl peroxides, and many other organic peroxides, are hazardous in their own right. It is important that the active oxygen is reacting in the intended way, and is not being converted into organic peroxides or other hazardous material; this can be prevented by understanding the chemistry of the system and by routine analysis of the reaction mixture. For example, *in situ* chlorine generation may oxidize nitrogenous materials to the potentially detonable nitrogen trichloride.

Reactions with hydrogen peroxide and peroxycarboxylic acids are exothermic. When hydrogen peroxide decomposes to oxygen and water, the heat generated is approx. 98 kJ mol^{-1}. In comparison, the oxidation of most organic compounds liberates 3–4 times as much heat. As a consequence, even when compositions are not in the detonable zone, appreciable temperature rises can occur. To obtain high yields, it is desirable to carry out reactions in a controlled manner and maintain peroxygen content as low as possible to prevent by-product formation. Temperature control on its own should be adequate but care should be taken to ensure over-cooling does not occur, as a slight exotherm is often required to ensure reaction is taking place. It is also advisable to pre-heat the reactor contents to about 5 °C below the proposed operating temperature before adding the peroxygen species. This will minimize percompound build-up at low temperatures before the reaction is initiated, thus limiting possible run-away reactions. Until a reaction is familiar, the reaction mixture should be analysed for peroxygen content during addition. Incidents have been caused by increasing reaction temperatures, which occur once all the percompound has been added, particularly if no reaction has taken place. In such circumstances, the reaction must be abandoned and a further experiment carried out at a higher operating temperature. It is advisable to have fast methods of cooling available to prevent any incidents occurring due to run-away reactions.

Reactions using active oxygen compounds must always be provided with adequate venting so that decomposition does not result in a pressure build-up. Release of oxygen can lead to oxygen enrichment of the atmosphere above the normal 21% oxygen in air, and can consequently greatly increase the susceptibility to ignition of flammable materials and vapours, and the intensity of any fire or explosion that results from it. Adequate precautions should be taken during work-up, particularly in the case of distillation or evaporative crystallization, to prevent concentration of peroxidic species. If formation of an additional phase occurs, whether liquid or precipitate, it must be investigated with appropriate precautions, as it may contain active oxygen. The absence of percompounds should be confirmed before commencing any purification operations. Destruction of residual peroxides will be discussed later.

In summary, the following check-list should be referred to when employing percompounds with organic materials:

(1) Wear adequate personal protection.
(2) Clean all glassware and the working area.
(3) Protect vessels from sources of contamination.
(4) Store active oxygen compounds away from sunlight and heat in ventilated containers.
(5) Vent all reactors adequately to ensure pressure relief if decomposition occurs.
(6) Carry out new reactions on a small scale, *i.e. ca.* 10 g.
(7) Use alcohol thermometers, stainless steel thermocouples and stainless steel ball hydrometers.
(8) Always plan reactions.
(9) Always add the percompound to the organic material (except with acid anhydrides).
(10) Control addition carefully, observing the reaction.
(11) Provide efficient agitation. Stop the peroxygen feed if the agitator fails.
(12) Ensure the content of hydrogen peroxide (or equivalent) does not exceed 20% *m/m* during the reaction.
(13) Pre-select the reaction temperature. Do not increase the temperature after addition if no reaction takes place.
(14) Supply adequate cooling to the reaction.
(15) Analyse the reaction mixture and remove or destroy any percompound present before distillation or crystallization.
(16) Never use acetone or other lower aliphatic ketones as a solvent for extraction or cleaning.
(17) Use nitrogen to render inert flammable atmospheres.

In the vast majority of cases the precautions quoted above will ensure no incidents occur.

Product solutions or effluents of peroxygen reactions may contain variable amounts of unreacted peroxide, usually in the form of hydrogen peroxide, percarboxylic acids, and/or organic peroxide. For reasons related to safety, waste treatment or product stability, it is usually necessary to destroy unreacted peroxide species in the product solution or effluent prior to discharge or work-up, and certainly before any product concentration process. The remainder of this chapter summarizes methods for removing residual peroxides which can be applied to commercial processes.

Most oxidation with peroxygen compounds requires a slight excess of oxidant to facilitate efficient conversion of the substrate. Frequently, the excess of peroxide is decomposed under the conditions of the reaction, *e.g.* at elevated temperatures, particularly in strongly basic or acidic conditions and/ or in the presence of metal catalysts. However, in many cases, unreacted peroxide persists when the reaction is terminated. In two-phase systems consisting of an aqueous and an organic layer, it is often sufficient to separate the aqueous phase and water-wash the organic phase until no residual peroxide is present. Alternative removal methods will be required where the peroxide is difficult to remove by washing, where recovery of material from the aqueous phase is required (*e.g.* solvent reclamation) or where safe disposal of the aqueous phase cannot be provided. Similarly, in a water-miscible medium where physical separation is not possible, a method for peroxide removal by chemical reduction or physical decomposition must be employed. A final level of $< 0.1\%$ m/m available oxygen will normally allow solutions containing volatile solvents to be concentrated to at least 10% of their initial volume before having to re-check the solution. If necessary, the solution can be retreated to complete the peroxygen decomposition. It is clear that it can be extremely difficult to remove the last few ppm residual peroxide from a solution. And finally, where no material recovery from aqueous solutions is required, consideration may be given to blending the solution with other process waste streams to effect a reduction in toxicity, biodegradability or oxygen demand of the other waste streams whilst simultaneously removing the peroxide.

Several methods of decomposing peroxide are employed commercially. Reactions with chemical reductants have the advantage of transferring oxygen from the peroxide to the reductant molecule, thereby avoiding enrichment of the atmosphere oxygen level. The most commonly used reducing agents for treatment of either water miscible or immiscible waste is solutions of sulfite or bisulfite. Where contact with water must be avoided, treatment with a non-aqueous reducing system may be required. Sulfur dioxide, hydrazine, tertiary phosphines and thioacids have all been employed in this capacity on a laboratory scale, whereas sulfite, and sulfur dioxide are more easily handled on a production scale. It is worth noting that the presence of excess of reducing agent in the treated solution may render subsequent testing negative, and it is therefore essential to ensure that peroxide is indeed removed by the reductive treatment. This is best achieved by adding the product solution to an agitated solution of excess reductant above ambient temperature. Although this mode of treatment is not always necessary (*e.g.* where the excess peroxide is in the form of hydrogen peroxide or alkylhydroperoxide under neutral conditions), it is required where organic peroxy acids are likely to be present. For example, it has been shown that the reaction of excess peracid with sulfite can lead to the formation of diacyl peroxide. The diacyl peroxide is an explosive species and is stable to reduction under normal conditions, but it is not detected under normal wet analytical techniques. Its formation can only be prevented by reducing the peracid with excess reductant. Where it is undesirable to introduce a reducing agent into a product solution, it is normally possible to use a peroxide decomposition agent in the form of a homogeneous or heterogeneous catalyst. These agents may be high surface area substances (*e.g.* active carbon or a transition/heavy metal or oxide). The main function of such a system is to transfer electrons to the peroxide molecule. The decomposition process, therefore, liberates oxygen. When employing catalytic decomposition systems, it is important to ensure the oxygen evolved is not allowed to mix with volatile organics. This problem is best avoided by working in an open system purged with nitrogen to prevent oxygen enrichment. In practice, this danger can be eliminated by operating below the flash point of the product mixture, and maintaining the atmospheric oxygen content below 10% *v/v* (by nitrogen dilution).

Homogeneous decomposition processes are best suited to aqueous solutions which are to be discarded. Frequently used catalytic agents include iron salts [iron(II) sulfate], copper salts [copper(II) sulfate] and catalase (limited to small-scale operations). The efficiency of such systems may show extreme pH dependence. For example, the metal salts are less effective in acid systems, whereas the enzyme methods are restricted to near-neutral pH. However, an enzyme called *Aspergillis niger* shows activity at a pH of 2–3. When the aqueous phase is to be recovered (*e.g.* by distillative or extractive processes), a solid decomposition catalyst is often preferable, since it may be readily removed by filtration. In this way, subsequent contamination of columns or stills involved in the product work-up is prevented. Further, rarely do heterogeneous decomposition processes affect the integrity of the product. Examples of catalysts known to be particularly effective in this area include platinum, platinum black, silver, cobalt or reduced palladium (either as gauzes or on supports), and the active manganese ore, pyrolusite. These systems are particularly unique in their high activity in acidic solutions. The major drawback with heterogeneous decomposition catalysts is their propensity to lose activity in the presence of oils, inorganic phosphates, colloidal tin and silica complexes, *etc.* Similarly, highly acidic liquors or those containing strong chelating agents tend to leach the catalysts from their supports.

The previously mentioned list of solid catalysts is not meant to exclude other transition/heavy metals or their oxides, most of which are known to be good decomposition catalysts for peroxygens. Notable exceptions include tantalum and tin compounds, the latter being used as stabilizers in certain grades of hydrogen peroxide. Non-metallic agents which provide high surface areas and contain Lewis acid sites have also been used as heterogeneous peroxide decomposition agents. These include activated carbons, calcined alumina, zeolites and aluminosilicates. Although such agents are less effective in acid solution compared with their metal counterparts for removing peroxides from alkali solution, they are generally preferred from a cost perspective. As is generally the case with solid decomposition catalysts, increasing the temperature substantially improves their catalytic activity. Figure 1.29 illustrates a

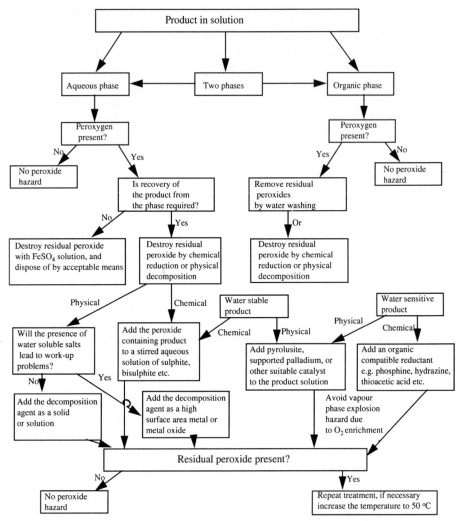

Figure 1.29 *Residual peroxides: removal and destruction.*

flowsheet which can be employed for choosing the best decomposition method for removal of residual peroxides from process liquors.

In general, peroxides are more difficult to remove from acidic solution. This trend arises due to two factors: the loss of activity of many catalytic agents (particularly homogeneous catalysts) and the inherently greater stability of peroxides in slightly acidic solution (this phenomenon relates to the so-called equivalence point of hydrogen peroxide solutions, which corresponds to a pH value of *ca.* 4.5). In fact, hydrogen peroxide can be considered a mild reducing agent at pH < 2–3. These factors are illustrated in Table 1.8, which surveys the performance of various decomposition agents for removal of residual peroxide (hydrogen peroxide and peracetic acid) from a simulated acidic process liquor.

Table 1.8 *Removal of residual peroxide in acidic pH[a] (simulated liquor: 10% glacial acetic acid; 3% hydrogen peroxide; 87% water)*

Reagent	Amount added (% m/m)	AvOx removed at 25 °C (%)	Reaction time (h)	AvOx removed at 60 °C (%)	Reaction time (h)
Blank	—	0.0	3.00	0.5	3.00
HCl	2.2	6.0	3.00	91.0	3.00
HCl	6.0	37.0	3.00	98.0	1.50
Reduced Pd[b]	0.02 as Pd	100	0.25	100	0.25
Oxidized Pd[b]	0.02 as Pd	38	3.00	96.0	3.00
Activated charcoal[c]	1.0	39	3.00	98.0	3.00
Montmorillonite	1.0	19.0	3.00	91.0	3.00
FeSO$_4$	2.5	46.0	3.00	98.0	3.00
MnO$_2$	1.0	21.0	3.00	64.0	3.00
Catalase[d] pH = 2	0.04	22.0	3.00	—	—
Catalase[d] pH = 4	0.04	35.0	3.00	—	—

[a] pH = 2 unless otherwise stated; [b] 2% on aluminosilicate (< 200 mesh); [c] Darco 6–60 (100–325 mesh); [d] Bovine liver catalase containing 11 000 units/mg (1 unit = 1 μmol hydrogen peroxide per min at pH = 7 and 25 °C).

As discussed earlier, heterogeneous catalysts using a supported metal such as palladium are most effective. It should be mentioned, however, that after six re-cycles, the supported palladium catalyst has lost 35% of the active palladium metal, due to leaching into the solution. Decomposition processes become considerably more effective as pH of the solution increases. In the near-neutral region, this is thought to be attributed to increased activity of many of the catalytic agents. As an example, Table 1.9 illustrates the relative performance of several decomposition agents in a simulated liquor containing hydrogen

Table 1.9 *Removal of residual peroxide in neutral pH[a] (simulated liquor: 10% methanol; 3% hydrogen peroxide; 87% water)*

Reagent	Amount added (% m/m)	AvOx removed at 25 °C (%)	Reaction time (h)	AvOx removed at 60 °C (%)	Reaction time (h)
Blank pH = 7	—	2.8	3.00	7.1	3.00
Reduced Pd[b]	0.02 as Pd	100	0.25	100	0.25
Oxidized Pd[b]	0.02 as Pd	100	1.00	100	0.50
Activated charcoal[c]	1.0	83	3.00	90.7	3.00
Montmorillonite	1.0	10.6	3.00	55.6	3.00
FeSO$_4$	2.5	70.4	3.00	99.1	1.00
MnO$_2$	1.0	97.3	1.50	98.9	1.00
Catalase[d] pH = 2	0.001	99.0	0.25	—	—

[a] pH = 7 adjusted by NaOH; [b] 2% on aluminosilicate (< 200 mesh); [c] Darco 6–60 (100–325 mesh); [d] Bovine liver catalase containing 11 000 units/mg (1 unit = 1 μmol hydrogen peroxide per min at pH = 7 and 25 °C).

Table 1.10 *Removal of residual peroxide in basic pH^a (simulated liquor: 10% methanol; 3% hydrogen peroxide; 87% water)*

Reagent	Amount added (% m/m)	AvOx removed at 25 °C (%)	Reaction time (h)	AvOx removed at 60 °C (%)	Reaction time (h)
Blank pH = 12	–	41.5	3.00	93.7	3.00
Reduced Pd^b	0.02 as Pd	100	0.25	100	0.25
Oxidized Pd^b	0.02 as Pd	99.7	0.50	99.0	0.50
Activated charcoal^c	1.0	98.9	1.00	99.6	1.00
Montmorillonite	1.0	98.0	3.00	99.8	1.00
FeSO$_4$	2.5	99.8	1.00	99.8	1.00
MnO$_2$	1.0	99.7	0.50	99.7	0.50
Catalase^d pH = 12	0.001	61.9	3.00	—	—

^{a} pH = 12 adjusted by NaOH; ^{b} 2% on aluminosilicate (< 200 mesh); ^{c} Darco 6–60 (100–325 mesh); ^{d} Bovine liver catalase containing 11 000 units/mg (1 unit = 1 μmol hydrogen peroxide per min at pH = 7 and 25 °C).

peroxide, methanol and trace of orthophosphoric acid. In contrast to acidic solutions, higher pH environments are much less aggressive towards leaching metals from their supports.

Continuing with the pH trend, peroxide solutions are readily decomposed in alkaline solution, whereas near-neutral decomposition processes rely heavily on catalytic activity of the decomposition agent, alkaline processes relying more on the inherent instability of the perhydroxyl anion under high pH conditions.

$$H_2O_2 + OH^{\ominus} \rightleftharpoons HOO^{\ominus} + H_2O$$
$$\downarrow$$
$$0.5 O_2 + OH^{\ominus}$$

Table 1.10 surveys several catalytic agents for removing residual peroxides from a liquor containing hydrogen peroxide, anionic surfactant and ethanol. Whilst general recommendations have been offered, these should serve only as a guideline.

5 Toxicology and Occupational Health Aspects of Hydrogen Peroxide

In humans, brief contact of hydrogen peroxide with the skin leads to irritation and whitening (cutaneous emphysema), the severity of which depends on the concentration of the hydrogen peroxide solution. Longer contact or higher concentration can lead to burns. Contact with the eyes can lead to serious

Table 1.11 *Acute toxicities of hydrogen peroxide in animals*

Test	Animal	Area	Concentration
LD_{50}	Mouse	Oral	2538 mg/kg
LD_{50}	Mouse	Oral	2000 mg/kg
LD_{50}	Rat	Oral	4060 mg/kg
LD_{50}	Mouse	Dermal	1.2×10^4 mg/kg
LC_{50}	Rat	Inhalation	2000 mg/m^3

injury. Hydrogen peroxide aerosol or vapour causes irritation and in severe cases damage to the upper respiratory tract and lungs.[53,54] The human reaction to the irritating effect of hydrogen peroxide on the mucous membrane and skin is far more sensitive than that of the rat. The threshold concentration for acute irritative effects of gaseous hydrogen peroxide on the respiratory tract is 60 mg m^{-3} in rats, but only 10 mg m^{-3} in humans; the corresponding values for skin are 110 mg m^{-3} for rats and 20 mg m^{-3} for humans.[55] Hydrogen peroxide has not been found to produce teratogenic or carcinogenic effects in humans; mutagenic or chromosomal effects have not been observed. The MAK and TLV-TWA values for hydrogen peroxide are 1.4 mg m^{-3}. Acute toxicities in animals are shown in Table 1.11.[56]

Acute dermal toxicity depends on hydrogen peroxide concentration. With 90% *m/m* hydrogen peroxide, the dermal LD_{50} in rabbit is 650 mg kg^{-1} and in rat 4800 mg kg^{-1}.[57] So far, no problems associated with reproduction in animals have been found.[58] Hydrogen peroxide has a mutagenic effect on fungi and bacteria (*e.g.* Ames test), but not on insects or mammalian cells in vitro.[59] Oral administration of hydrogen peroxide produced tumours in the small intestine of mice.[59] However, when hydrogen peroxide was given orally to rats, no significant differences occurred between the two test groups and the control group.[60] No evidence of tumour formation was found one year after exposure of mouse skin to 5% *m/m* hydrogen peroxide.[61]

6 Conclusion

It is hoped that this chapter has given the reader an understanding of the production of hydrogen peroxide over the last two centuries and why its application in the early years was limited strictly to bleaching. The major breakthrough was made when the autoxidation of hydroquinones (AO process) was discovered. The AO process allows for the production of more stable and higher concentrations of aqueous hydrogen peroxide, thus extending the range of applications for the reagent, particularly in the field of fine organic synthesis.

This chapter also illustrates that by observing some simple guidelines, aqueous hydrogen peroxide can be used commercially in a safe manner. It is hoped that the reader will now contemplate the use of such a versatile reagent within their particular area of expertise.

References

1 L.J. Thenard, *Ann. Chim. Phys.*, 1818, **8**, 306.
2 W.C. Schumb, C.N. Satterfield and R.L. Wentworth, in 'Hydrogen Peroxide', Rheinhold Publ. Co., New York, 1955, p. 18.
3 H. Meidinger, *Ann. Chem. Pharm.*, 1853, **88**, 57.
4 H. Berthelot, *C.R. Hebd. Seances Acad. Sci.*, 1878, **86**, 71.
5 W.M. Weigart, H. Delle and G. Kabish, *Chem.-Ztg.*, 1975, **99**, 101.
6 W. Manchot, *Liebigs Ann. Chim.*, (a) 1901, **314**, 377; (b) 1901, **316**, 331.
7 J.H. Walton and G.W. Filson, *J. Am. Chem. Soc.*, 1932, **54**, 3228.
8 BASF, DRP 649234; Pfleiderer, DRP 6583767.
9 Chemical Econonmics Handbook, 1997.
10 Degussa DE, 1273499.
11 BASF DE, 801840.
12 Degussa, DE 1542089.
13 Laporte Chemicals Ltd., GB 718306.
14 Laporte Chemicals Ltd., GB 718307.
15 FMC Corp., US 3009782.
16 J.A. Williams, *Chem. Ind. (London)*, 1959, 76.
17 Allied Chemical Corp., US 3902347.
18 Degussa, DE 1273499.
19 E.I. Dupont de Nemours & Co., US 2902347.
20 Laporte Chemicals Ltd., GB 991413.
21 E.I. Dupont de Nemours & Co., CA 649850.
22 Columbia Southern Chemical Corp., GB 795433.
23 Pittsburgh Plate Glass Co., GB 924625.
24 Laporte Chemicals Ltd., GB 841323.
25 Shell Development Corp., US 2871104.
26 N.V. De Bataafse Petroleum Maatschappij, BE 1002295.
27 E.J. Pneva, *Chimitscheskoja Promyschlenost*, 1976, **9**, 19.
28 Arco Chemical Technology L.P., CA 2099173.
29 Arco Chemical Technology L.P., US 5268160.
30 A.C. Lee and T.C. Choi, *Chem. Eng. Commun.*, 1993, **123**, 179.
31 P. Tatapudi and J.M. Fenton, *J. Electrochem. Soc.*, 1993, **140**, 55.
32 P. Tatapudi, These et Dissertation Academique (THESE), US 1993, University Microfilms Int., *Dissertation Abstract Int.*, 1994, **55**, 1549.
33 Arco Chemical Technology L.P., US 5463090.
34 Arco Chemical Technology L.P., EP 568337.
35 Elf Atochem, EP 179699.
36 Pechiney Ugine Kuhlmann, BE 780470.
37 A. Frusteri, E.N. Savinov, V.N. Parmon and N. Giordino, *Catal. Lett.*, 1994, **27**, 355.
38 A. Itoh, Y. Kuroda, T. Kitano, Z.H. Guo and K. Sasaki, *J. Mol. Catal.*, 1991, **69**, 236.
39 A.C. Lee and T.C. Chou, *Chem. Eng. Commun.*, 1993, **123,** 179.
40 J.S. Do and C.P. Chen, *J. Appl. Electrochem.*, 1994, **24**, 936.
41 E.S. Shanley, E.M. Roth, G.M. Nicholls and M. Kilpatrick, *J. Am. Chem. Soc.*, 1956, **78**, 5190.
42 J.R. Kolczynski, E.M. Roth and E.S. Shanley, *J. Am. Chem. Soc.*, 1957, **79**, 531.
43 A.G. Mitchell and W.F.K. Wynne-Jones, *Trans. Faraday Soc.*, (a) 1955, **51**, 1690; (b) 1956, **52**, 824.
44 W.R. Busing and H.A. Levy, *J. Chem. Phys.*, 1965, **42**, 3054.
45 G. Scotchard, G.M. Karanagh and L.B. Ticknor, *J. Am. Chem. Soc.*, 1952, **74**, 3715.
46 Gmelins 'Handbuch der Anorganischen Chemie', 8 Auflage, Sauerstoff, Lieferung 7, System-Number 3, Verlag Chemie, 1966.
47 J. MacKenzie, in 'Plant/Operations Progress', 1991, **10**, 164.

48 Shell Chemical Corp., 'Concentrated Hydrogen Peroxide', Summary of Research Data on Safety Limitations, 1961.
49 Graviner Co., 'Explosion Protection Seminar', January 1986.
50 Schumb, Satterfield and Wentworth, in 'Hydrogen Peroxide', Reinhold Publishing, 1955.
51 E.S. Shanley and F.P. Greenspan, *Ind. Eng. Chem.*, 1947, **39**, 1536.
52 'Safety Limits for Hydrogen Peroxide/Organics/Strong Acid Systems', Solvay Interox Technical Information.
53 Merkblatt der Berufsgenossenschaft der Chemischen Industrie, Wasserstoffperoxid, 11009, 7184 FRG.
54 F. Hauschild, R. Ludwig and M. Muhlberg, 'Corrosive Action of Hydrogen Peroxide', *Naunyn-Schmie de Bergs Arch. Exp. Pathol. Pharmakol.*, 1958, **235**, 51.
55 M.T. Budagovskiya, M.K. Vadachikoriya and A.I. Desyatov, 'Poisoning with Hydrogen Peroxide', *Otravl. Peridrolem Voen. Med. Zh.*, 1971, **9**, 79.
56 P.P. Ljarskij, 'Toxikologisch-Hygienische Charakteristik von Disinfektionsmitteln auf der Basis von Wasserstoffperoxide und Seinen Derivaten', *Gig. Sanit.*, 1983, **6**, 28.
57 Food and Drug Administration, 'Evaluation of Health Aspects of Hydrogen Peroxide as a Food Ingredient', US Department of Commerce, National Technical Information Service, 1979, PB 80–104 607.
58 IARC Monogr. Eval. Carcinog. Risk Chem. Man., 1985, **36**, 285.
59 A. Ito, H. Watanabe, M. Naito and Y. Naito, in 'Induction of Duodenal Tumors in Mice by Oral Administration of Hydrogen Peroxide', *Gann.*, 1981, **72**, 174.
60 T. Takayama and S. Takizawa, 'Carcinogenicity Study of Hydrogen Peroxide in Rats', Annual Report of Cancer Research, Ministry of Health and Welfare, Japan 1980.
61 Y. Kurokawa, 'Studies on Promoting and Complete Carcinogenic Activity of some Oxidizing Chemicals in Skin Carcinogenisis', *Cancer Lett.*, 1984, **24**, 299.

CHAPTER 2

Activation of Hydrogen Peroxide Using Inorganic and Organic Species

1 Introduction

The following chapter will give an account of the general principles of hydrogen peroxide activation in the presence of a variety of chemical entities. The basic chemistry of hydrogen peroxide will be discussed, followed by a discord on its activation in the presence of inorganic and organic materials. Finally, the chapter will conclude with a brief treatment as to how hydrogen peroxide can be stabilized against the loss of its active oxygen.

2 Basic Chemistry of Hydrogen Peroxide

The E^0 (redox potential) value of hydrogen peroxide is 1.76 V measured for the half-reaction shown (Figure 2.1).

On the face of the E^0 value alone, one would expect the compound to be a relatively powerful oxidant (Table 2.1). However, hydrogen peroxide is, in fact, a relatively weak oxidizing agent. It can achieve some oxidations unaided, $e.g.$ it reacts slowly with substrates such as olefins, aromatic hydrocarbons, alkanes, $etc.$, but for the majority of applications, it requires activation in some way or other. This is reflected by the fact that hydrogen peroxide has been employed for bleaching purposes for over a century[1] and some activation methods such as Fenton's reagent are almost as old.[2]

By far the bulk of useful peroxygen chemistry has been discovered in the last 50 years, and many catalytic methods, described later, are even more recent. The low intrinsic reactivity of hydrogen peroxide is actually an advantage, since it

$$H_2O_2 + 2\overset{\oplus}{H} + 2\overset{\ominus}{e} \longrightarrow 2H_2O \qquad E^0 = 1.76 \text{ V}$$

Figure 2.1 *Half-reaction of hydrogen peroxide to water.*

Table 2.1 *Oxidation potentials of a range of oxidants*

Oxidant	E^0/V
F_2	3.00
$HO^\bullet$	2.80
$^1O_2$	2.42
O_3	2.01
H_2SO_5	1.81
H_2O_2	1.76
$KMnO_4$	1.70
HO_2^-	1.70
$HOCl$	1.49
Cl_2	1.27
ClO_2	1.27
O_2	1.20

means that it can be selectively activated to perform a given oxidation. Figure 2.2 provides a summary of the various active oxidants derived from hydrogen peroxide. The remainder of the chapter will expand upon each of these methods.

Although hydrogen peroxide is a weak oxidant, it does have both electro-

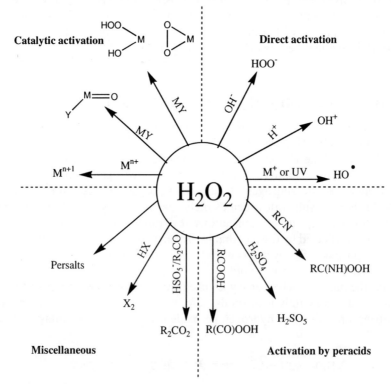

Figure 2.2 *Activation of hydrogen peroxide.*

$$H_2O_2 \quad \rightleftharpoons \quad \overset{\delta^+}{HO}\text{-----}\overset{\delta^-}{OH}$$

Figure 2.3 *Polarization of hydrogen peroxide.*

philic and nucleophilic properties. The electrophilic character arises from the fact that the O–O bond is easily polarized (Figure 2.3).

Undissociated hydrogen peroxide behaves, to some extent, as a nucleophile, being about 10^4 times more nucleophilic than water. For example, hydrogen peroxide readily adds to carbonyl bonds giving rise to hydroxyhydroperoxides (peracetals and perketals). Such compounds are often used as polymerization initiators on account of their radical decomposition at moderate temperatures (O–O bond homolysis). Neutral hydrogen peroxide can also react with activated acyl compounds such as anhydrides to give peroxyacids.

In alkaline solution, hydrogen peroxide dissociates (Figure 2.4).

The perhydroxyl anion, HO_2^-, is a powerful nucleophile (see Chapter 3) and will attack substrates such as electron-deficient olefins (*e.g.* α,β-unsaturated ketones) and aldehydes. The perhydroxyl anion is also of value in bleaching and product purification, particularly of natural products. In addition, it can be used to generate more powerful oxidants by mixing with electron-deficient acyl compounds (giving peroxyacids) or with nitriles (Figure 2.5).

Under strongly acidic conditions (normally non-aqueous), hydrogen peroxide can be protonated or converted to the equivalent of the hydroxyl cation (Figure 2.6).

The resulting strong electrophile can be used in some oxidations, notably phenol hydroxylation, however the extreme acidic conditions required limit the

$$H_2O_2 \quad \rightleftharpoons \quad \overset{\oplus}{H} + \overset{\ominus}{HO_2} \qquad pK_a = 11.6$$

Figure 2.4 *Dissociation of hydrogen peroxide under alkaline pH.*

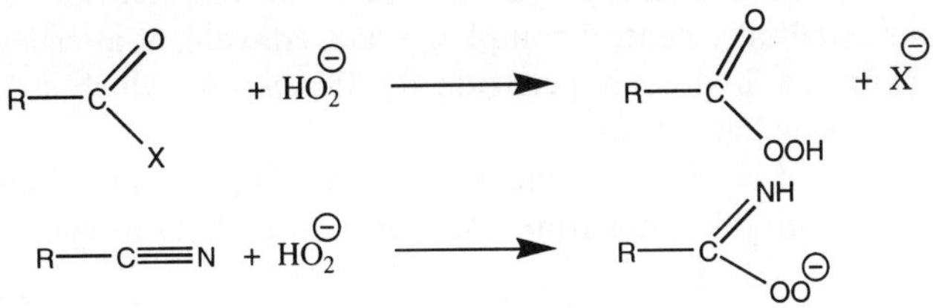

Figure 2.5 *Generation of more powerful oxidants via the use of perhydroxyl anion.*

$$H_2O_2 + \overset{\oplus}{H} \rightleftharpoons H_3\overset{\oplus}{O_2}$$

$$H_2O_2 + HA \rightleftharpoons H_3\overset{\oplus}{O_2} + \overset{\ominus}{A}$$

Figure 2.6 *Action of strongly acidic conditions on hydrogen peroxide.*

applicability of this oxidant species; many products tend to react further by oxidation or acid catalysed condensation or rearrangement.

The O–O bond in hydrogen peroxide is relatively weak,[3] approx. 213 kJ mol^{-1}, and is susceptible to homolysis by a variety of methods including thermal, photolytic, radiolytic and metal-redox. The active species produced is the hydroxyl radical; when generated by UV irradiation at 254 nm, two hydroxyl radicals per mol of hydrogen peroxide are produced. In terms of oxidizing power, it is second only to fluorine. This high oxidizing power correlates to a relative lack of selectivity as an oxidant, and hence the hydroxyl radical has a limited role in synthesis. The use of hydrogen peroxide–UV systems is, however, known in water disinfection, and is growing in effluent treatment, where the power to degrade refractory organics to products treatable by biological processes is of value.

The hydroxyl radical can also be generated from several one-electron reducing metal ions,[4] of which the most common is iron(II), this combination being known as Fenton's reagent (see Section 3).

The following two sections will discuss the activation of hydrogen peroxide in the presence of inorganic/organometallic species and organic compounds respectively.

3 Activation of Hydrogen Peroxide in the Presence of Inorganic and Organometallic Species

There are two solid peroxygens which dominate the area of inorganic hydrogen peroxide chemistry, namely sodium percarbonate (PCS) and sodium perborate (PBS). It is also worth noting that the urea hydrogen-bonded complex of hydrogen peroxide, known as urea–hydrogen peroxide (UHP) is also an important solid peroxygen which will be discussed here for the sake of completeness.

UHP can be represented by the structure in Figure 2.7. UHP is readily crystallized from aqueous strong hydrogen peroxide and urea solutions.[5] Being an anhydrous essentially neutral complex, it is a convenient alternative to very highly concentrated hydrogen peroxide ($>85\% \, m/m$), which is difficult to obtain and relatively hazardous.

The application of UHP in organic synthesis has been explored[6] and has been proven to be a suitable substitute for concentrated hydrogen peroxide in

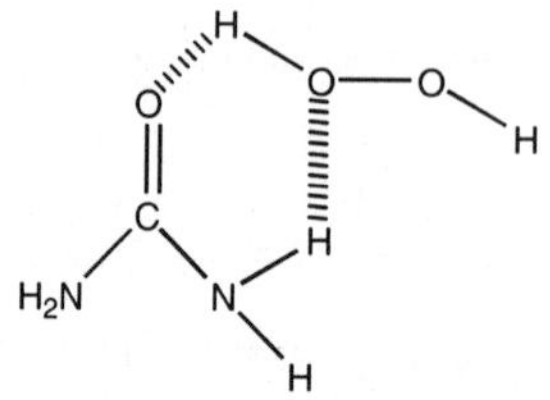

Figure 2.7 *Hydrogen-bonded complex of urea and hydrogen peroxide.*

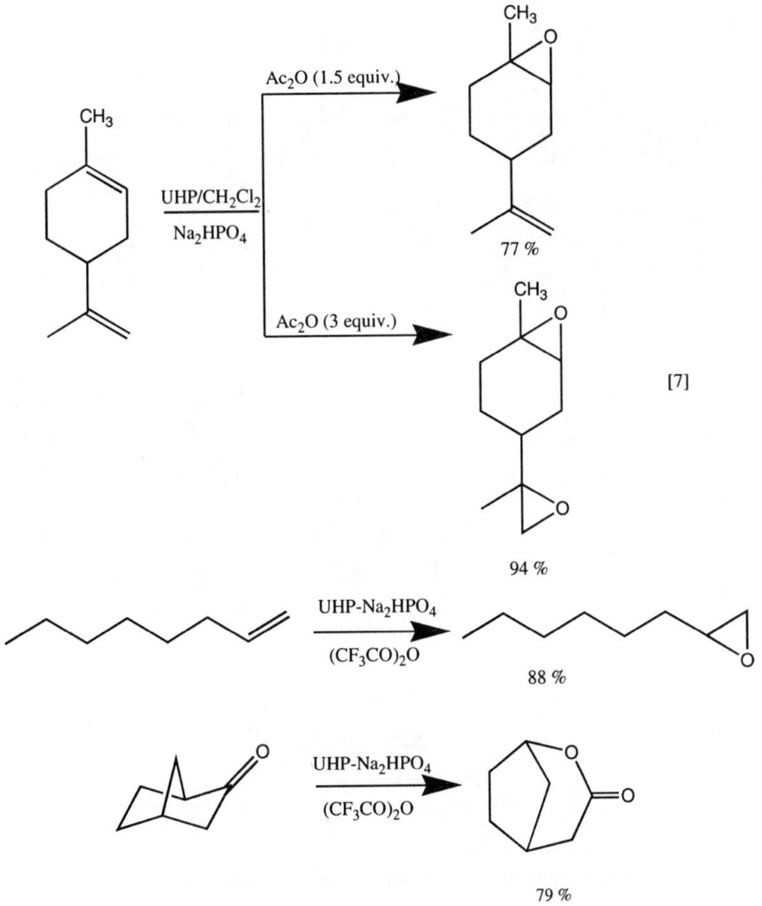

Figure 2.8   *Use of UHP to oxidize a number of organic materials to industrially useful intermediates.*

generating trifluoroperacetic acid from the anhydride. A number of reactions employing UHP to generate peroxyacids *in situ* are outlined in Figure 2.8.

Sodium perborate and sodium percarbonate are peroxygen compounds available at a relatively low cost. This is a result of their extensive use in the detergent industry as precursors to the bleaching agents which are generated *in situ*.[8–10] The structures of the two materials are very different (Figure 2.9). PBS is a true persalt and can be described as $NaBO_3 \cdot nH_2O$, where $n = 1$–4. Depending on the value of n, the PBS is called either PBS-1 or PBS-4.

The structure of PBS corresponds to a six-membered heterocyclic dianion.[11] PCS, on the other hand, is not a true percarbonate, but is a perhydrate $(Na_2CO_3 \cdot 1.5H_2O_2)$, as shown in Figure 2.9. The integrity of the adduct is due to hydrogen peroxide bonding between carbonate anions and hydrogen peroxide molecules.[12] Its rather confusing name has arisen because of historical uncertainties over its structure[13] and PCS should in fact be considered as a solid

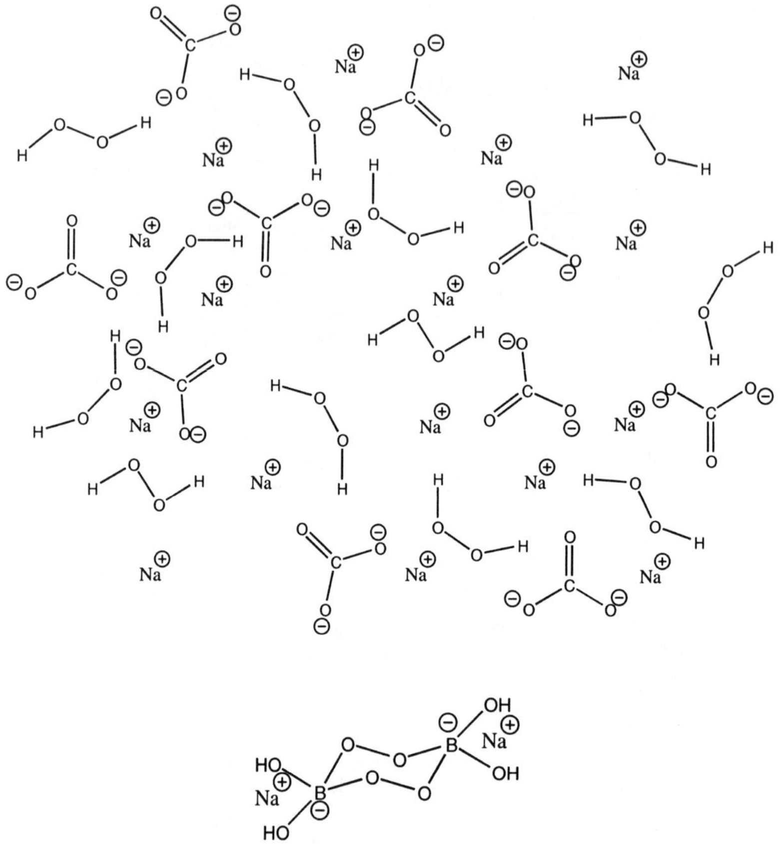

Figure 2.9   *Structures of sodium percarbonate and sodium perborate.*

form of hydrogen peroxide like UHP.[14] PBS and PCS are easy to handle and are relatively stable at room temperature, their thermal decomposition occurring well above 100 °C.[14] In water, both PCS and PBS yield a mildly alkaline solution of hydrogen peroxide[15] although the borate exerts a buffering and mild stabilizing effect. In non-aqueous media, however, the chemistry of PBS is quite distinct,[16] especially when in combination with acetic acid. For example, synthetically important oxidative transformations, previously only carried out with stoichiometric metal oxidants, can be performed with sodium perborate inconjunction with acetic acid.

As alluded to earlier, concentrated hydrogen peroxide, as well as not being commercially available, is also difficult to handle safely. Consequently, the ability of PBS and PCS to release oxidative species into an organic medium has made them extremely useful synthetic tools. Figure 2.10 summarizes a number of examples where PBS and PCS have been employed in this capacity.

PBS and PCS have both been shown to be effective reagents for the oxidation of organic species. It is therefore likely that their employment with metal-centred catalysts, in place of hydrogen peroxide, will grow in the future.

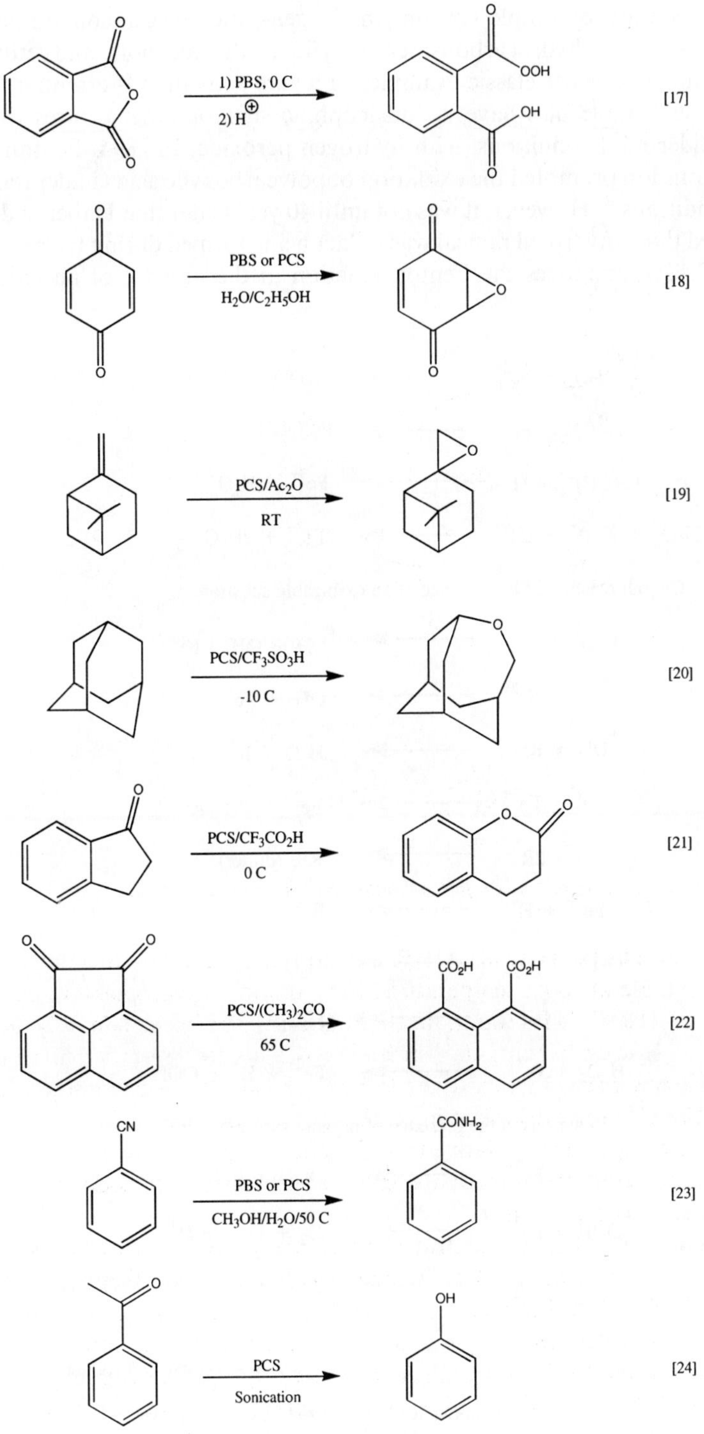

Figure 2.10 *Some of the oxidative transformations effected by PBS and PCS.*

The oxidation of simple carbon-, hydrogen- and oxygen-containing compounds, such as hydrocarbons, carboxylic acids, alcohols and ethers by hydrogen peroxide are classic examples of where activation is essential. This is because such molecules have no nucleophilic sites or other centres that can react, under mild conditions, with hydrogen peroxide. In 1894, Fenton found that ferrous ion promoted the oxidation of polycarboxylic acids under relatively mild conditions.[25] However, it was not until 40 years later that Haber and Weiss suggested that a hydroxyl radical was in fact being formed during the reaction.[26] Figure 2.11 summarizes the Fenton reaction in the absence of an oxidizable

$$H_2O_2 + Fe^{2+} \longrightarrow Fe(OH)^{2+} + {}^{\bullet}OH$$

$${}^{\bullet}OH + Fe^{2+} \longrightarrow Fe(OH)^{2+}$$

$$Fe(OH)^{2+} + H^+ \longrightarrow Fe^{3+} + H_2O$$

$$H_2O_2 + 2Fe^{2+} + 2H^+ \longrightarrow 2Fe^{3+} + 2H_2O$$

Overall reaction in the absence of an oxidizable substrate

$$H_2O_2 + Fe^{2+} \longrightarrow {}^{\bullet}OH + OH^- + Fe^{3+}$$

$${}^{\bullet}OH + Fe^{2+} \longrightarrow OH^- + Fe^{3+}$$

$${}^{\bullet}OH + RH \longrightarrow H_2O + R^{\bullet}$$

$$R^{\bullet} + Fe^{3+} \longrightarrow Fe^{2+} + \text{'Product'}$$

$$2R^{\bullet} \longrightarrow RR \text{ (dimer)}$$

$$R + Fe^{2+} + H^+ \longrightarrow RH + Fe^{3+}$$

$${}^{\bullet}OH + H_2O_2 \longrightarrow H_2O + {}^{\bullet}OOH$$

$$2\overset{\bullet}{O}H \longrightarrow H_2O_2$$

$$H_2O_2 + Fe^{3+} \longrightarrow Fe^{2+} + H^+ + {}^{\bullet}OOH$$

Reactions in the presence of organic substrates (RH)

$${}^{\bullet}OOH + Fe^{2+} \longrightarrow {}^-OOH + Fe^{3+}$$

$${}^{\bullet}OOH + Fe^{3+} \longrightarrow O_2 + Fe^{2+} + H^+$$

$$R^{\bullet} + O_2 \longrightarrow RO_2^{\bullet} \longrightarrow RO_2^- + Fe^{3+}$$

$$\longrightarrow \text{'Product'}$$

Competing reactions leading to loss of oxidative power

Figure 2.11   *The Fenton reaction.*

$$H_2O_2 + Fe^{2+} \longrightarrow FeO^{2+} + H_2O$$

$$FeO^{2+} + Fe^{2+} + H^+ \longrightarrow Fe(OH)^{2+} + Fe^{3+}$$

$$Fe(OH)^{2+} + H^+ \longrightarrow Fe^{3+} + H_2O$$

Figure 2.12 *Possible formation of ferryl ion during Fenton-type chemistry.*

substrate, in the presence of organic substrates and suggests the competing reactions which lead to the loss of the oxidative power of the system.

It is interesting to note that several research groups propose the oxidizing intermediate to be an iron(IV) species such as the ferryl ion (FeO^{2+}),[27] especially when the iron centre is attached to a porphyrin ring. Figure 2.12 illustrates a possible alternative mechanism involving the ferryl ion to that outlined above.

Unfortunately, it is not very easy to distinguish between the two mechanisms since they both fit the rate law:

$$-d[H_2O_2]/dt = k[H_2O_2][Fe(II)]$$

It should be pointed out that even if one starts with a single oxidation state of iron in the presence of hydrogen peroxide, both states are soon produced. The ferric ion can oxidize perhydroxyl radical $(HO_2^{\bullet})$ and/or its conjugate ion $(O_2^{\bullet-})$, as well as hydrogen peroxide. The chemistry of iron(III) with hydrogen peroxide has been comprehensively reviewed by Brown, Jones and Suggett.[28]

The facile oxidation of organic substrates in the presence of a Fenton system is believed to be due to the production of the hydroxyl radical, as mentioned earlier. Consequently, a number of techniques have been employed to generate the species independently of any iron centres. Two techniques worthy of a mention are photolysis and radiolysis. Photolytic activation can be used to cause homolysis of the peroxygen bond because peroxides have a relatively broad absorption band above 300 nm (Figure 2.13).[29]

Other peroxygen species can also be photolytically cleaved to yield the hydroxyl radical and another radical centre. For example, homolysis of peroxymonosulfate $(HOOSO_3)$ generates $OH^{\bullet}$ and $SO_4^{\bullet-}$. The concentration of the generated hydroxyl radical can be controlled by variation of the wavelength and the intensity used. The photolysis of hydrogen peroxide in the presence of alcohol produces EPR (electron paramagnetic resonance) spectra which indi-

$$H_2O_2 \xrightarrow{h\nu} 2HO^{\bullet}$$

$$HO^{\bullet} + H_2O_2 \longrightarrow H_2O + {}^{\bullet}OOH$$

$$2{}^{\bullet}OOH \longrightarrow O_2 + H_2O_2$$

Figure 2.13 *Photocatalysis of hydrogen peroxide.*

cate attack of the hydroxyl radical at positions α and β to the alcohol functional group. The ratio of attack at the two positions is often solvent dependent.

Radiolysis has also been employed to generate the hydroxyl radical. However, because very energetic particles are used (x-rays, γ-rays, electron beams, *etc.*) aqueous solutions are used instead of hydrogen peroxide. Water molecules can be cleaved homolytically and heterolytically to produce three radical species: hydroxyl radical, hydrated electron and hydrogen atom.[30] A great many rate constants of hydroxyl radical with reductants, especially alcohols, have been measured using radiolysis combined with EPR or electronic spectroscopy.[31]

On the face of it, the Fenton system seems ideally suited for the oxidation of organic species. As a consequence of the high reactivity of the hydroxyl radical (approx. 10^9 M^{-1} s^{-1}), there is very little selectivity and this limits the application of Fenton chemistry in fine organic synthesis. It is, however, efficacious as an oxidation system for the removal of organic effluents from water courses. These systems will be further discussed in Chapter 5. Some organic syntheses have been carried out in the presence of Fenton's reagent and a few illustrative examples are outlined in Figure 2.14.

It should be noted that in practice, low conversions of substrate are deliberately aimed for, in an attempt to reduce by-product formation.

Iron compounds are not the only one-electron reductants capable of carrying

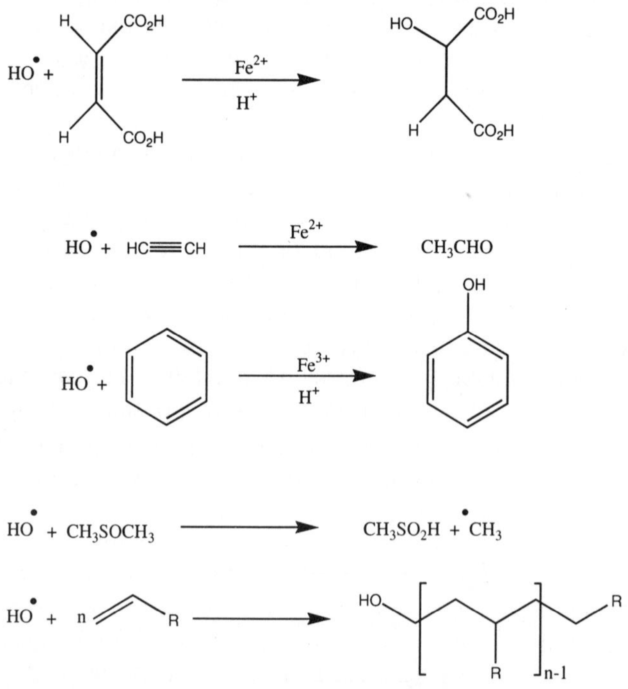

Figure 2.14   *Application of Fenton chemistry in the preparation of oxidized organic compounds.*

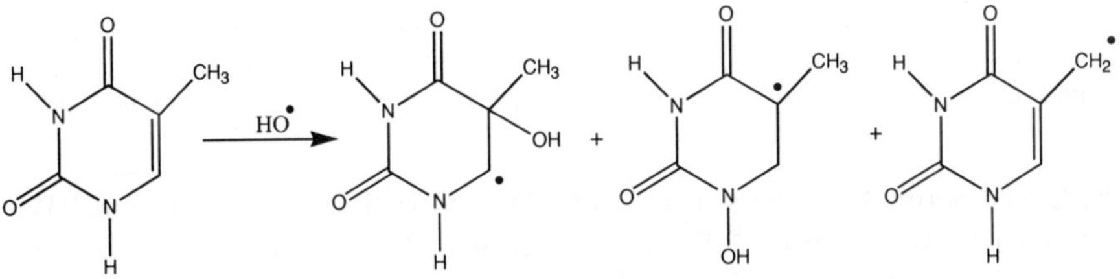

Figure 2.15 *Hydroxyl radical attack on guanine.*

out the chemistry described above, and not all one-electron reductants have hydroxyl radical as the active oxidant. For example, on the basis of spin-trapping and kinetic experiments, Johnson and co-workers have concluded that when $Cu(I)_{aq}$ and $Cu(Phen)_2^{2+}$ react with hydrogen peroxide, the active oxidant is not the hydroxyl radical, but possibly a Cu(III) species formed from the breakdown of a $CuOOH^-$ peroxide species.[32] Vanadium(IV),[33] vanadium(V),[34] titanium(III),[35] chromium(II)[36] and cobalt(II)[37] are all one-electron reductants which have been extensively studied.

Finally, Fenton's reagent induced hydroxyl radical formation can be extremely toxic to biological systems,[38] and it is now believed that these radicals are the driving force behind the initiation of cascading cytoxic free radical reactions. The long-chain molecule of DNA can be cleaved or damaged by the hydroxyl radical, with the deoxyribose ring moiety and the amine bases being the most susceptible to attack (Figure 2.15).[39] The C–H bonds in deoxyribose are very reactive towards hydroxyl radical. Proteins and lipids can also be attacked leading to severe cell damage in some cases.

Unfortunately, it is not possible to give comprehensive coverage to Fenton's chemistry here, however, the author hopes that enough justice has been afforded to the subject to whet the appetite of those wishing to study the area further.

Hydrogen peroxide and other peroxygen species can function as primary or stoichiometric oxidants when in combination with other elements present in catalytic quantities. Hypervalent forms of the element are produced and these are the active species of the system. Some of these oxidants are well established, for example, the use of hydrogen peroxide with osmium tetroxide for alkene hydroxylation.[40] More recently, hydrogen peroxide has been used with ruthenium compounds, under phase-transfer conditions, for alcohol oxidation, and alkene cleavage.[41] Hydrogen peroxide in combination with cobalt or cerium compounds has also been employed for the side-chain oxidation of aromatic materials.[42] Such systems have been shown to have advantages over the more traditional air/oxygen systems in terms of product selectivity.

One of the limitations of employing hydrogen peroxide in oxidations with hypervalent metals is its ability to act as a reducing agent as well as an oxidizing agent. The titration of hydrogen peroxide using cerium(IV), permanganate or dichromate are well established examples of such a phenomenon, whereby hydrogen peroxide is oxidized to oxygen. However, two of these systems also illustrate the pH dependence of such a reaction; in alkaline solution,

$$M^{n+} + H_2O_2 \longrightarrow {}^{(n+2)+}M{=}O + H_2O$$

Figure 2.16 *Formation of a metal 'oxene' species in the presence of hydrogen peroxide.*

hydrogen peroxide oxidizes chromium(III) to chromium(IV) and cerium(III) to cerium(IV), the reverse of the acid reactions. Peroxymonosulfates and per-carboxylic acids do not exhibit reducing properties to the same extent and are in general able to oxidize a wider range of elements. Caro's acid (H_2SO_4 + $H_2O_2 \leftrightarrow H_2SO_5 + H_2O$) in particular can be used in the metals industry to recover or separate metals using redox chemistry (Chapter 6). Caro's acid can oxidize manganese(II) to manganese(III) or manganese(VII), depending on the pH. Peracetic acid can be used to generate lead(IV) and iodide(III) reagents for organic synthesis applications.

A relatively recent field of study, which, will be expanded upon later, is that of transition metal substituted polyoxometallates.[43] The fact that they are, in most cases, completely inorganic means that they are inherently robust to oxidative degradation, hence, offering good prospects for industrial application.

Metal oxene compounds are formed from the interaction of a metal centre with a peroxygen and this reaction is classed as a two-electron transfer oxidation (Figure 2.16).

The oxene species formed can, under certain conditions, be a much more powerful oxygen transfer agent than hydrogen peroxide. It should be noted that there is a link between this system and the Fenton's chemistry discussed earlier; both go through an intermediate in which hydrogen peroxide or its anion is ligated to the metal. In fact, under scrupulously anhydrous conditions, iron salts with hydrogen peroxide exhibit epoxidation and regioselective hydroxylation typical of oxenes but inconsistent with hydroxyl radical chemistry.[44]

The most important oxene chemistry driven by peroxygens is that of the metalloporphyrins of iron and manganese. The manganese systems operate in the presence of a co-catalyst such as imidazole to carry out oxygen transfer and insertion reactions, notably epoxidation,[45] amine oxidation[46] and hydroxyla-tion.[47] The reactions can be carried out in a two-phase system with a phase-transfer agent or in a single-phase system with a co-solvent for the hydrogen peroxide, catalyst, co-catalyst, and substrate. These systems are mimics for peroxidase enzymes, including cytochrome P-450, which uses iron porphyrin oxene chemistry (Figure 2.17).[48] Although metal catalysts are involved, the levels of such catalysts are often very small, and recovery can be simple. Unfortunately, the lifetime of such catalysts is often limited due to self-attack by the oxene or radical side reactions. Current work is directed towards supported forms of metalloporphyrins, where site isolation is expected to improve stability towards oxidation (the immobilization of such catalysts will be discussed further in Chapter 4).

A number of workers have shown that a small amount of nitrogenous base with manganese metalloporphyrins greatly increases the stereospecificity of alkene epoxidation.[49–51] Figure 2.18 illustrates the dual behaviour of imidazole during the epoxidation of alkenes by hydrogen peroxide.[52]

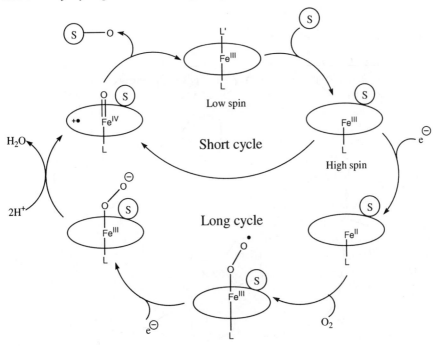

Figure 2.17 *Catalytic cycle of cytochrome P-450.*

It has recently been suggested that addition of a carboxylic acid further enhances the catalytic activity of manganese(III) porphyrins with nitrogenous bases in the epoxidation of alkenes by hydrogen peroxide.[53–54] This has been attributed to the formation of a peracid ligated to the metal, Mn^{III}–OOCOR, with a weaker peroxide bond than Mn^{III}–OOH, encouraging heterolytic cleavage. This is, however, highly unlikely. A much more reasonable explanation is that the mode of action is similar to that of imidazole in that the acid–base catalysis plays a pivotal role; such chemistry is effected by the carboxylic acid. A major problem with the employment of a nitrogenous base such as imidazole is that they are readily oxidized to N-oxides by hydrogen peroxide. This has promoted studies into the use of oxidatively stable co-catalysts such as amine N-oxides[55] and ammonium acetate.[56]

The last decade has greatly increased our knowledge on how metalloporphyrins are able to interact with hydrogen peroxide without producing the unspecific hydroxyl radical. It would therefore appear likely that in the near future, industrial applications of such metalloporphyrins with hydrogen peroxide will be started in earnest.

Before leaving the area of oxene chemistry, there is one further system worthy of mention: the manganese Schiff-base complexes. The Schiff-base complexes were prepared in response to the Katsuki–Sharpless system for stereospecific epoxidation (Figure 2.19).[57] The Katsuki–Sharpless system consists of titanium(IV) isopropoxide and (+)- or (−)-diethyl tartrate with

H2O2
MnIII
MnIII
MnV
MnIII
H2O
OH
HN
N
H
O

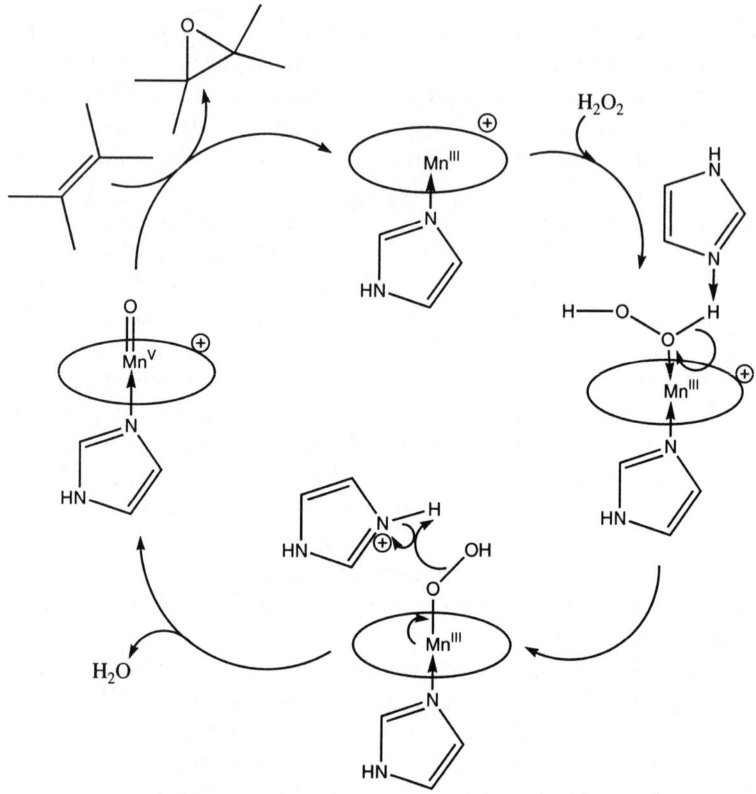

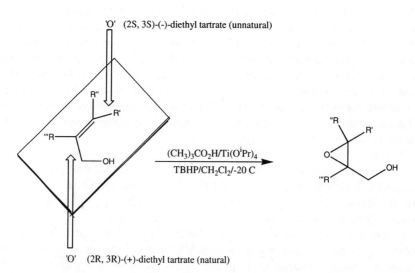

Figure 2.18 *Dual function of imidazole in the epoxidation of alkenes by hydrogen peroxide catalysed by manganese(III) porphyrins.*

'O' (2S, 3S)-(-)-diethyl tartrate (unnatural)
R"
R'
'''R
OH
(CH3)3CO2H/Ti(OiPr)4
TBHP/CH2Cl2/-20 C
"R
R'
O
OH
'''R
'O' (2R, 3R)-(+)-diethyl tartrate (natural)

Figure 2.19 *Oxygen transfer to the alkene using the Katsuki–Sharpless asymmetric epoxidation system.*

t-butyl hydroperoxide as the oxidant, all of which are commercially available. The mechanism is believed to involve attack on the allylic alcohol substrate by a compound formed from the titanium(IV) isopropoxide and the diethyl tartrate. A complex that contains the substrate and the *t*-butyl hydroperoxide is produced.[58] Which enantiomer of the epoxide is formed is determined by whether (+)- or (−)-diethyl tartrate is used.

Since its discovery, the Katsuki–Sharpless system for asymmetric epoxidation has been used in the synthesis of a huge number of complex carbohydrates.[59] The Katsuki–Sharpless method, however, requires the presence of a hydroxyl group adjacent to the double bond, to anchor the alkene in the active centre of the catalyst. In recent years, catalytic systems based on manganese complexes of chiral Schiff-bases have been developed by Jacobson[60] and by Katsuki.[61] These catalysts control the enantioselectivity of the epoxidation via non-bonded interactions, hence, the number of substrates which can be sterospecifically epoxidized is practically unlimited. Epoxidation is normally carried out at ambient temperature, and commonly employs sodium hypochlorite as the oxidant. Hydrogen peroxide[62] and triplet oxygen[63] have also been employed, but to date the results have been disappointing. The active oxygen species is believed to be an oxene species ($Mn^V=O$). Two of Jacobsen's complexes are illustrated in Figure 2.20.

Whilst the chiral manganese complexes can epoxidize alkenes with high enantioselectivity (> 90% e.e.), they are not particularly stable. This instability is probably due to the easily oxidizable imine and phenoxide ligands on the complex. Attempts are currently being made to immobilize Schiff-bases in order to increase their stability in a similar manner to the metalloporphyrins discussed earlier.

Metal peroxo and hydroperoxo systems are the largest and probably the most important of the oxidant classes within this category. Metal peroxo complexes are formed rapidly in water over a wide pH range from a variety of d^0 metal compounds, mainly of groups IVb, Vb and VIb, notably titanium(IV), vanadium(V), molybdenum(VI) and tungsten(VI). These complexes are electrophilic in nature and many of them have oxygen transfer properties. Their re-formation from the parent compound and hydrogen peroxide is rapid enough for them to be used catalytically in oxidation systems. As such they can be regarded as inorganic catalytic analogues of percarboxylic acids (see

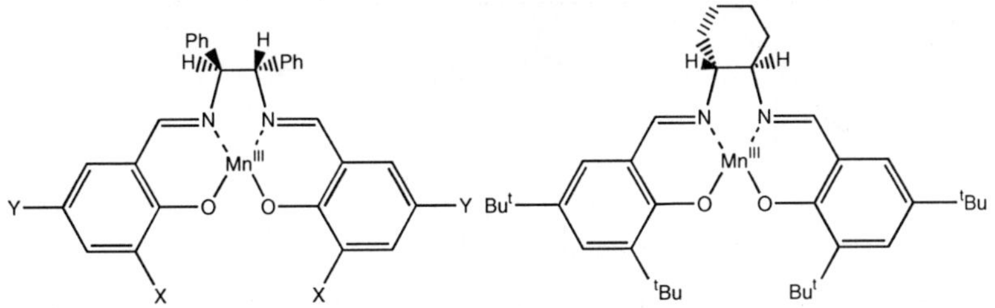

Figure 2.20   *Jacobsen's manganese Schiff-base complexes.*

Section 4) which immediately implies a large number of potential applications in oxidation.

Titanium complexes in aqueous solution are used to determine hydrogen peroxide in analytical chemistry but are not very active as oxidants. There is, however, a very important heterogeneous catalyst system based on titanium peroxo chemistry, which will be more fully discussed in Chapter 4, and is only briefly referred to here. The catalyst is called titanium silicalite (TS-1) and was discovered by Enichem.[64–66] The silicalite has a ZSM-5 structure containing small amounts (*ca.* <2% *m/m*) of titanium substituted for the silicon in the framework. It is produced hydrothermally from tetraethylorthosilicate and titanium(IV) isopropoxide in the presence of a tetrapropylammonium hydroxide templating agent. The material catalyses a wide range of oxidations (Figure 2.21).[67–72] Applications are, however, limited by the size of the zeolite channel (5.5 Å), since the substrate must pass down these in order to reach the active titanium sites. Work on analogues of TS-1 using larger pore zeolites is an active area of current research.[73]

The synthesis of peroxo metal complexes with hydrogen peroxide is easily accomplished, normally by one of two methods. Firstly, using early transition

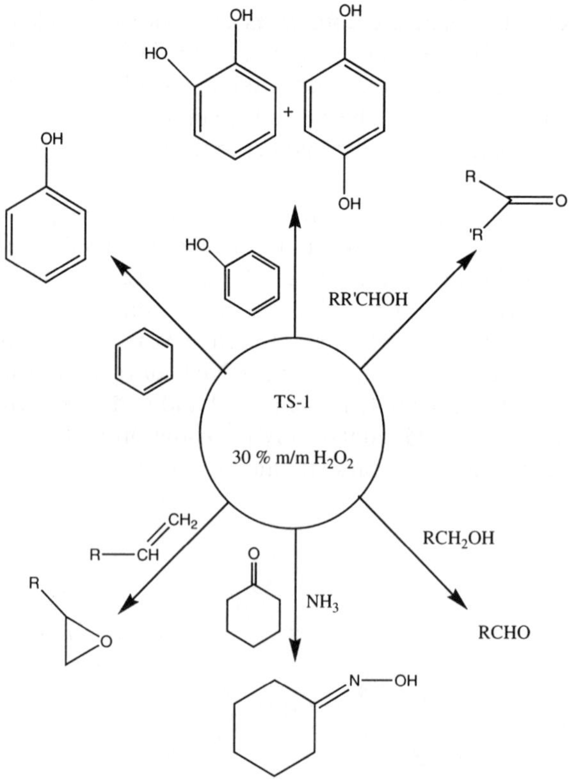

Figure 2.21   *Oxidative transformations catalysed by TS-1.*

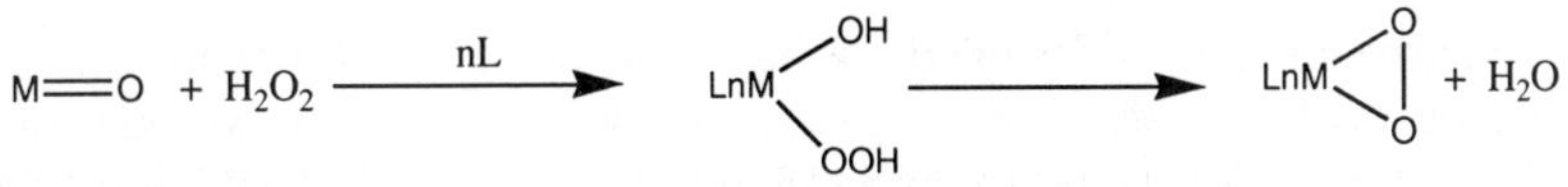

Figure 2.22 *Preparation of peroxometal complexes.*

metals; the metal oxides are dissolved in aqueous hydrogen peroxide, often in the presence of ligands (Figure 2.22). The ligands can be organic, inorganic or a mixture of both.

This method of preparation was originally reported by Mimoun and co-workers.[74] The intermediate hydroperoxy complex is rarely isolated, especially in the presence of hydrogen peroxide, however if an alkylperoxide is employed, then the ring closure reaction does not take place and alkylperoxo species are formed.[75] The triangular metal oxirane species formed is of immense importance for the purpose of oxygen transfer to organic substrates.

The second method often employed for the preparation of peroxo complexes involves the use of the late transition metals[76] and exploits the mild acidic properties of hydrogen peroxide, *i.e.* the reaction is an acid–base exchange (Figure 2.23).

The latter method has the drawback that it can only be applied successfully to a limited number of compounds, such as $[(RCO_2)Pd(OOH)_2]$ ($R = CF_3, CH_3$)[77] and $[(CH_3CO_2)Cu_2(OH)_2(OOH)]_n$.[78] Alternative synthetic procedures leading to group VIII metal hydroperoxy complexes involve either oxygen insertion into a metal hydride bond[79] or protonation of easily accessible metal oxirane complexes with strong acids.[80]

The vast majority of the d^0 complexes formed from the metal oxides and hydrogen peroxide exhibit pentagonal bipyrimidial geometry around the central metal with the peroxo group co-ordinated side-on. However, the two M–O bond distances are slightly different, which is believed to be one reason why oxygen transfer reactions are relatively facile. Figure 2.24 illustrates the geometry for a vanadium dipicolinc acid complex.

$$LnM\text{---}X + H_2O_2 \rightleftharpoons LnM\text{---}OOH + HX$$

Figure 2.23 *Preparation of hydroperoxy complexes.*

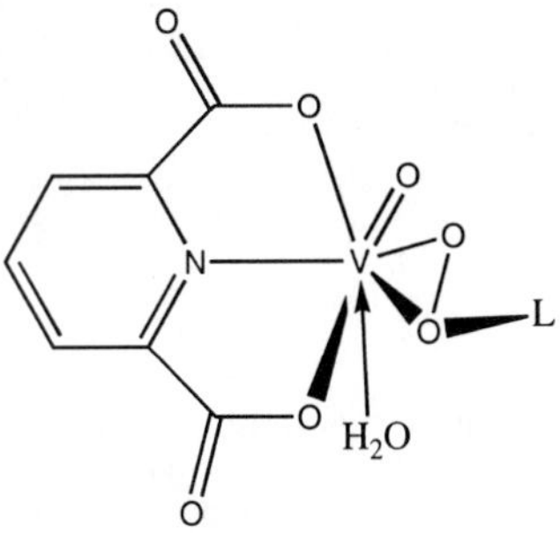

Figure 2.24 *Geometry of $VO(dipic)(O_2)(H_2O)$.*

The peroxo complexes of vanadium have not, by comparison with the other three elements (Ti, W, Mo) cited, been extensively employed for oxygen transfer reactions. The ease of the redox step vanadium(V) to vanadium(IV) introduces a mixture of two-electron and one-electron character into vanadium peroxo chemistry, which in the case of alkene epoxidation leads to side reactions of the substrate and products.[81]

The most typical peroxo complexes showing clear electrophilic behaviour in the oxidation of a variety of organic substrates are those of molybdenum and tungsten, the list of which is extensive.[82] Many of these can be isolated and used stoichiometrically as oxidants in organic solvents. However, such isolation can be hazardous, and stoichiometric use on a large scale would require a high metal inventory even though re-cycle is relatively easy. Of more industrial relevance is the catalytic use of these complexes, usually in a two-phase system, with a phase transfer agent. In such a situation, the peroxo complex is formed in the aqueous phase from hydrogen peroxide and a catalytic amount of the metal(VI) compound; the peroxo complex is then taken by the phase transfer agent into the organic phase in which the substrate is dissolved. After oxidation, the oxo complex is regenerated in the aqueous phase. Such systems epoxidize alkenes and oxidize organo-nitrogen and sulfur centres. Simple molybdate and tungstate salts can be the metal source, but more powerful systems are formed by the addition of phosphate or arsenate or by the use of pre-formed polyoxometallate species such as the Keggin dodecatungstophosphate ion (Figure 2.25).[83]

Both of these species are believed to function through an intermediate known as the Venturello complex (Figure 2.26);[84] one of the peroxo oxygens is

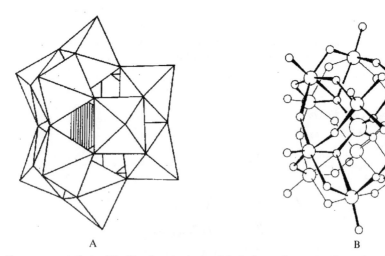

A   B

Two representations of the Keggin unit: A = polyhedral, metal atoms are located at the centres of the octahedra. B = ball and stick, where the large circles represent the metal atoms, and the heteroatom resides in the centre. The small circles denote the oxygen atoms. The structure contains essentially four types of oxygen i.e. those bordering the heteroatom, those at the outside of the structure, those forming bridges in M_3O_{10}, and those forming bridges between M_3O_{10} groups.

Figure 2.25 *Keggin structure of heteropolyperoxometallates.*

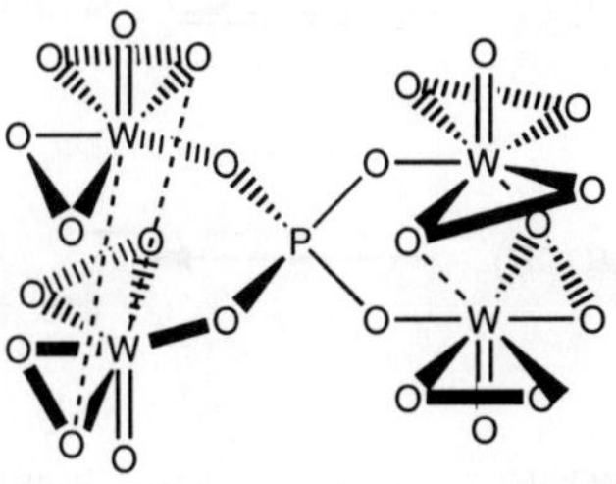

Figure 2.26 *Venturello complex.*

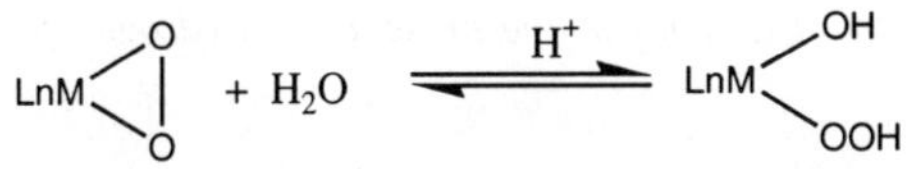

Figure 2.27 *Addition of water across a peroxo metal complex.*

activated towards electrophilic transfer by the non-bonded interaction with an adjacent metal centre.

The metal peroxo complexes are equivalent through addition of water across the O–O bond to form hydroxo hydroperoxo species (Figure 2.27).

In the cases of tungsten and molybdenum, the complexes are believed to be intermediates in the oxidation of alcohols to carbonyl compounds (a hydride abstraction reaction).[85] Figure 2.28 illustrates some of the oxygen transfer

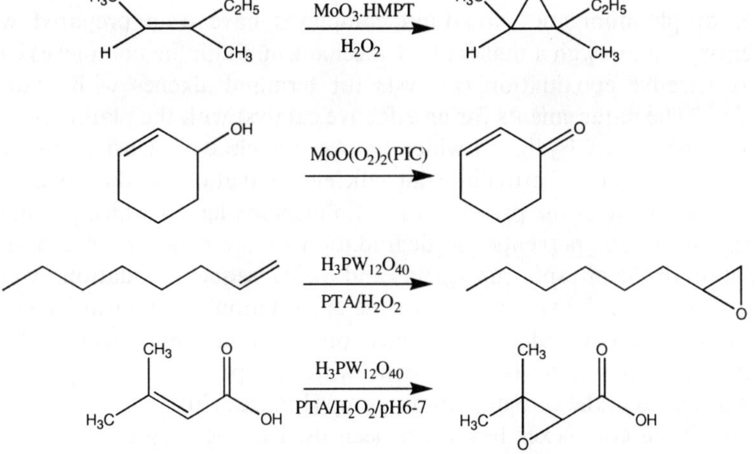

Figure 2.28 *Examples of oxygen transfer reactions of hydrogen peroxide in the presence of molybdenum(VI) and tungsten(VI) complexes.*

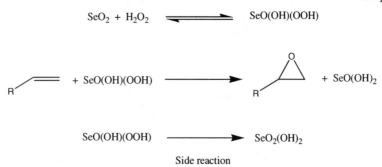

Figure 2.29   *Epoxidation of alkenes in the presence of hydrogen peroxide and selenium oxide.*

reactions which can be carried out with hydrogen peroxide in the presence of tungsten and molybdenum catalysts.

It is well known that other non-metal oxides can react with hydrogen peroxide to form similar compounds which can be viewed as inorganic peracids. Such species include boron(III), arsenic(III) and selenium(IV). For example, selenium dioxide can be used as a catalyst for epoxidations or amine oxidations through perselenous acid (Figure 2.29).[86]

These catalytic oxidations, unlike those with molybdenum and tungsten, require the continuous removal of water by azeotrope to regenerate the active oxidant. In addition, selenium and arsenic are highly toxic, and their total containment within the process presents some difficulties. Although organic resin-bound analogues have been reported,[87] it is still believed that the molybdenum and tungsten catalytic chemistry is much more amenable to current and future use in clean industrial processes.

Before leaving this section, it is worth mentioning an important class of peroxo complexes which are produced from d^8 transition metal species. A number of platinum and palladium complexes have been prepared which transfer oxygen through a nucleophilic mechanism. Platinum complexes can be used as selective epoxidation catalysts for terminal alkenes with hydrogen peroxide.[88] The requirements for an effective catalyst with the platinum system is the presence of a CF_3 ligand which reduces the electron density around the platinum centre, hence providing an efficient activation of the alkene. The complex also requires the presence of a diphosphine ligand which, because of the chelating effect, prevents the degradation of the catalyst, thus avoiding dissociation of the phosphorus ligands and the subsequent oxidation by excess hydrogen peroxide.[89] Investigation of the epoxidation mechanism by kinetics has found a second order dependence on platinum concentration.[90] This behaviour is believed to be consistent with the presence of two different platinum centres in the rate-determining step (Figure 2.30).

The platinum complexes have also been used in the presence of hydrogen peroxide to prepare cyclic lactones from cyclic ketones (Figure 2.31).[91]

It should be noted that the d^8 systems being nucleophilic in nature comple-

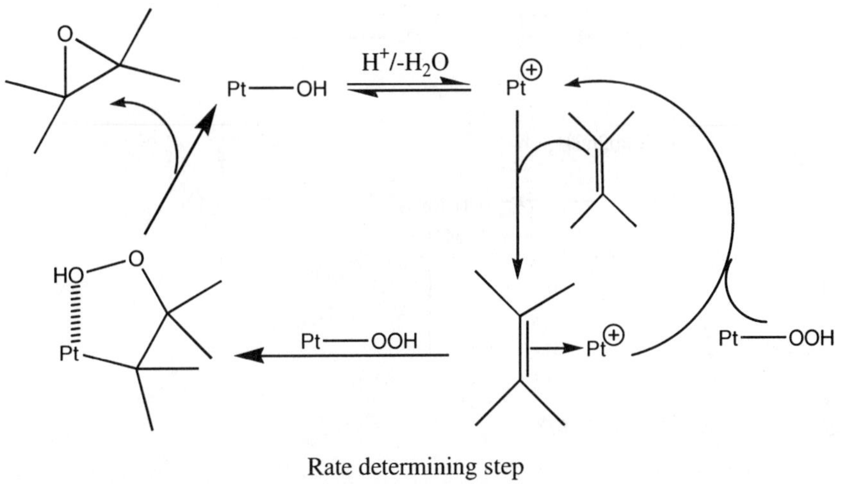

Rate determining step

Figure 2.30 *Possible mechanism of epoxidation with hydrogen peroxide in the presence of Strukel catalysts.*

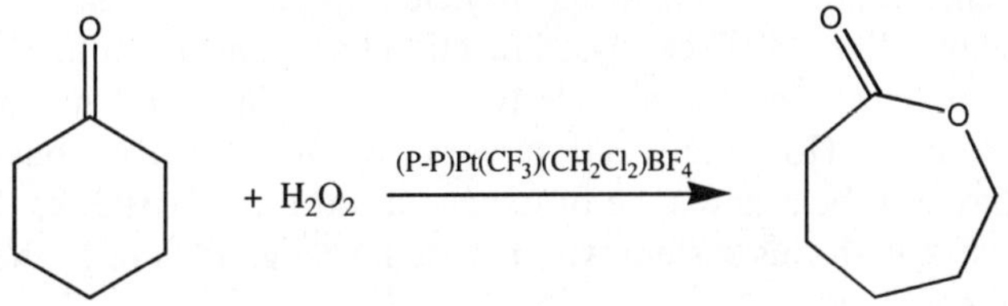

Figure 2.31 *Oxidation of cyclohexanone in the presence of a Strukel catalyst.*

ment the d^0 peroxo metal complexes which are electrophilic oxygen transfer agents. Consequently, the d^8 systems work better in the presence of electron deficient substrates, and the d^0 better with electron-rich materials.

Finally, a mention should be made about the one peroxo system which will become more and more dominant: the organometallic oxides of rhenium(VII). Such compounds have been found to be of outstanding catalytic activity for a number of oxygen transfer reactions with hydrogen peroxide.[92] The best studied complex is methyltrioxorhenium(VII) (MTO) and its congeners. Figure 2.32 illustrates its synthesis. Epoxidation, aromatic oxidation and halide oxidation with these complexes have been studied with hydrogen peroxide and shown to be remarkably efficacious.

A remarkable oxygen transfer reaction of MTO to alkenes in the presence of hydrogen peroxide was recently discovered by Sharpless and co-workers at Scripps Research Institute in La Jolla, California.[93] The reaction is the first catalytic reaction for efficient epoxidations under non-acidic conditions. The reaction uses aqueous hydrogen peroxide as the oxidant and MTO in the presence of pyridine ligands as the catalyst. The reaction converts concentrated solutions of alkenes in dichloromethane to epoxides at room temperature with

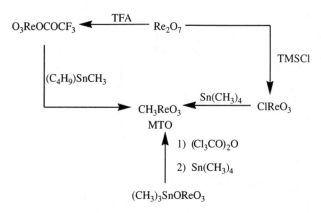

Figure 2.32   *Preparation of MTO.*

yields up to 99%. The new reaction efficiently epoxidizes non-terminal alkenes. Even the extremely sensitive epoxides from α-methylstyrene, indene and phenylcyclohexene are isolated in high yields (92–99%, Figure 2.33), although terminal alkenes are only epoxidized in yields up to 80% even after days of reaction. The pyridine–MTO catalysed reaction has many advantages including reactivity, robustness, inexpense, relatively safe oxidant and no undesirable secondary products. How the reaction works is still a mystery, but Sharpless believes it may well be a new type of oxygen atom transfer species. It is likely that future work with this system will include investigations in to the preparation of chiral epoxides.

The subject of catalytic oxidation with hydrogen peroxide is an area in which many exciting new developments are taking place. The driving force behind these investigations is the need for cleaner processes, particularly in fine chemical manufacture, and shorter routes with higher product selectivity. In

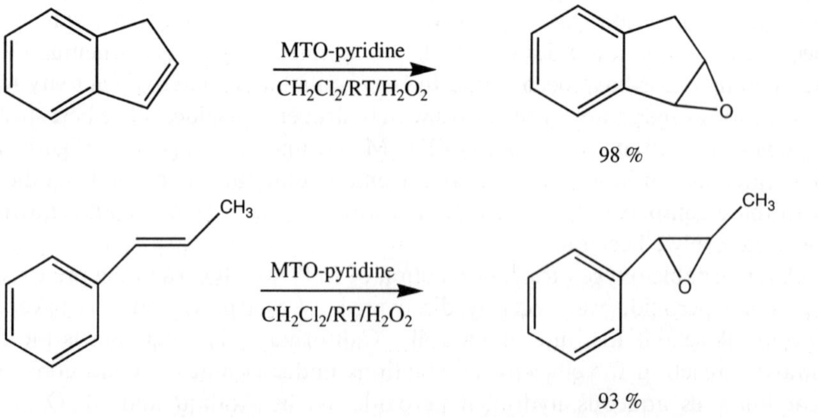

Figure 2.33   *Non-acidic oxygen transfer reactions of MTO–pyridine in the presence of hydrogen peroxide.*

$$H_2SO_4 + H_2O_2 \rightleftharpoons H_2SO_5 + H_2O$$

Caro's acid

$$2HX + H_2O_2 \longrightarrow X_2 + 2H_2O \qquad X = Cl, Br, I$$

Halogen acid activation

Figure 2.34 *Activation of hydrogen peroxide in the presence of sulfuric acid and halogen acid.*

particular, the design of new heterogeneous catalysts for liquid-phase oxidation will play an important role in these developments. This is not only because they offer the advantage of ease of recovery, but because they also offer the possibility of designing site-isolated redox metal catalysts, similar to TS-1: unique substrate chemio-, regio- and stereo-selectivities.

Two other hydrogen peroxide activation routes which need to be mentioned in this section are Caro's acid and the generation of halogens with hydrogen peroxide. The general schemes for their manufacture are shown in Figure 2.34.

Caro's acid is the name given to products obtained when strong sulfuric acid and hydrogen peroxide are mixed. The active oxygen transfer species in the system is peroxymonosulfuric acid (H_2SO_5). Equilibration is rapid at concentrations in the g dm^{-3} region or greater, and the reaction that occurs is exothermic (largely due to the heat of dilution of sulfuric acid and water). The amount of peroxomonosulfuric acid present in any mixture can be readily predicted from the initial composition.

Peroxomonosulfate for synthesis applications can be derived from the triple salt ($2KHSO_5 \cdot KHSO_4 \cdot K_2SO_4$). While this remains a convenient, though relatively expensive, solid source for small-scale use, technology for controlled generation of Caro's acid is now available.[94] In general, the properties of peroxomonosulfate systems in synthesis are similar to a mixture of percarboxylic acids and mineral acid (see Section 4). One important difference is their use in the generation of dioxiranes, where they are uniquely suitable. A drawback of the peroxomonosulfate system is the co-production of sulfate. The sulfate may, however, be re-cycled in large scale application.

In strong acidic aqueous solutions, halogen acids, except for hydrogen fluoride, are converted stoichiometrically to the free halogen by hydrogen peroxide. In halogen substitution reactions, only half of the free halogen is used, and the other half ends up as waste ($X_2 + RH \rightarrow RX + HX$). However, when hydrogen peroxide is employed, the HX by-product can be re-cycled and reacted with hydrogen peroxide to produce the free halogen. Such recycle has three benefits: avoidance of acid/salt effluent from the process, protection of acid sensitive products and, for bromine and iodine, a cost saving. Figure 2.35 shows how the halogen generation process can be operated either via *in situ* generation or *ex situ* generation.[95] The hydrogen peroxide/hydrogen bromide system is particularly convenient, and its uses are described more fully in Chapter 3.

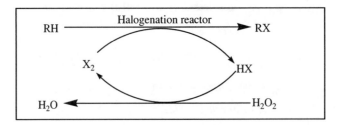

In-situ halogenation

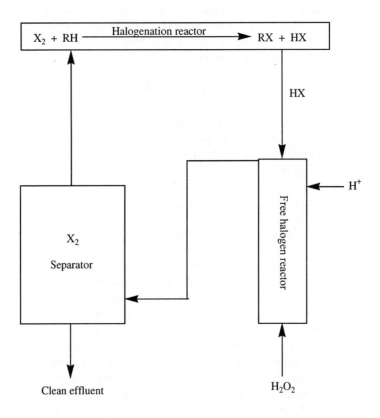

Ex-situ generation of X_2

Figure 2.35 In situ *and* ex situ *generation of free halogen for organic halogenation applications.*

In addition, the use of hydrogen peroxide and the halogen acid as primary reactants avoids the need for free halogen storage on-site, which is not always allowed or desirable. For greatest efficiency, the system must be operated at low pH, as these conditions minimize any side-reactions (outlined in Figure 2.36).

$$X_2 + H_2O \xrightleftharpoons[H^+]{OH^-} HOX + HX$$

$$H_2O_2 + HOX \rightleftharpoons H_2O + HX + O_2$$

Figure 2.36 *Competing side reactions during the preparation of free halogen from hydrogen peroxide and hydrogen halide.*

4 Activation of Hydrogen Peroxide in the Presence of Organic Compounds

Activation of hydrogen peroxide can occur by interaction with an organic compound. The following section will discuss this topic by reference to the following system types:

- Percarboxylic acids, diacyl peroxides and peresters.
- Alkyl hydroperoxides and dialkyl peroxides.
- Percarboximidic acid.
- Dioxiranes.
- Miscellaneous.

Currently, percarboxylic acid compounds are generally the main species employed for activating hydrogen peroxide towards electrophilic oxidation, such as epoxidation, Baeyer–Villiger and heterocyclic *N*-oxidation. Percarboxylic acids are usually prepared via an equilibrium reaction (Figure 2.37).

For most carboxylic acids, a strong acid must be added to achieve an acceptable equilibrium rate. Sulfuric, sulfonic or phosphoric acids can be used, although there are a few carboxylic acids strong enough to catalyse their own peracid formation, notably formic and trifluoroacetic acid. Peracids themselves are weakly acidic, with most pK_a values in the range 7.5–8.5.[96] For example, peracetic acid has a pK_a of 8.2 compared with acetic acid at 4.8. The two peracids most relevant to the chemical industry are performic (HCO_3H) and peracetic (CH_3CO_3H). Performic acid is always generated *in situ* via the reaction shown in Figure 2.37, as it is neither sufficiently stable nor safe to isolate as an equilibrium mixture. Peracetic acid can, however, be used in a variety of forms including *in situ* generation, and pre-formed equilibrium mixtures, distilled aqueous solutions and solvent-extracted products.

In situ generation is a convenient method used for substrates which can tolerate the acidity required to catalyse the equilibrium. If a faster reaction is required, this can be achieved by using the strongest pre-formed products, which typically contain 35–40% *m/m* peracetic acid with about 45% *m/m*

$$CH_3CO_2H + H_2O_2 \rightleftharpoons CH_3CO_3H + H_2O$$

Figure 2.37 *Equilibrium present when hydrogen peroxide reacts with a carboxylic acid.*

acetic acid and $< 15\%$ *m/m* water, $< 5\%$ *m/m* residual hydrogen peroxide and
1% *m/m* sulfuric acid or less. The acidity of the sulfuric acid can be suppressed
by addition of sodium acetate to the reaction mixture, and the effects of the free
acetic acid can be moderated by addition of bases such as sodium carbonate,
sodium hydrogenphosphate, or magnesium hydroxide. For highly acid-sensitive
reactions (such as epoxidations where the epoxide product is prone to ring
opening or Baeyer–Villiger oxidation of cyclic ketones where the lactone is
prone to ring opening), distilled peracetic acid may be used. Under a range of
controlled conditions, equilibrium mixtures can be distilled to give a product
containing essentially only peracetic acid ($> 30\%$ *m/m*) and water. This technology has been proven on a plant scale[97] and lends itself to the re-cycle of acetic
acid. Peracetic acid can also be extracted into organic solvents such as ethyl or
isopropyl acetates to give organic solutions containing over 20% *m/m* peracid.
These solutions are useful in the oxidation of water-sensitive substrates.

The *in situ* generation of performic and peracetic acids has an important
application in producing epoxidized soya bean oil (ESBO), a plasticizer and
stabilizer. The acetic acid is often re-cycled. Technology for the manufacture of
propylene oxide[98] and of epichlorohydrin[99] using internal cycle systems for
peracetic or perpropionic acids has also been developed.

Peracids can also be prepared from reaction of hydrogen peroxide with acyl
halides, anhydrides, amides, dialkyl phosphates, N-acylimidazoles, aromatic
aldehydes, lipase catalysis and esters (Figure 2.38).[100–107]

Tetraacetylethylenediamine (TAED) and pentaacetylglucose (PAG) are
employed as bleach pre-cursors of peracetic acid; this allows the bleach to
occur at a temperature lower than when hydrogen peroxide is used. Perhydrolysis of the sulfonated aromatic ester, shown in Figure 2.38, gives superb bleach
performance on textiles, probably due to the surface active properties of the
compound. The acyl dialkyphosphates and the N-acylimidazoles are also
effective peracetic acid precursors at low temperature.

Lipase has been employed to prepare peracids, usually *in situ*.[108] The lipase
catalysed peracid production from carboxylic acids has been used for the mild
epoxidation of alkenes.[109] A number of immobilized lipases exist, including one
from *Candida Antarctica* (Novozym® 435), which catalyses the conversion of
fatty acids to peroxy fatty acids.[110]

Pre-formed peracids used in the laboratory include 3-chloroperbenzoic
acid,[111] which is long established in synthesis, and magnesium monoperoxyphthalate.[112] Polymer-supported percarboxylic acids have also been prepared
for organic synthesis purposes (Figure 2.39).[113]

Percarboxylic acids can be used for the epoxidation of alkenes,[114]
acetylenes,[115] the hydroxylation of alkanes,[116] and the oxidation of alcohols,[117]
and of sulfur[118] and nitrogen compounds.[119]

Figure 2.40 illustrates some of the methods which can be employed
for the preparation of diacyl peroxides [$(RCO_2)_2$], both symmetrical and
asymmetrical.[120–122]

Under normal circumstances, it is best to avoid the formation of diacyl
peroxides, because they are particularly hazardous. It is important to note that

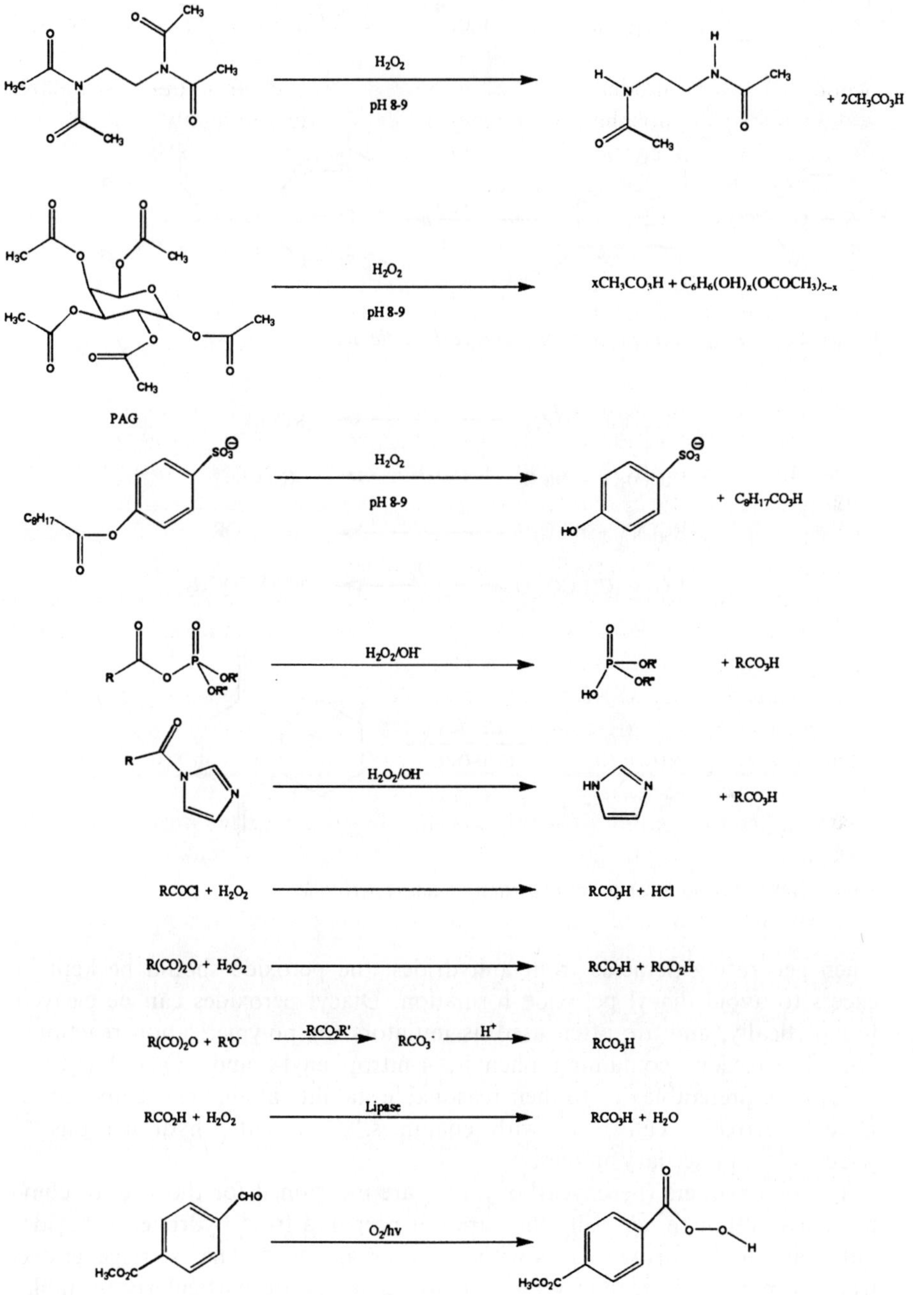

Figure 2.38 *Methods for the preparation of percarboxylic acids.*

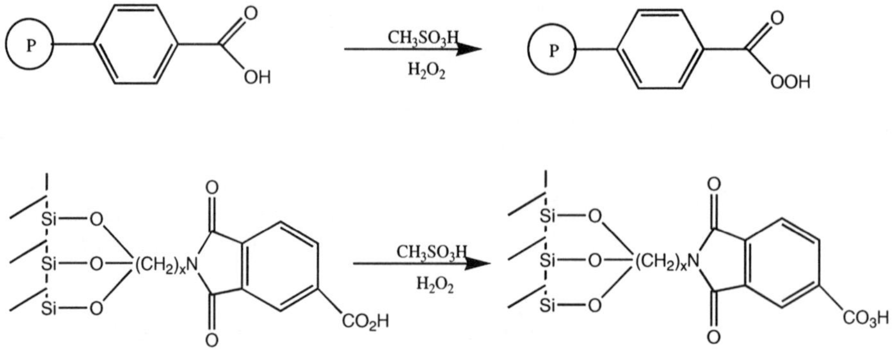

Figure 2.39 *Preparation of supported percarboxylic acids.*

$$2RCOCl + H_2O_2 \longrightarrow (RCO_2)_2$$

$$(RCO)_2O + Na_2O_2 \longrightarrow (RCO_2)_2$$

$$RCOCl + R'CO_3H \longrightarrow RCO_3COR'$$

$$RCHO + (CH_3CO)_2O \xrightarrow{O_2} RCO_3COCH_3$$

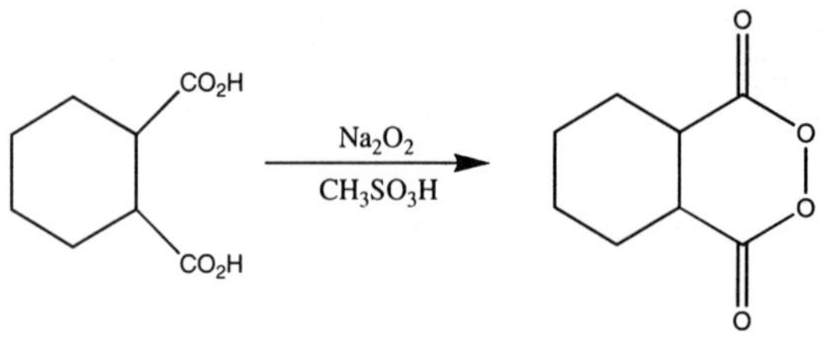

Figure 2.40 *Methods for the preparation of diacyl peroxides.*

when generating peracids from anhydrides, the peroxide should be kept in excess to avoid diacyl peroxide formation. Diacyl peroxides can be cleaved homolytically, and are often used as initiators for polymerization reactions. Diacyl peroxides containing phenyl-, 4-nitrophenyl-, and $-OC(CH_3)_2CCl_3$ groups are preferable due to their reasonable stability at ambient temperature. Diacyl peroxides can react with enamines,[123] aromatic hydrocarbons,[124] phenols,[125] and tertiary amines.[126]

The peroxyesters (peroxycarboxylates) are mentioned for the sake of completeness, although generally they are not prepared from hydrogen peroxide, but from alkyl hydroperoxides with acylating agents.[127] The esters generated from primary and secondary alkyl hydroperoxides are particularly unstable, and decompose easily to give carbonyl compounds and carboxylic acids. The tertiary alkyl peroxycarboxylates are relatively more stable. However, again it is advisable to avoid their generation. Their reaction chemistry is similar to that of diacyl peroxides.

$$H_2O_2 + OH^- \longrightarrow HOO^- + H_2O$$

$$HOO^- + R_3CX \longrightarrow R_3CO_2H + X^-$$

Figure 2.41 *Preparation of alkyl hydroperoxides.*

Dialkyl peroxides, like diacyl peroxides and peroxyesters, are characterized by homolysis of the O–O bond, which is promoted thermally, photochemically or by transition metal catalysis. The combination of steric factors, and the poor leaving group ability makes simple dialkyl peroxides (ROOR) almost unreactive in heterolytic oxygen atom transfer.[128] Consequently, no further mention of these species will be made.

Alkyl hydroperoxides (ROOH), on the other hand, are more distinctly reactive. They are generated via nucleophilic displacement of suitable substrates by hydrogen peroxide in alkaline media (Figure 2.41).

The above method can also be used for the preparation of primary and secondary hydroperoxides.[129] In addition, the tertiary alkyl hydroperoxides can be generated from the perhydrolysis of alcohols using concentrated ($> 50\% \, m/m$) hydrogen peroxide.[130] Figure 2.42 summarizes a range of transformations which can be carried out in the presence of alkyl hydroperoxides.

Generally, most oxygen transfer reactions employing alkyl hydroperoxides require transition metal activation, since the alkoxide anions are relatively poor leaving groups, even poorer than the hydroxide anion. The reasons why they are sometimes employed in preference to hydrogen peroxide are as follows:

- Some alkyl hydroperoxides, such as TBHP (*t*-butyl hydroperoxide), are less sensitive to contamination by metals compared to hydrogen peroxide and percarboxylic acids. However, with the increased knowledge of hydrogen peroxide stabilization, this point is somewhat negated.
- Alkyl hydroperoxides are readily soluble in hydrocarbon solvents. This is an important advantage, as many of the solvents employed with hydrogen peroxide can retard the catalyst's activity.
- Oxidations with alkyl hydroperoxides are carried out at close to neutral conditions as possible, which makes them particularly useful towards the derivatization of acid-sensitive substrates.
- The alcohol formed after the oxygen transfer is complete may be readily removed by distillation.
- Some stereospecific epoxidations such as the Katsuki–Sharpless system only function with alkyl hydroperoxides.

When alkaline hydrogen peroxide is added to organonitriles, the hydrogen peroxide decomposes; this dismutation is known as the Radzizsewski reaction.[136] The initial product formed is a percarboximidic acid [RC(NH)OOH], which can react either with more hydrogen peroxide to liberate oxygen and the amide, or oxidize a substrate such as an olefin (Figure 2.43) in a similar way to a percarboxylic acid.

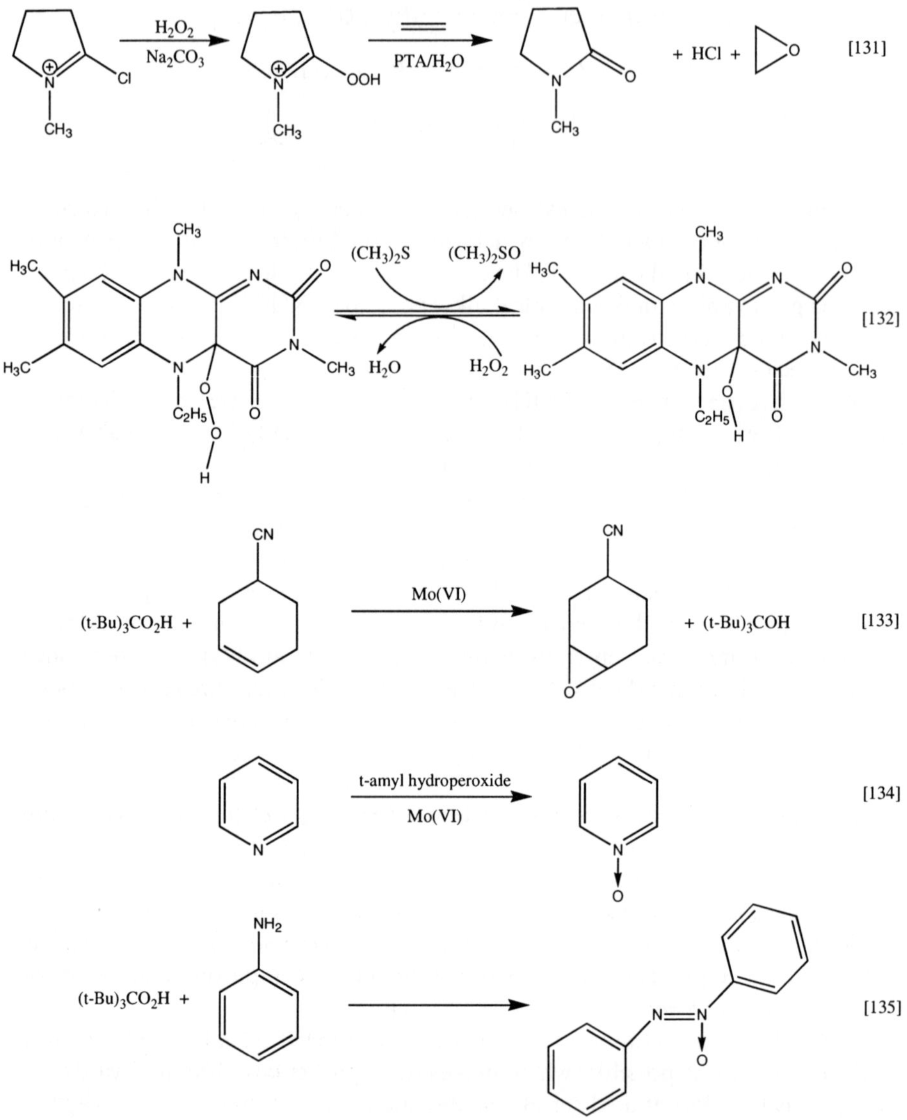

[131]

[132]

[133]

[134]

[135]

Figure 2.42   *Oxygen transfer to organic substrates using alkyl hydroperoxides.*

This system was discovered around 1960, and was first described by Payne.[137] The method can be used for the mild epoxidation of a range of substrates.[138] It is a relatively safe and inexpensive route, and avoids the use of peroxycarboxylic acids. Trichloroacetonitrile (Cl$_3$CCN) and hydrogen peroxide afford a highly efficient oxygen atom transfer agent for epoxidations.[139] Substrates containing carbonyl groups will not undergo Baeyer–Villiger oxidation with Payne's reagent.[140] The amide by-product of these reactions, which is usually precipitated and hence easily separated, makes the system convenient for small-scale

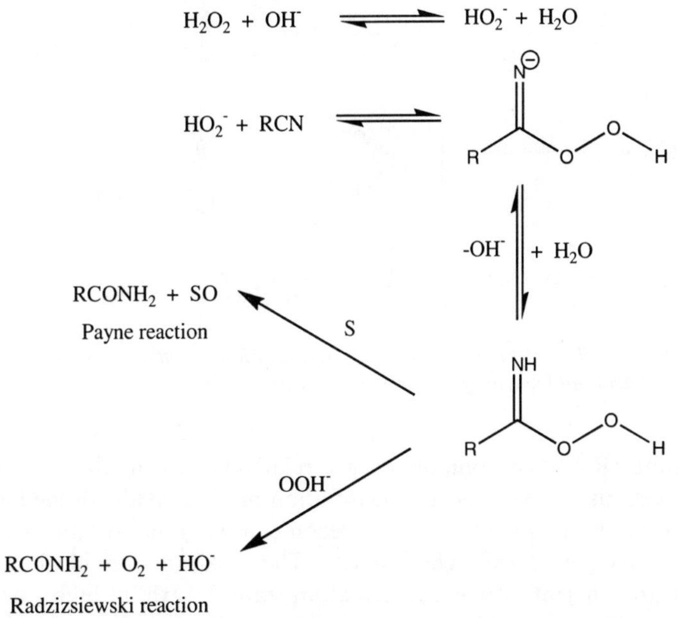

$$H_2O_2 + OH^- \rightleftharpoons HO_2^- + H_2O$$

$$HO_2^- + RCN \rightleftharpoons$$

RCONH$_2$ + SO

Payne reaction

S

-OH$^-$ + H$_2$O

RCONH$_2$ + O$_2$ + HO$^-$

OOH$^-$

Radzizsiewski reaction

Figure 2.43 *Possible mechanism for the formation of peroxyimidic acid from the addition of hydrogen peroxide to organic nitriles.*

synthesis. However, the stoichiometric production of the amide is a disadvantage which has limited the industrial use of the system. In principle, acetamide can be dehydrated back to acetonitrile using acid catalysts, but this is relatively difficult. In spite of the drawbacks, it is believed that a hydrogen peroxide/acetonitrile process for epoxidation has been used on an industrial scale.[141]

Isocyanates (R—N=C=O) as well as dialkylcarbodiimides (R—N=C=N—R) can also serve in activating hydrogen peroxide to perform epoxidations under mild conditions.[142,143]

The Payne system has been employed in the presence of antibodies to effect catalytic enantioselective epoxidation of unfunctionalized alkenes.[144] Interestingly, Chen and co-workers[145] have discovered that acetamide, at a slightly acidic pH in the presence of hydrogen peroxide, can be employed for the epoxidation of alkenes (Figure 2.44).[145]

The researchers proposed mechanisms for the reaction which involved either *in situ* formation of performic acid or peroxyimidic acid (Figure 2.45).

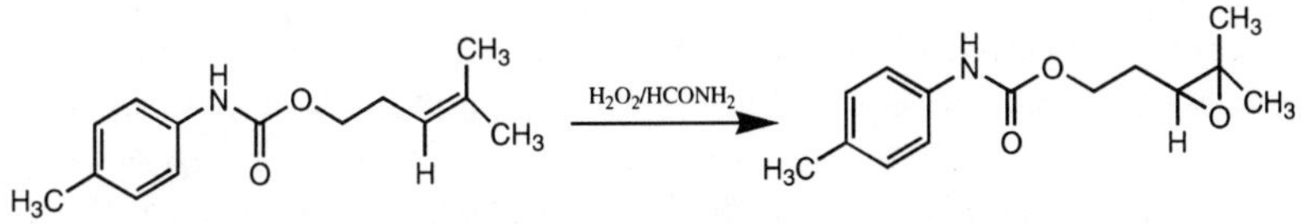

Figure 2.44 *Epoxidation of alkenes in the presence of acetamide–hydrogen peroxide.*

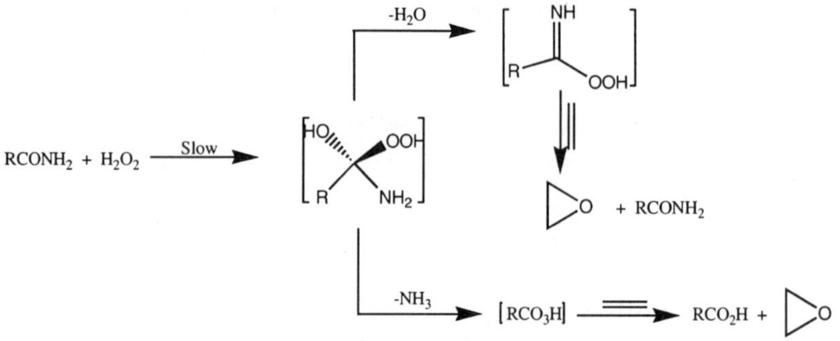

Figure 2.45   *Possible mechanisms for the epoxidation of alkenes in the presence of acetamide–hydrogen peroxide systems.*

Dioxirane ($RR'CO_2$) compounds are relatively new in the arsenal of the synthetic chemist, however since the isolation of dimethyldioxirane by Murray and Jeyaraman in 1985,[146] it has become a very important oxidant for preparative oxygen transfer chemistry.[147] The dioxiranes are ideal oxidants in that they are efficient in their oxygen atom transfer, exhibit high chemo- and regio-selectivities, act catalytically, are mild towards the substrate and oxidized product, and perform under strictly neutral conditions. The compounds are prepared from peroxymonosulfate and ketones under neutral to mildly alkaline conditions (Figure 2.46).

The most commonly employed ketone is acetone, however a more powerful alternative is methyl trifluoromethyl ketone.[148] The source of the peroxymonosulfate is either the triple salt ($2KHSO_5 \cdot KHSO_4 \cdot K_2SO_4$) or neutralized Caro's acid (nominally $NaHSO_5$).[149] The scope of the method is limited by the tendency of many ketones to undergo Baeyer–Villiger oxidation in the presence of dioxirane formation (Figure 2.47).

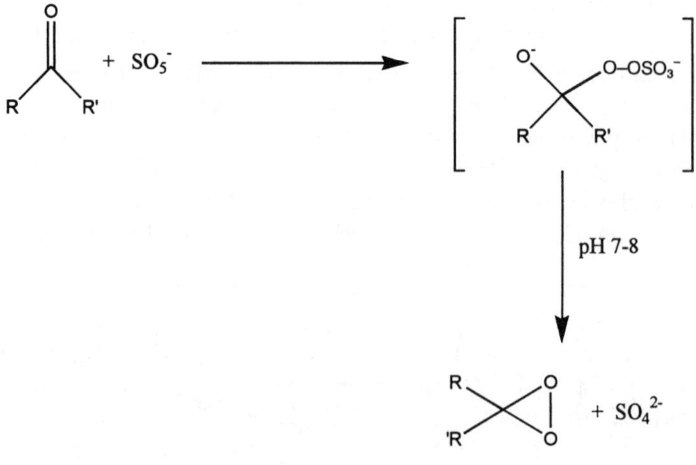

Figure 2.46   *Preparation of dioxiranes from peroxymonosulfate and ketones.*

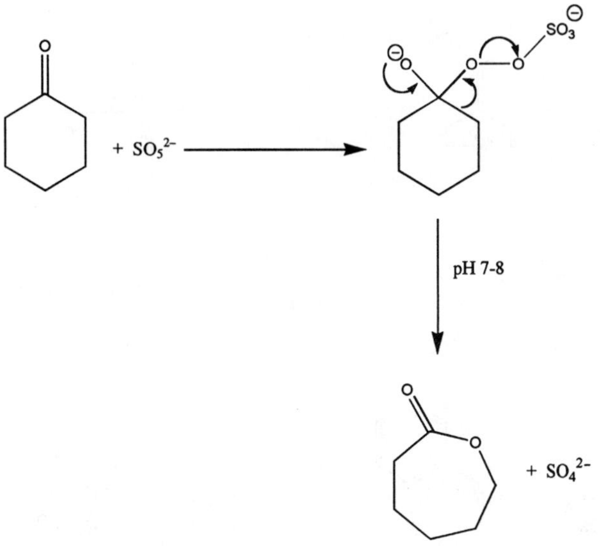

Figure 2.47 *Baeyer–Villiger oxidation of cyclohexanone in the presence of peroxomonosulfate system.*

Dioxiranes can either be used *in situ* or *ex situ*. The peroxymonosulfate used to generate the dioxirane forms a stoichiometric amount of sulfate, which is clearly a problem for re-cycle or disposal and has hindered its application on a large scale. Despite this problem, dioxiranes have attracted interest as selective delignification agents for pulp and paper, and it is claimed that the sulfate problem, on this very large scale, can be solved.[150] Figure 2.48 demonstrates some of the oxygen atom transfer transformations which can be carried out using dimethyldioxirane.

Recently, the question as to what the active oxidant is in ketone–peroxymonosulfate epoxidation has been raised by Armstrong.[151] Armstrong carried out a set of ^{18}O-labelled experiments in an attempt to elucidate this question. The work indicated that no transfer of the [18]O label to the epoxide was observed (Figure 2.49).

The explanation for these experimental results, *i.e.* the lack of label transfer, is that the tetrahedral species (A) resulting from the addition of HSO_5^- to the carbonyl group is capable of epoxidation. Ring closure of (A) is likely to be the rate-determining step in dioxirane formation. This work is important from a synthetic viewpoint, since it is crucial in the development of chiral ketones for the catalytic asymmetric epoxidation and the design of probes of transition state stereoselectivities that the nature of the oxidizing species is understood.

The discussion so far has been in relation to bimolecular electrophilic oxidation by hydrogen peroxide via a two-electron donor substrate. However, the oxidation process can also be envisaged as a single electron transfer process (SET). The two mechanisms are not that dissimilar[152] and can be viewed as two sides of the same coin. The bimolecular mechanism can be visualized as an

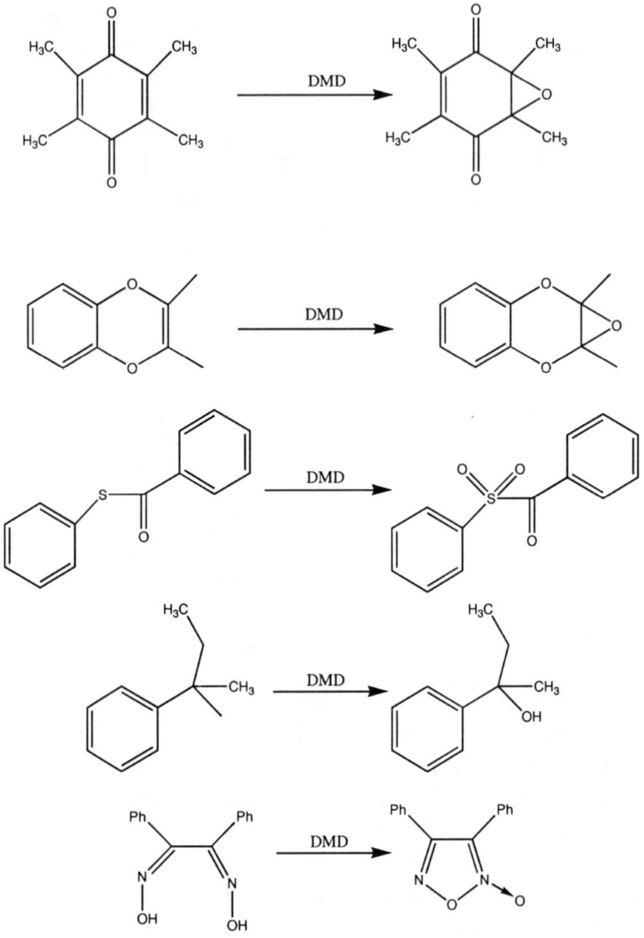

Figure 2.48 *Oxygen atom transfer ability of DMD with different substrates.*

inner-sphere mechanism displacing the peroxygen bond, accompanied by ligand transfer, with one of the OH groups of the peroxide having the role of the bridging ligand, whilst the SET mechanism is substantially an outer-sphere process. The concept of SET mechanisms playing a role in a number of bimolecular processes is a popular one, and those interested in the subject should read the book by Eberson.[153] There are some reactions of organic substrates and hydrogen peroxide which clearly show SET-type mechanisms. For example, the oxidation of the radical cation 1,1-dimethyl-4,4'-bipyridinium dichloride is one such reaction.

Four-membered peroxides are an interesting group of compounds, and will be discussed briefly. Such compounds have the ability to yield excited state carbonyl fragments upon thermolysis;[154] in such cases, chemiluminesence is often observed. A classic illustration of this type of chemistry is the reaction of hydrogen peroxide with oxalyl derivatives, for example, when the 2,4-dinitro-

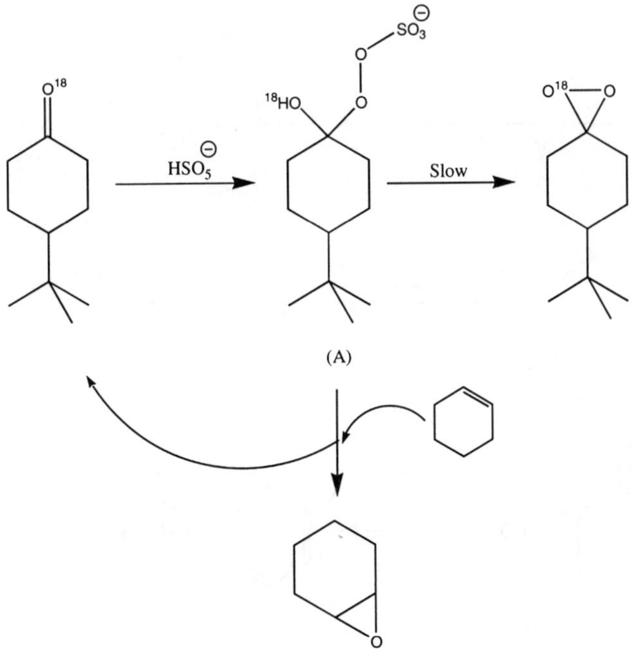

Figure 2.49 *Possible non-involvement of dioxiranes for the* in situ *epoxidation of alkenes.*

phenyl diester of oxalic acid is reacted with concentrated alkaline peroxide and 9,10-bis(phenylethylynl)anthracene. This system has some commercial application in the manufacture of 'chemical' torches.[155]

The three-membered peroxide systems produced from ketones in the presence of peroxymonosulfate have been discussed separately above because of their importance. There is, however, another three-membered ring peroxide which needs noting: the reaction of alkoxysulfuranes with anhydrous hydrogen peroxide. Such systems have been employed for the low temperature epoxidation of alkenes (Figure 2.50).[156]

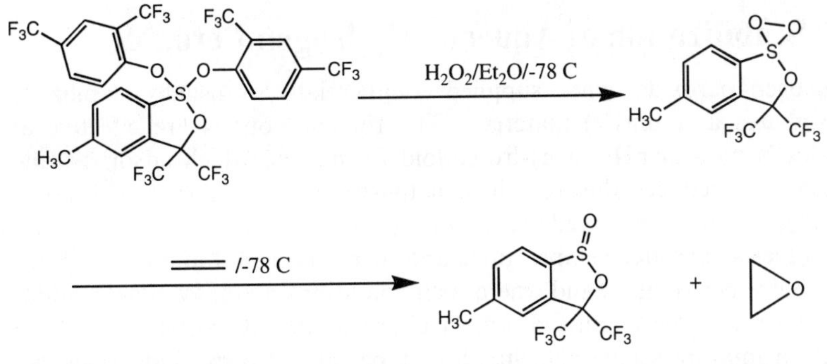

Figure 2.50 *Activation of hydrogen peroxide in the presence of alkoxysulfuranes.*

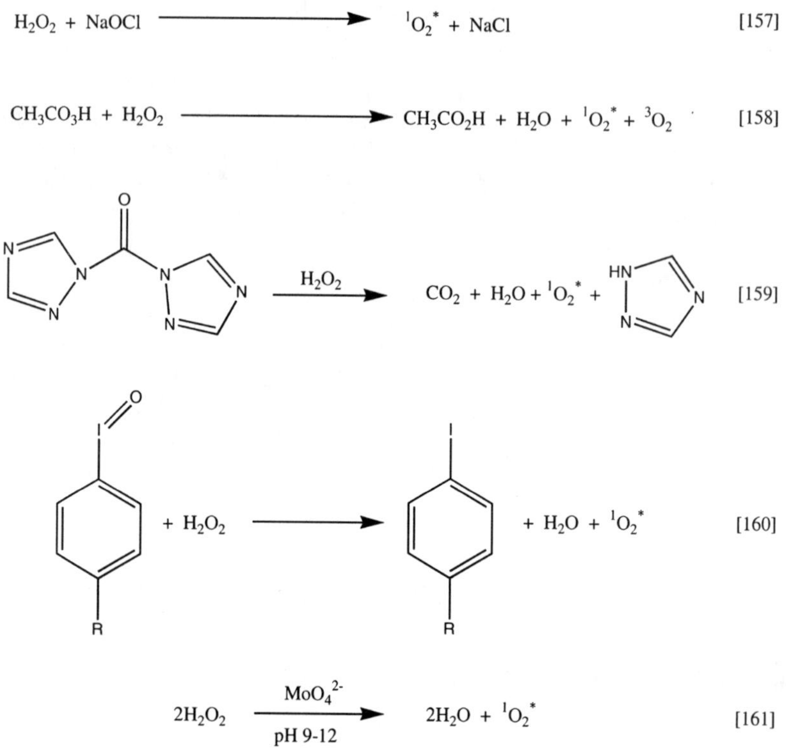

$$H_2O_2 + NaOCl \longrightarrow {}^1O_2{}^* + NaCl \qquad [157]$$

$$CH_3CO_3H + H_2O_2 \longrightarrow CH_3CO_2H + H_2O + {}^1O_2{}^* + {}^3O_2 \qquad [158]$$

$$\xrightarrow{H_2O_2} CO_2 + H_2O + {}^1O_2{}^* + \qquad [159]$$

$$+ H_2O_2 \longrightarrow \qquad + H_2O + {}^1O_2{}^* \qquad [160]$$

$$2H_2O_2 \xrightarrow[\text{pH 9-12}]{MoO_4{}^{2-}} 2H_2O + {}^1O_2{}^* \qquad [161]$$

Figure 2.51 *Methods for the generation of singlet molecular oxygen from hydrogen peroxide.*

Finally, the decomposition of hydrogen peroxide can, under non-radical pathways, give molecular oxygen in a singlet state $({}^1O_2, {}^1\Delta_g)$. There are a number of ways in which singlet molecular oxygen can be generated, these are illustrated in Figure 2.51.

Singlet oxygen can be used to oxidize a range of substrates, such as phenols, acrylics, steroids and terpenes.[162]

5 Stabilization of Aqueous Hydrogen Peroxide

Hydrogen peroxide, when supplied commercially, is usually stabilized with phosphates and tin(IV) materials. The tin compounds are effective at the product's natural pH via hydro-colloid formation, which adsorbs transition metals and reduces their catalytic activity. In the majority of cases, extra stabilization is not required when hydrogen peroxide or its derivatives are used in synthesis. Elevated temperatures and increased metal impurities all tend to destabilize peroxygens, and where such conditions are unavoidable, additional stabilizers may be employed, added either to the hydrogen peroxide or the reaction mixture separately. Stabilizer type falls into two categories: sequestrants and radical scavengers.

Polydentate ligands, usually aminocarboxylic acids (EDTA, DTPA, CDTA) and phosphonic acids (EDTMP, DTPMP, CDTMP), will sequester many transition metals and reduce their activity as decomposition catalysts, especially in alkaline systems. They work by both co-ordinative saturation, and by stabilizing the higher oxidation state of the metal, thus preventing its cycling back to the lower oxidation state. The aminopolycarboxylates are good short-term stabilizers for hydrogen peroxide, though DTPA is much better than EDTA against iron, since the latter does not co-ordinatively saturate iron(II). All tertiary nitrogen-containing sequestrants undergo N-oxidation with hydrogen peroxide over a period of days if added to 35% or 50% *m/m* products.[163] In the case of the aminocarboxylates, this drastically reduces the strength of the binding to the metals, hence stabilization is impaired, and the sequestrant itself is degraded. The phosphonates, on the other hand, remain strong chelating agents even after N-oxidation. Relative effects of sequestrants in alkaline hydrogen peroxide have also been examined.[164] A useful nitrogen-free sequestrant for hydrogen peroxide is 1-hydroxyethylidene diphosphonic acid (HEDP). In general, solutions containing both percarboxylic acids and free hydrogen peroxide are more prone to catalytic decomposition than either component alone, since the two peroxides themselves behave as a redox couple.[165]

The sequestrant approach to stabilization is usually better than the scavenger approach, as it is based on prevention rather than cure. It can, however, only act on metal-based catalytic decomposition, and is poor in strongly acidic systems.

Radical scavengers can be used in addition to the sequestrant approach, or as an alternative where sequestrants are not appropriate. Their use is based on conversion of very reactive radicals, produced during chain decomposition processes, to very stable radicals, hence stopping the chain mechanism. Simple alcohols can be effective stabilizer components, as are almost all aromatic compounds (commonly used as anti-oxidants in food or plastics), *e.g. p*-hydroxybenzoates, butylated hydroxytoluene, anisole, *p*-butylcatechol, gallates. All of these materials give relatively stable radicals on one-electron oxidation. Sequestrant N-oxides act as scavengers through one-electron oxidation to stable nitroxides.[166]

6 Conclusion

The author hopes that the chapter has shown how the reactivity, and selectivity towards oxygen transfer reactions of hydrogen peroxide can be utilized for synthetic purposes. Generally, for laboratory and industrial application, activation of hydrogen peroxide by transition metal ions is the method of choice, although in a number of cases, the transition metal route can lead to disposal and environmental problems. The reaction of hydrogen peroxide with organic compounds can, therefore, provide a viable alternative to metal ion activation.

The subject is highly relevant in the search for more effective, safe and economical activation methods, employing hydrogen peroxide as the primary oxidant source.

References

1 L.A. Beeman and J.S. Reichert, 'The Bleaching of Pulp', TAPPI Monograph, Series No. 10, New York, 1953; A.H. Gilbert, 'Detergent Age', 1967, June 18.

2 H.J. Fenton, *Proc. Chem. Soc. London*, 1893, **9**, 113; H.J. Fenton, *J. Chem. Soc.*, 1894, **65**, 899.

3 D.F. McMillen and D.M. Golden, *Ann. Rev. Phys. Chem.*, 1982, **33**, 493.

4 A. Samani, D. Meisel and G. Czapski, *J. Chem. Soc., Dalton Trans.*, 1972, 1273.

5 C. Lu, E.W. Hughes and P. Giguere, *J. Am. Chem. Soc.*, 1941, **63**, 1507.

6 M.S. Cooper, H. Heaney, A.J. Newbold and W.R. Sanderson, *Synlett.*, 1990, 553.

7 H. Newman and R.B. Angier, *Tetrahedron*, 1970, **26**, 825.

8 J. Remound, *Rev. Prod. Chim.*, 1954, **57**, 401.

9 T.K. Das, A.K. Mandavawalla and S.K. Datta, *Colourage*, 1983, **30**, 15.

10 P. Kuzel, T. Lieser and M. Dankowski, *Chim. Oggi.*, 1986, October 60.

11 A. Manson, *Acta. Chem. Scand.*, 1961, **15**, 934.

12 M.A.A.F. Corrondo, W.P. Griffith, W.P. Jones and D.P. Shapski, *J. Chem. Soc., Dalton Trans.*, 1977, 2323.

13 C.C.R. Rocchiccioli, *Acad. Sci. Paris*, 1965, **261**, 361.

14 Kirk-Othmer, 'Encyclopedia of Chemical Technology', 3rd Edition, Wiley, New York, Vol. 17, 1982, p. 6.

15 A.K. Galway and W.J. Hood, *J. Chem. Soc., Faraday Trans. 1*, 1982, **78**, 2815.

16 G.W. Kabalka, T.M. Shoup and N.M. Goudgaon, *J. Org. Chem.*, 1989, **54**, 5930.

17 Y. Ogata and H. Shimizu, *Bull. Chem. Soc. Jpn.*, 1979, **52**, 635.

18 A. Rashid and G. Read, *J. Chem. Soc., C*, 1967, 1323.

19 G. Xie, L. Xu, J. Hu, S. Ma and F. Tao, *Tetrahedron Lett.*, 1988, **29**, 2967.

20 G.A. Olah, Q. Wang, N. Krass and G.K.S. Prakash, *Rev. Roum. Chim.*, 1991, **36**, 567.

21 G.A. Olah, Q. Wang, N.J. Trivedi and G.K.S. Prakash, *Synthesis*, 1991, 739.

22 D.T.C. Yang and T.T. Evans, *Synth. Commun.*, 1993, **22**, 2587.

23 A. McKillop and D. Kemp, *Tetrahedron*, 1989, **45**, 3299.

24 G.W. Kabalka, N.K. Reddy and C. Narayana, *Tetrahedron Lett.*, 1992, **33**, 865.

25 H.J.H. Fenton, *J. Chem. Soc.*, 1894, **65**, 899.

26 N. Uri, *Chem. Rev.*, 1952, **50**, 375.

27 J.T. Groves and M. van den Puy, *J. Am. Chem. Soc.*, 1976, **98**, 5290.

28 S.B. Brown, P. Jones and A. Suggett, *Prog. Inorg. Chem.*, 1970, **13**, 170.

29 J.H. Baxendale and J.A. Wilson, *Trans. Faraday Soc.*, 1957, **53**, 344.

30 M. Masarwa, H. Cohen, D. Meyerstein, D.L. Hickman, A. Bakac and J.H. Espenson, *J. Am. Chem. Soc.*, 1988, **110**, 4293.

31 T. Farataziz and A.B. Ross, NSRDS-NBS, No. 59, U.S. Department of Commerce, U.S. Government Printing Office, Washington D.C. 1977.

32 G.R.A. Johnson and N.B. Nazhat, *J. Am. Chem. Soc.*, 1987, **109**, 1990.

33 H.B. Brooks and F. Sicillo, *Inorg. Chem.*, 1971, **10**, 2530.

34 M. Anderson, J.O. Edwards and A.A. Green, *Inorg. Chim. Acta.*, 1969, **3**, 655.

35 R.O.C. Norman and P.R. West, *J. Chem. Soc., B*, 1969, 389.

36 S. Rahhal and H.W. Richter, *J. Am. Chem. Soc.*, 1988, **110**, 3126.

37 C.P. Moorhouse, B. Halliwell, M. Grootveld and J.M.C. Gutteridge, *Biochim. Biophys. Acta.*, 1985, **843**, 261.

38 D.C. Borg and K.M. Schaich, in 'Oxygen Radicals in Tissue Damage', ed. B. Halliwell, Faseb, Bethesda, MD, 1988.

39 W.M. Garrison, *Chem. Rev.*, 1987, **87**, 381.

40 N.A. Milas and S. Suismann, *J. Am. Chem. Soc.*, 1936, **58**, 1302; 1937, **59**, 2345.

41 G. Barak, J. Dakka and Y. Sasson, *J. Org. Chem.*, 1988, **53**, 3553.

42 Solvay Interox, W.O. 93/00319.

43 A.M. Khenkin and C.L. Hill, *Mendeleev Commun.*, 1993, 140; R. Neumann and M. Gara, *J. Am. Chem. Soc.*, 1994, **116**, 5509.

44 J.T. Groves and G.A. McClusky, *J. Am. Chem. Soc.*, 1976, **98**, 859; H. Sugimoto and D.T. Sawyer, *J. Am. Chem. Soc.*, 1985, **107**, 5712.
45 J.P. Renaud, P. Battioni, J.F. Bartoli and D. Mansuy, *J. Chem. Soc., Chem. Commun.*, 1985, 888.
46 P. Battioni, J.P. Renaud, J.F. Bartoli and D. Mansuy, unpublished results.
47 P. Battioni, J.P. Renaud, J.F. Bartoli and D. Mansuy, *J. Chem. Soc., Chem. Commun.*, 1986, 341.
48 D. Mansuy, *Pure Appl. Chem.*, 1987, **59**, 759.
49 R. Labeque and L.J. Marnett, *J. Am. Chem. Soc.*, 1989, **111**, 6621.
50 D. Mansuy, P. Battioni and J.P. Renaud, *J. Chem. Soc., Chem. Commun.*, 1985, 888.
51 D. Mansuy, P. Battioni and J.P. Renaud, *J. Chem. Soc., Chem. Commun.*, 1984, 1255.
52 D. Mansuy, P. Battioni, J.P. Renaud, J.F. Battioni, M. Reina-Artiles and M. Fort, *J. Am. Chem. Soc.*, 1988, **110**, 8462.
53 P.L. Anelli, S. Banfi, F. Montanari and S. Quici, *J. Chem. Soc., Chem. Commun.*, 1989, 779.
54 S. Banfi, A. Maiocchi, A. Moggi, F. Montanari and S. Quicci, *J. Chem. Soc., Chem. Commun.*, 1990, 1794.
55 A.M. d' A. Rocha Gonsaves, R.A.W. Johnstone, M.M. Pereira and J. Shaw, *J. Chem. Soc., Perkin Trans. 2*, 1991, 645.
56 A. Thelland, P. Battioni and D. Mansuy, *J. Chem. Soc., Chem. Commun.*, 1994, 1035.
57 T. Katsuki and K.B. Sharpless, *J. Am. Chem. Soc.*, 1980, **102**, 5974.
58 B.K. Sharpless, *J. Am. Chem. Soc.*, 1991, **113**, 106.
59 J.G. Hill, B.E. Rossiter and K.B. Sharpless, *J. Org. Chem.*, 1983, **48**, 3607.
60 W. Zhang, J.L. Loebach, S.R. Wilson and E.N. Jacobsen, *J. Am. Chem. Soc.*, 1990, **112**, 2801.
61 T. Katsuki, *Coord. Chem. Rev.*, 1995, **140**, 185.
62 T. Schwenkreis and A. Berkessel, *Tetrahedron Lett.*, 1993, **34**, 4785.
63 T. Yamada, K. Imagawa and T. Nagata, *Chem. Lett.*, 1992, 2231.
64 U. Romato, A. Esposita, A. Maspero, C. Neri and M.G. Clerici, *Chim. Ind.*, 1990, **72**, 610.
65 Enichem Sintesi, EP 226257.
66 B. Notari, in 'Structure Activity and Selectivity Relationships in Heterogeneous Catalysis', ed. R.K. Grasselli and A.W. Sleight, Elsevier, Amsterdam, 1991, 243.
67 Exxon Chemical, WO 90/05126.
68 Anic, DE 3309669.
69 Enichem Anic, EP 522634.
70 Montedipe SpA, EP 208311.
71 Enichem Sintesi, EP 190609.
72 Anic SpA, EP 102097.
73 M.A. Camblor, A. Corma, A. Martinez and J. Perez-Pariente, *Zeolites*, 1993, **13**, 83.
74 H. Mimoun, I. Seree de Roch and L. Sayas, *Bull. Soc. Chim. France*, 1969, 1481.
75 H. Mimoun in 'Comprehensive Co-ordination Chemistry', eds. G. Wilkinson, R.D. Gillard and J. McCleverty, Pergamon, Oxford, 1987, Chapter 61, p. 317.
76 F. Di Furia and G. Modena, *Rev. Chem. Intermediates*, 1985, **6**, 51.
77 M. Roussel and H. Mimoun, *J. Org. Chem.*, 1980, **45**, 5381.
78 E.I. Ochiai, *Inorg. Nucl. Chem. Lett.*, 1973, **9**, 987.
79 M.T. Atley, M. Preece, G. Strukel and B.R. James, *Can. J. Chem.*, 1983, **61**, 1332.
80 H. Mimoun, *Pure Appl. Chem.*, 1981, **53**, 2389.
81 Solvay Interox, unpublished results.
82 M.H. Dickman and M.T. Pope, *Chem. Rev.*, 1994, **94**, 569.
83 C.L. Hill, in 'Catalytic Oxidations with Hydrogen Peroxide as the Oxidant', ed. G. Strukel, Kluwer Academic Publishers, Dordrecht, 1992.
84 C. Venturello and R.D. Aloisio, *J. Org. Chem.*, 1988, **53**, 1553.

85 S.E. Jacobsen, D.A. Muccigrosso and F. Mares, *J. Org. Chem.*, 1979, **44**, 921.
86 Interox S.A., US 5086189.
87 R.T. Taylor and L.A. Flood, *J. Org. Chem.*, 1983, **48**, 5160.
88 G. Strukel and R.A. Michelin, *J. Chem. Soc., Chem. Commun.*, 1984, 1358.
89 G. Strukel and R.A. Michelin, *J. Am. Chem. Soc.*, 1985, **107**, 7563.
90 A. Zanardo, F. Pinna, R.A. Michelin and G. Strukel, *Inorg. Chem.*, 1988, **27**, 1966.
91 G. Strukel, M. del Todesco Frixstone, R. Giovanetti and F. Pinna, 'Studies in Surface Science and Catalysis', Vol. 66, ed. L. Simandi, Elsevier, Amsterdam, 1991, p. 405.
92 W.A. Hermann, *J. Organomet. Chem.*, 1995, **500**, 149.
93 K.B. Sharpless, J. Rudolph, K.L. Reddy and J.P. Chiang, *J. Am. Chem. Soc.*, 1997, **119**, 6189.
94 Solvay Interox, WO 93/08144.
95 C.W. Jones, 'Speciality Chemicals', February 1996.
96 A.J. Everett and G.J. Minkoff, *Trans. Faraday Soc.*, 1953, **49**, 410.
97 Solvay Interox, unpublished results.
98 Propylox S.A., US 4160778; 4193929; 4267124; 4294776.
99 Propylox S.A., US 4304639; 4325888; 4358609.
100 F. Burzio, *Chimica Oggi*, 1990, **8**, 23.
101 J. Rebek, *Heterocycles*, 1981, **15**, 517.
102 P.R.H. Speakman, *Chem. Ind.*, 1978, 579.
103 D. Swern, in 'Organic Peroxides', ed. D. Swern, Vol. I, Wiley-Interscience, New York and London, 1970, p. 313.
104 J.Y. Nedelec, J. Sorba and D. Lefort, *Synthesis*, 1976, 821.
105 R. Yoshizawa and T. Inukai, *Bull. Chem. Soc. Jpn.*, 1969, **42**, 3238.
106 R.D. Bach, W.M. Klein, R.A. Ryntz and J.W. Holubka, *J. Org. Chem.*, 1979, **44**, 2569.
107 N. Kawabe, K. Okada and M. Ohmo, *J. Org. Chem.*, 1972, **37**, 4210.
108 F. Bjorkling, S.B. Godtfredson and O. Kirk, *J. Chem. Soc., Chem. Commun.*, 1990, 1301.
109 F. Bjorkling, H. Frymann, S.B. Godtfredson and O. Kirk, *Tetrahedron*, 1992, **48**, 4587.
110 M. Rush gen. Klass and S. Warwel, Proceedings 21st. World Congress of the International Society for Fat Research (I.S.F.), Den Haag, Oct 1–6, 1995.
111 L.F. Fieser and M. Fieser, in 'Reagents for Organic Synthesis', Vol. 1, Wiley-Interscience, New York, 1967, p. 135.
112 P. Brougham, M.S. Cooper, D.A. Cummerson, H. Heaney and N. Thompson, *Synthesis*, 1987, 1015.
113 C.R. Harrison and P. Hodge, *J. Chem. Soc., Chem. Commun.*, 1974, 1009.
114 J.G. Traynham, G.R. Franzen, G.A. Knesel and D.J. Northington, *J. Org. Chem.*, 1967, 3285.
115 J. Ciabattoni, R.A. Campbell, C.A. Renner and P.W. Concannon, *J. Am. Chem. Soc.*, 1970, **92**, 3826.
116 U. Frommer and V. Ullrich, *Z. Naturforsch*, 1971, **26B**, 322.
117 J.A. Cella and J.P. McGrath, *Tetrahedron Lett.*, 1975, 4115.
118 G. Kreze, W. Schramm and G. Cerc, *Chem. Ber.*, 1961, **94**, 2060.
119 R.F. Evans, M. van Ammers and H.J. den Hertog, *Rec. Trav. Chim.*, 1959, **78**, 408.
120 H. Kleinfeller and K. Rastader, *Angew. Chem.*, 1953, **75**, 5426.
121 H. Wieland, T. Ploetz and K. Indest, *Justus Liebigs. Ann. Chem.*, 1937, **532**, 179.
122 A.H. Alberts, H. Wynberg and J. Strating, *Synth. Commun.*, 1973, **3**, 237.
123 S.O. Lawrenson, H.J. Jacobsen and E.H. Larson, *Acta. Chem. Scand.*, 1963, **17**, 1188.
124 P. Kovaic, C.G. Reid and M.J. Brittain, *J. Org. Chem.*, 1970, **35**, 2152.
125 C. Walling and R.B. Hodgson, *J. Am. Chem. Soc.*, 1958, **80**, 228.
126 R. Huisgen and W. Kolbeck, *Tetrahedron Lett.*, 1965, 783.

127 R.C. Schnur, *Chem. Eng. News*, 1981, **59**, 19.
128 R.A. Sheldon, in 'The Chemistry of Functional Groups, Peroxides', ed. S. Patai, Wiley, New York, 1983, Chapter 6, p. 161.
129 H. Williams and S. Mosher, *J. Am. Chem. Soc.*, 1954, **76**, 3495.
130 H. Kropf and C.R. Bernert, *Justus Liebigs. Ann. Chem.*, 1971, **751**, 109.
131 T.C. Bruice, *J. Chem. Soc., Chem. Commun.*, 1983, 14.
132 J.D. Rodrieguez and J.P. Dulcere, *J. Org. Chem.*, 1991, **56**, 469.
133 G.A. Tolstikov, V.P. Yurev and U.M. Dzhemilev, *Russ. Chem. Rev.*, 1975, **44**, 319.
134 M.N. Sheng and J.G. Zajacek, *J. Org. Chem.*, 1968, **33**, 588.
135 P.A. Bartlett, F.R. Green and T.R. Webb, *Tetrahedron Lett.*, 1977, 331.
136 B. Radzizsewski, *Chem. Ber.*, 1884, **17**, 1289; K.B. Wiberg, *J. Am. Chem. Soc.*, 1953, **75**, 3961.
137 G.B. Payne, P.H. Deming and P.H. Williams, *J. Org. Chem.*, 1961, **26**, 659.
138 R.D. Bach and J.W. Knight, *Org. Synth.*, 1981, **60**, 63.
139 L.A. Arias, S. Adkins, C.J. Nagel and R.D. Bach, *J. Org. Chem.*, 1983, **48**, 888.
140 G.B. Payne, *Tetrahedron*, 1962, **18**, 763.
141 Monsanto Co., EP 161234.
142 N. Matsumura, N. Sonoda and S. Tsutsumi, *Tetrahedron Lett.*, 1970, 2029.
143 S. Krishnan, D.G. Kuhn and G.A. Hamilton, *Tetrahedron Lett.*, 1977, 1369.
144 A. Koch, J.L. Reymond and L.J. Learner, *J. Am. Chem. Soc.*, 1994, **116**, 803.
145 Y. Chen and J.L. Reymond, *Tetrahedron Lett.*, 1995, **36**, 4015.
146 R.W. Murray and R. Jeyaraman, *J. Org. Chem.*, 1985, **50**, 2847.
147 R.W. Murray, *Chem. Rev.*, 1989, **89**, 1187; A.W. Curci and J.O. Edwards, *Acc. Chem. Res.*, 1989, **22**, 205.
148 R. Mello, M. Fiorento, O. Sciacovelli and R. Curci, *J. Org. Chem.*, 1988, **53**, 3890.
149 C.W. Jones, J.P. Sankey, W.R. Sanderson, M.C. Rocca and S.L. Wilson, *J. Chem. Res. (S)*, 1994, 114.
150 C.L. Lee, R. Hodikjan, A. Skothos, G. Sacciadis and J.T. Wearing, 'Proc. Int. Pulp Bleaching Conference', 1994, 273.
151 A. Armstrong, P.A. Clarke and A. Wood, *J. Chem. Soc., Chem. Commun.*, 1996, 849.
152 A. Pross, *Acc. Chem. Res.*, 1985, **18**, 212.
153 L. Eberson, in 'Electron Transfer Reactions in Organic Chemistry', Springer-Verlag, Berlin, 1987.
154 K.R. Kopecky, P.A. Lockwood, J.E. Filby and R.W. Reid, *Can. J. Chem.*, 1973, **51**, 468.
155 M.M. Rauhut, *Acc. Chem. Res.*, 1969, **2**, 80.
156 L.D. Martin and J.C. Martin, *J. Am. Chem. Soc.*, 1977, **99**, 3511.
157 C.S. Foote, S. Wexler, W. Ando and R. Higgins, *J. Am. Chem. Soc.*, 1968, **90**, 975.
158 D.F. Evans and M.W. Upton, *J. Chem. Soc., Dalton Trans.*, 1985, 2525.
159 J. Rebek, R. McCready and S. Wolf, *J. Org. Chem.*, 1979, **44**, 1485.
160 D.F. Evans and M.W. Upton, *J. Chem. Soc., Dalton Trans.*, 1985, 1141.
161 L.J. Csanyi, *Trans. Met. Chem.*, 1990, **15**, 371.
162 J.M. Aubery, *Membrane Lipid Oxidation*, 1985, **2**, 65.
163 A. Coeur and J. Alary, *Tenside*, 1967, **4**, 65.
164 D.F. Evans and M.W. Upton, *J. Chem. Soc., Dalton Trans.*, 1985, 1146.
165 D.F. Evans and M.W. Upton, *J. Chem. Soc., Dalton Trans.*, 1985, 1151.
166 S. Croft, B.C. Gilbert, J.R. Lindsay-Smith, J.K. Stell and W.R. Sanderson, *J. Chem. Soc., Perkin Trans. 2*, 1992, 153; D.M. Davis and R.M. Jones, *J. Chem. Soc., Perkin Trans. 2*, 1989, 1323.

CHAPTER 3

Application of Hydrogen Peroxide for the Synthesis of Fine Chemicals

1 Introduction

The use of hydrogen peroxide within the manufacture of bulk chemicals is limited, mainly due to economic constraints, and consequently molecular oxygen tends to be the oxidant of choice. Whilst the use of air and/or oxygen methods of oxidation are outside the scope of this book, it is worth mentioning that over the past decade a large body of research has been conducted on selective oxidation of organic functional groups with such reagents.[1-7] Generally, however, liquid-phase oxidations with molecular oxygen are radical chain processes,[8] and the intermediate alkylperoxy and alkoxy radicals are largely indiscriminate in their reactivity. Selective oxidation is normally only observed with relatively small molecules containing one reactive group. Thus, although molecular oxygen is a relatively cheap oxidant with limited waste problems, its scope is usually limited to a small number of simple petrochemicals.

The economics of fine chemical and pharmaceutical production, in contrast, allows a broader range of primary oxidants to be employed. Indeed, even though hydrogen peroxide is more expensive than oxygen, it can be the oxidant of choice because of its simplicity of operation, *i.e.* in fine chemical and pharmaceutical use, the total cost of equipment and raw material may be lower for oxidation employing hydrogen peroxide over oxygen.[9]

This chapter will introduce the reader to the following areas of synthetic organic chemistry using peroxygen reagents:

- Epoxidation of alkenes.
- Hydroxylation of alkenes.
- Cleavage of double bonds.
- Alcohol oxidation.
- Carbonyl oxidation

- Aromatic side chain oxidation.
- Organo-nitrogen oxidation.
- Organo-sulfur oxidation.
- Halogenation.
- Reactions at aromatic nuclei.

The chapter will be mainly concerned with homogeneous liquid-phase oxidation of organic substrates. For a discussion on heterogeneous liquid-phase oxidation, refer to Chapter 4.

2 Epoxidation of Alkenes

Epoxidation is an oxygen-transfer reaction for which hydrogen peroxide and its derivatives are particularly well suited. This is related to the fact that a wide range of peroxygen systems have been successfully applied to a series of alkene types, and some of these applications are amongst the longest established industrial uses of peroxygens. Epoxides are valuable and versatile commercial intermediates owing to the large range of reactions they undergo. The majority of transformations occur with active hydrogen compounds (Figure 3.1) such as

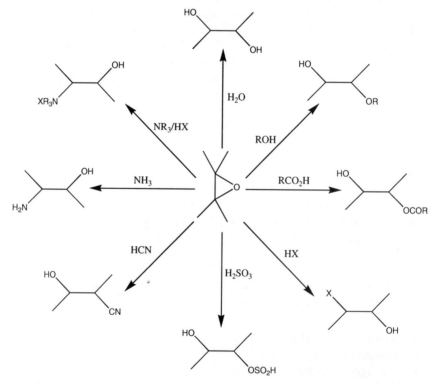

Figure 3.1 *Ring opening reactions of epoxides in the presence of active hydrogen compounds.*

ammonia, amines, organic acids, alcohols and sulfur compounds.[10] Epoxides undergo re-arrangement to carbonyl compounds[11] or allyl alcohols.[12]

Polymerization of short chain diepoxides is an important industrial route to cross-linked polymers.[13]

The reaction of alkenes with peroxycarboxylic acids to produce epoxides was discovered by Prilezhaev over 80 years ago.[14] It is still the most widely used method for epoxidation, and considerable work has been carried out to elucidate the mechanism. The commonly accepted explanation for oxirane formation involves a cyclic polar process where the proton is transferred intramolecularly to the carbonyl oxygen, with simultaneous attack by the alkene π-bond. This concerted process was suggested by Bartlett,[15] and because of the unique planar transition structure it is referred to as the 'butterfly' mechanism (Figure 3.2).

The oxygen atom transfer from a peroxy acid to an alkene is facilitated by electron-donating substituents on the carbon–carbon double bond and electron-withdrawing groups on the peroxy acid.[16] Further, conclusions which can be made from the experimental data collected over the years, include:

- *cis*-Alkenes react faster than corresponding *trans*-alkenes.
- Cycloalkenes generally react much faster than dialkyl-substituted alkenes.
- The effect of aryl substituents is minimal giving only small increases in rate.
- In a conjugated diene, the first double bond will be epoxidized at a faster rate than the second double bond.
- Strained alkenes such as norbornene react faster relative to unstrained alkenes such as cyclohexene.
- Electron-withdrawing groups present in the alkene markedly reduce the reaction rate, *e.g.* ketones, carboxylic acids or esters.

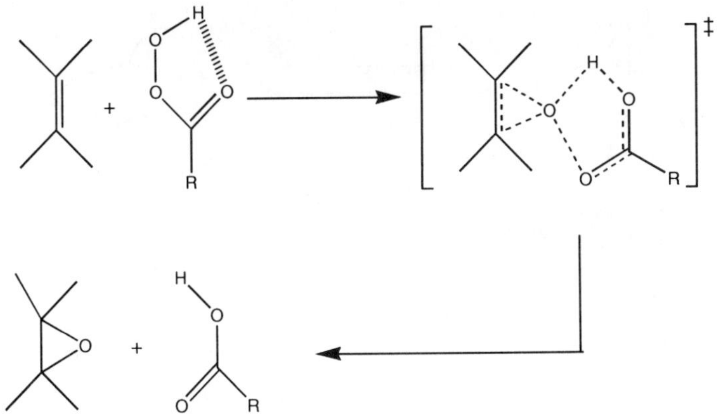

Figure 3.2 *Bartlett 'butterfly' mechanism for the epoxidation of alkenes with peroxycarboxylic acids.*

The order of reactivity of the peroxy acid increases with the pK_a value, *i.e.* peroxytrifluoroacetic > monopermaleic > *p*-nitroperbenzoic > *m*-chloroperbenzoic > performic > perbenzoic > peracetic.

On the basis of theoretical studies by Bach and co-workers,[17] it was found that the nucleophilic π-bond of the alkene attacks the O–O σ-bond in an S_N2 fashion with displacement of a neutral carboxylic acid. There are, however, some mechanistic anomalies. For example, a protonated peracid should be a much more effective oxygen transfer agent over its neutral counterpart, but experiments have shown only modest rate enhancements for acid catalysed epoxidation. Early attempts to effect acid catalysis in alkene epoxidation where relatively weak acids such as benzoic acid were employed proved unsuccessful.[18] The picture is further complicated by contradictory data concerning the influence of addition of acids on epoxidation rates.[19] Trichloroacetic acid catalyses the rate of epoxidation of stilbene with perbenzoic acid, but retards the rate of a double bond containing an ester constituent such as ethyl crotonate.[20] Recent work has shown that a seven-fold increase in the rate of epoxidation of *Z*-cyclooctene with *m*-chloroperbenzoic acid is observed upon addition of the catalyst trifluoroacetic acid.[21] Kinetic and theoretical studies suggest that the rate increase is due to complexation of the peroxy acid with the undissociated acid catalyst (HA) rather than protonation of the peroxy acid. *Ab initio* calculations have shown that the free energy of ethylene with peroxyformic acid is lowered by about 3 kcal mol^{-1} upon complexation with the catalyst.[21]

In certain circumstances, neighbouring functional groups may alter the direction of the peroxy acid attack. For example, with norbornene the *exo*-epoxide is formed, whereas under identical conditions, 7,7-dimethylnorbornene reacts to give predominantly the *endo*-epoxide (Figure 3.3).[22]

The rate of epoxidation is also affected by solvent. The reaction is particularly facilitated by chlorinated solvents. More polar solvents, particularly hydrogen bonding types, greatly decrease the rate of reaction. This considerable reduction

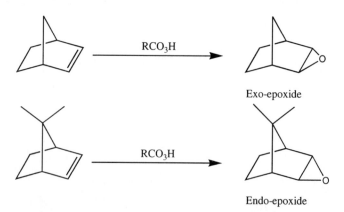

Exo-epoxide

Endo-epoxide

Figure 3.3 *Effect of neighbouring functional groups on the epoxidation of norbornene derivatives.*

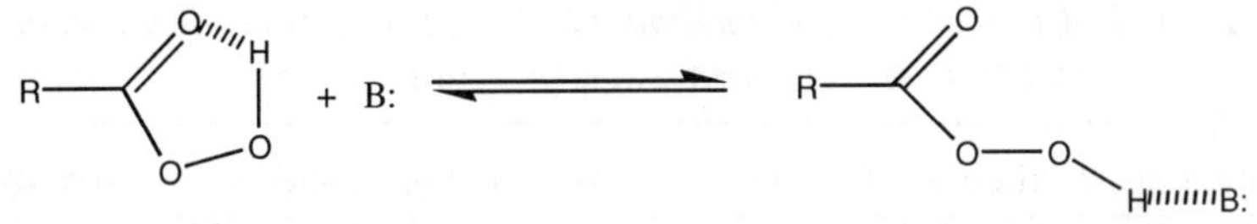

Figure 3.4 *Disruption of intramolecular bonding of a peracid species via a hydrogen bonding solvent.*

in epoxidation rate is due to disruption of the percarbonyl ring to form intermolecularly hydrogen bonded adducts (Figure 3.4).

The selection of a peracid for epoxidation is not solely dependent on the alkene and peracid structures. The stability of the resulting epoxide ring produced during oxidation has a significant effect on product yield, since stability is affected by electronic factors, *i.e.* electron-withdrawing groups reduce the oxirane stability. For example, epichlorohydrin is less stable than propylene oxide to ring-opening with organic acids. Carboxylic acids attack epoxides to produce mono-esters of diols. Mineral acids have a marked catalytic effect on ring cleavage. The ability of a carboxylic acid to give ring opening increases with decreasing pK_a value. Loss of epoxide in the presence of acids is the major cause of low yield. However, losses can be minimized with careful choice of peracid and reaction conditions.

The simplest and generally most economic means of epoxidation is to generate the percarboxylic acid in the reaction medium (Figure 3.5).

The alkene substrate is first dispersed in the reactor with the carboxylic acid in the presence of a solvent if necessary. Addition of hydrogen peroxide forms the peracid which facilitates epoxidation, and re-generates the carboxylic acid for further reaction. In consequence only low levels of carboxylic acid are required (0.2–0.3 mol per mol of double bond). This also has the advantage of reducing epoxide loss to acid-catalysed ring-opening. The application of two phases with or without the presence of solvent also improves the efficiency of the epoxidation. Peracids most suited to epoxidation *in situ* are performic and peracetic acid.

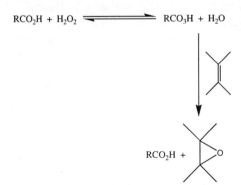

$$RCO_2H + H_2O_2 \rightleftharpoons RCO_3H + H_2O$$

$$RCO_2H +$$

Figure 3.5 *Epoxidation of an alkene via the* in situ *generation of a peracid.*

Table 3.1 *Typical procedures for the epoxidation of soyabean oil using* in situ
            performic and peracetic acid

In situ *performic acid*	Parameter	In situ *peracetic acid*	Parameter
Soyabean oil	1 unit	Soyabean oil	1 unit
Formic acid	0.14 units	Acetic acid	0.2 units
H_2O_2 (70% *m/m*)	0.3 units	H_2O_2 (70% *m/m*)	0.3 units
Temperature (°C)	60	H_2SO_4 (98% *m/m*)	0.01 units
Reaction time (h)	8–10	Temperature (°C)	60
		Reaction time (h)	10–15

Formic acid is a relatively strong acid and can self-catalyse the formation of
the peracid. In contrast, peracetic acid may only be generated by the addition of
a strong acid catalyst. The acids normally employed are sulfuric, phosphoric or
a cationic exchange resin. Some processes also use these catalysts in conjunction
with performic acid to decrease reaction time and limit the quantity of formic
acid used. This is economically important as formic acid cannot be recovered to
reduce costs.

Performic acid is an unstable, hazardous percarboxylic acid, and must always
be generated *in situ*. Epoxidation with *in situ* performic and peracetic acid are
well established commercial processes. They find application in the epoxidation
of alkenes, particularly those of high molecular weight. Many such epoxides are
produced on a large scale, and can be classified as vegetable oils, unsaturated
esters, unsaturated acids, α-alkenes, natural polymers and synthetic polymers.
The most important vegetable oil which is epoxidized commercially is soyabean
oil. World production of epoxidized soyabean oil (ESBO) exceeds 150 000
metric tons per annum. Epoxidized linseed oil is also important, but produced
at a lower rate than ESBO. Both products are formed by usual *in situ* performic
and peracetic acid techniques.[23,24] Typical procedures are outlined in Table
3.1.[25]

Synthetic polymers from alkene monomers are also a major target for
epoxidizers. Polyisoprenes,[26] polybutenes and polybutadienes have all been
epoxidized by general *in situ* peracid methods. In the case of liquid polybutadienes, a wide range of types is available. Products with varying contents of 1,4-
cis, 1,4-*trans*, and 1,2-vinyl alkene groups are available (Figure 3.6).

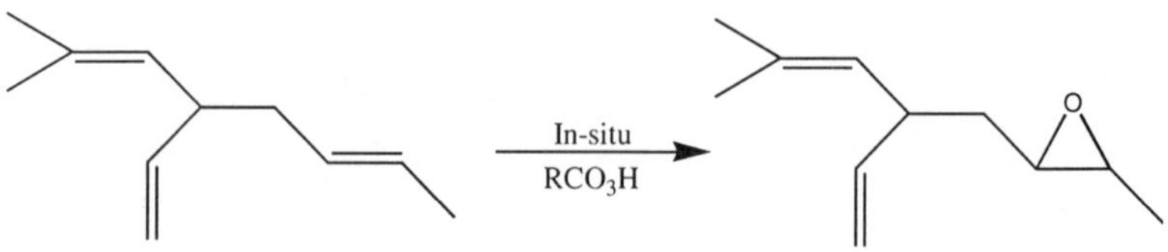

Epoxidation order: 1,4-trans > 1,4-cis > 1,2-vinyl

Figure 3.6 *Epoxidation of polyalkene materials with* in situ *formed peracids.*

Table 3.2 *Epoxidation of olefins employing* in situ *peracids from a UHP/acid anhydride system*

Substrate	Product	Time (h)	Isolated yield (%)	Reference
Styrene	Styrene oxide[a]	17	60	28
α-Methylstyrene	1-Phenyl-1-methyloxide[a]	10	75	29
Oct-1-ene	1,2-Epoxyoctane[b]	0.5	88	30
Phenyl allyl ether	1,2-Epoxy-3-phenoxypropane[c]	20	57	31
Chloresterol	3-β-Trifluoroacetoxy-5,6α-epoxycholestrane[c]	16	62	
Methyl methacrylate	Methyl 2-methyl-2,3-epoxypropanoate[d]	1	56	28
Geranyl acetate	6,7-Epoxygeranyl acetate[b]	5	82	32
Linalool	2-Methyl-5-(2′-propyl-2′-hydroxy)-2-vinyl tetrahydrofuran[c]	4	64	33

[a] $(CH_3CO)_2O$, UHP, Na_2HPO_4, CH_2Cl_2, R.T.; [b] $(CF_3CO)_2O$, UHP, Na_2HPO_4, CH_2Cl_2, reflux; [c] $(CF_3CO)_2O$, UHP, Na_2HPO_4, CH_2Cl_2, R.T.; [d] $(CH_3CO)_2O$, UHP, Na_2HPO_4, CH_2Cl_2, reflux.

On a laboratory scale, one particular method of producing *in situ* peracids for epoxidation of a wide range of substrates under mild conditions is via the use of urea hydrogen peroxide (UHP) in the presence of organic anhydrides.[27] The anhydride must be added slowly to the UHP, solvent and substrate to generate the peracid. If the UHP is added to the substrate, solvent and anhydride, the unstable and potentially explosive diacyl peroxides can be formed. Table 3.2 illustrates a number of substrates which have been epoxidized using this reagent system.

Lipases over the last decade have become more prevalent in organic synthesis. As a consequence of such activity, the lipases are now recognized as efficient, and useful catalysts for the modification of fats and oils by acidolysis of triglycerides, and for the synthesis or hydrolysis of carboxylic acid esters.[34] Lipases have also been used to catalyse the formation of peroxycarboxylic acids from the parent carboxylic acid and hydrogen peroxide. Bjorkling and co-workers have shown that immobilized lipases, such as *Candida antarctica* and *Candida cylindracea*, could perform exceptionally well in toluene or hexane as solvents.[35] The results were in accordance with the observation that lipases generally perform better for synthetic purposes in water-immiscible organic solvents compared with water-miscible ones.[36] The oxidation using lipases is therefore carried out in a two-phase system. For liquid alkenes, the immobilized enzyme is simply dispersed in the substrate with a catalytic amount of carboxylic acid (generally octanoic), and the hydrogen peroxide is added slowly. Close to quantitative yields of the epoxide are obtained, however, terminal alkenes react relatively slowly with the system (Figure 3.7).

The lipase method of *in situ* peracid formation is relatively safe, and the selectivity to epoxide, *i.e.* the amount of epoxide formed compared with the

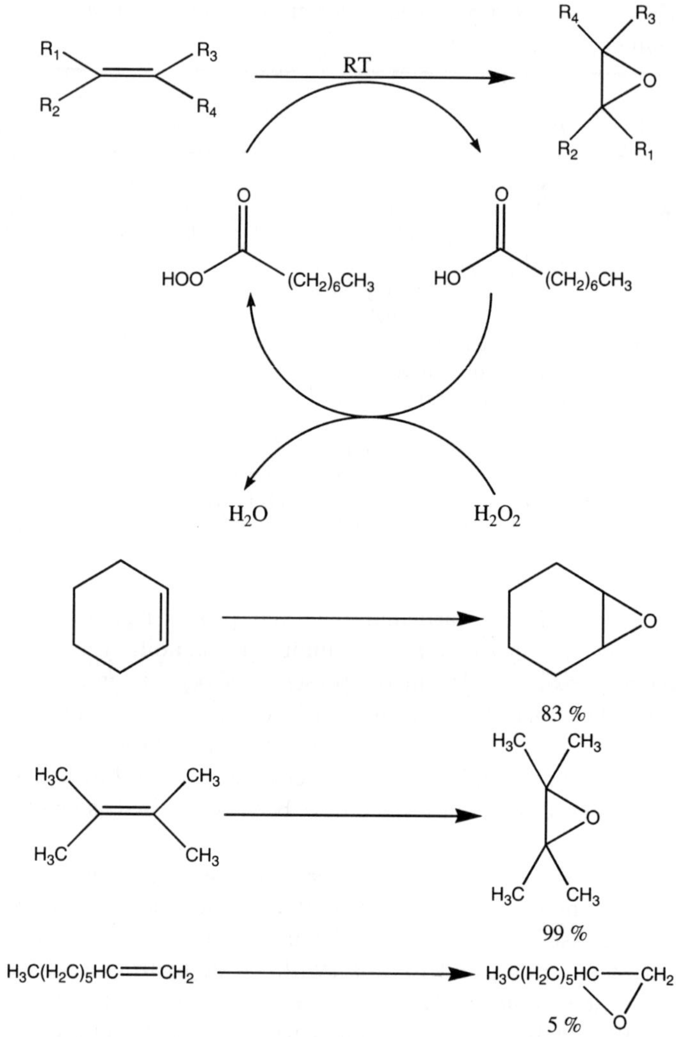

Figure 3.7 *Epoxidation of alkenes by lipase catalysed peracid formation in a two-phase system.*

conversion of the substrate (*e.g.* when yield = conversion, the selectivity is 100%) is generally greater than 90%. The method in the near future may be employed for the large scale manufacture of epoxides, even sensitive ones such as those based on terpene structures.

Where unsaturated compounds cannot be epoxidized via *in situ* techniques then equilibrium peracids or solvent extracted peracids are frequently used. The equilibrium peracetic acid is often used stoichiometrically and is the favoured method for unreactive alkenes. Kinetics favour the use of equilibrium peracids over *in situ* techniques. Ring opening is still, however, a major side-reaction but

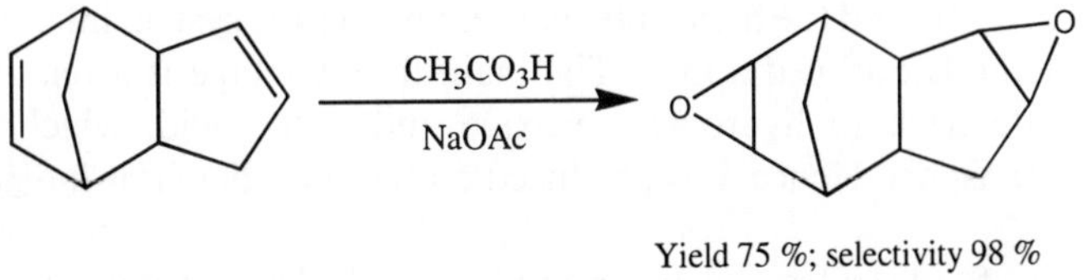

Yield 75 %; selectivity 98 %

Figure 3.8 *Epoxidation of dicyclopentadiene in the presence of buffered peracetic acid.*

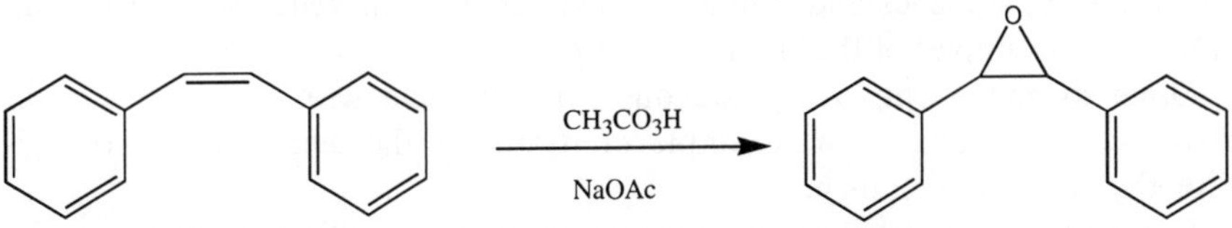

Figure 3.9 *Epoxidation of trans-stilbene with buffered peracetic acid.*

is generally more easily controlled by the use of aprotic solvents. The sulfuric acid content in the equilibrium peracetic acid is normally neutralized with sodium acetate. Buffered peracetic acid prepared in this way should only be used fresh. The solvent-free epoxidation of dicyclopentadiene is an example of the use of buffered peracetic acid (Figure 3.8).

trans-Stilbene is an example of the epoxidation of an electron-rich double bond using buffered peracetic acid (Figure 3.9).

Acid-labile epoxides undergo epoxide ring opening reactions even with weak acids, including acetic acid. Therefore, an inorganic base in the solid form is used sometimes with a solvent to minimize yield loss. The base is often sodium carbonate, disodium dihydrogen orthophosphate or magnesium hydroxide, and is normally added in large excess based on the total acid present in the peracid mixture.[39] α-Pinene is an industrially important product used in the flavour and fragrance field, which is epoxidized using equilibrium peracetic acid in the presence of sodium acetate and sodium carbonate (Figure 3.10).

Distilled aqueous peracetic acid is a high purity source of the peracid.[40] The major advantages of the distilled aqueous peracetic acid are that it contains no mineral acid, it has a low level of free acetic acid and it has a high water content. The lack of mineral acid and low acetic acid content can have advantages in

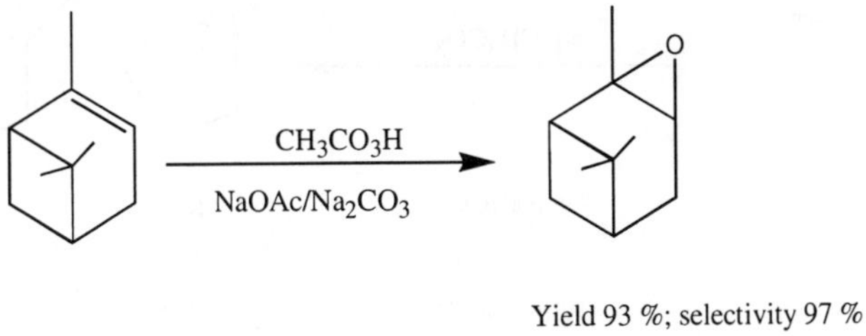

Yield 93 %; selectivity 97 %

Figure 3.10 *Epoxidation of α-pinene with buffered peracetic acid.*

terms of reactivity, and the high water content makes the system inherently safer than the methods outlined above. The main disadvantage is a relatively high reversion rate back to hydrogen peroxide and acetic acid, which makes it difficult to transport. Hence, it is produced on the site where it is intended to be used.

A variety of substrates can be epoxidized with distilled aqueous peracetic acid. It can be used directly with more robust substrates or for more sensitive alkenes in conjunction with a buffer. Examples of the former group include soyabean oil, stilbenes and synthetic polymers. Cyclohexene, isoamylene and styrene are examples of the latter class (Figure 3.11).

Other examples employing aqueous peracetic acid with alkene substrates have been reported.[41] One example of its use is the diepoxidation of 1,5-dimethyl-1,5-cyclooctadiene.[42]

In summary, peracetic acid in aqueous solution is an efficient reagent for the epoxidation of a variety of substrates, however, it should be used relatively quickly after its preparation. Sensitive substrates like isoamylene also require a buffer such as sodium hydrogen carbonate to affect high yield of the epoxide. Solvent extraction of an appropriate source of a peracid can lead to organic solutions of the oxidation species. Such methods have been available for many years,[43] however, very few are amenable to more than a small scale operation.

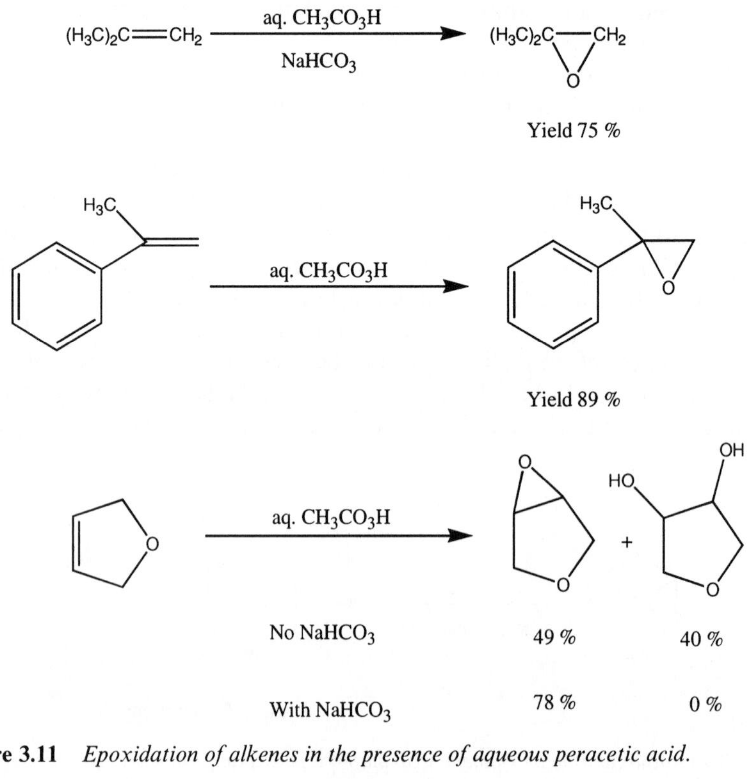

Figure 3.11 *Epoxidation of alkenes in the presence of aqueous peracetic acid.*

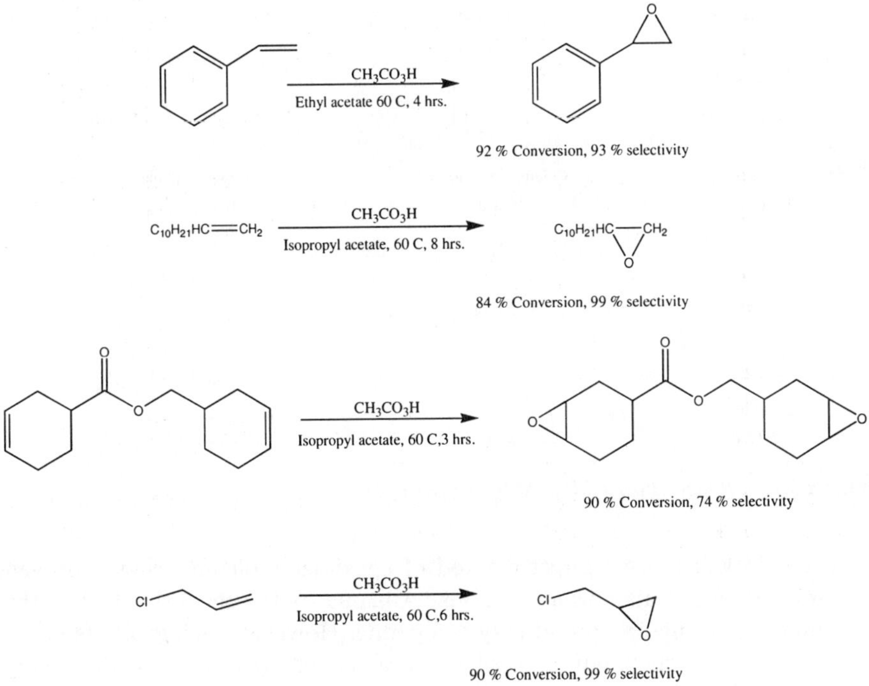

92 % Conversion, 93 % selectivity

84 % Conversion, 99 % selectivity

90 % Conversion, 74 % selectivity

90 % Conversion, 99 % selectivity

Figure 3.12 *Application of solvent-extracted peracetic acid for epoxidation of a range of alkenes.*

The approaches to anhydrous or essentially anhydrous solutions include reaction of the acid anhydride with hydrogen peroxide in the presence of a solvent,[44] oxidation of the analogous aldehyde[45] and azeotropic removal of water during peracid formation.[46] By far the easiest and safest method is to simply extract the equilibrium peracid into an appropriate solvent.[47] The solvents usually employed are ethyl acetate or isopropyl acetate. A range of substrates has been epoxidized using such extracted peracids (Figure 3.12).

Whilst the bulk of epoxidations are carried out with *in situ* performic or peracetic acids, a range of other peracids have been employed (Figure 3.13).

PTFAA is generally prepared by adding trifluoroacetic anhydride to very concentrated aqueous hydrogen peroxide or to UHP. The peracid is much more reactive than other peracids usually used for epoxidation.[48] Buffers have to be used with such a system together with low temperatures.[49] MMPP[50] is a non-shock-sensitive and non-deflagrating peroxygen compound, which has been used effectively for a range of epoxidations. MCPBA was first employed around 1951 for epoxidation[51] and has been widely used on a small scale. MCPBA will epoxidize a wide range of olefins and heterocyclic-containing olefins.[52] The relatively hazardous nature of MCPBA makes it a less desirable oxidant against MMPP. Amongst other peracids used in epoxidation are monoperphthalic,[53] *p*-nitroperbenzoic acid[54] and permaleic acid (Figure 3.14).[55]

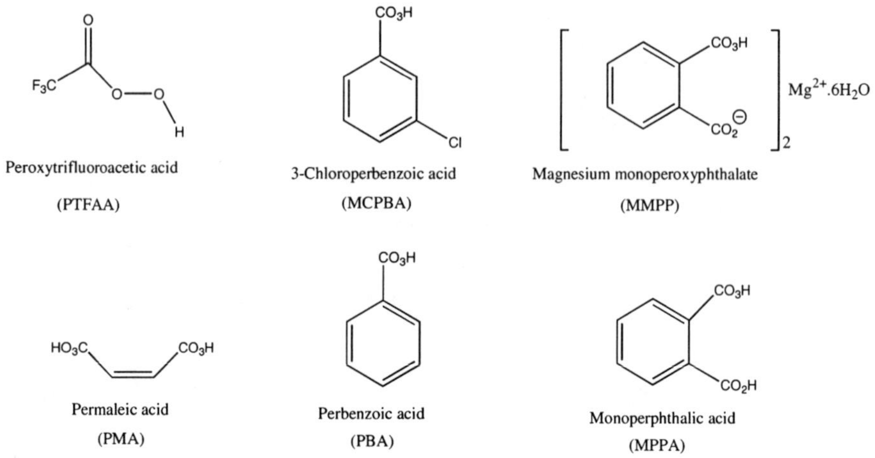

Peroxytrifluoroacetic acid

(PTFAA)

3-Chloroperbenzoic acid

(MCPBA)

Magnesium monoperoxyphthalate

(MMPP)

Permaleic acid

(PMA)

Perbenzoic acid

(PBA)

Monoperphthalic acid

(MPPA)

Figure 3.13   *Range of peracids used to prepare epoxides.*

The only widely used direct method of epoxidizing olefins using hydrogen peroxide is under alkaline conditions. Alkaline hydrogen peroxide has the advantage of being cheap and easy to operate. However, chemically its use is restricted to the epoxidation of olefins deactivated by electron-withdrawing groups. Industrial processes must be designed with care as hydrogen peroxide rapidly decomposes at high pH, and organic peroxides can be formed as by-products of the reaction. Alkaline hydrogen peroxide can be used to epoxidize stereoselectively. For example, when the two isomers of 3-methylpent-2-en-4-one are epoxidized, the same isomer is obtained in each case.[56] Other substrates epoxidized using alkaline hydrogen peroxide include acrolein[57] and isophorone,[58] unsaturated sulfones,[59] nitro compounds,[60] vinyl phosphonates[61] and fluorine compounds.[62] Alkaline hydrogen peroxide is used for the preparation of 3-methyl-2-epoxycyclopentanone[63] and epoxy-α-damascone.[64] Epoxy-α-damascone is an important intermediate in the production of the flavouring corylone (Figure 3.15).

Mildly basic conditions can be provided using quaternary ammonium fluorides with aqueous hydrogen peroxide in dimethyl sulfoxide solution. For example, carvone gives 82% epoxide in six hours at room temperature. The method is particularly suited to base-sensitive substrates such as cinnamaldehyde.[65]

Alkaline hydrogen peroxide can be reacted with organic nitriles (Payne's reagent) to produce an effective epoxidizing system. The active species is a peroxyimidic acid, which is similar in structure to percarboxylic acids but is not isolated.[66] Acetonitrile or benzonitrile are generally used as the nitrile source. The method is useful for the preparation of acid-labile epoxides such as glycidylaldehyde diethylacetal[67] and can also be used to epoxidize unsaturated ketones without affecting the carbonyl function. The main drawback with the system, which hinders its wider use industrially, is the formation of the co-product amide, which has to be disposed of, sold, used captively or dehydrated

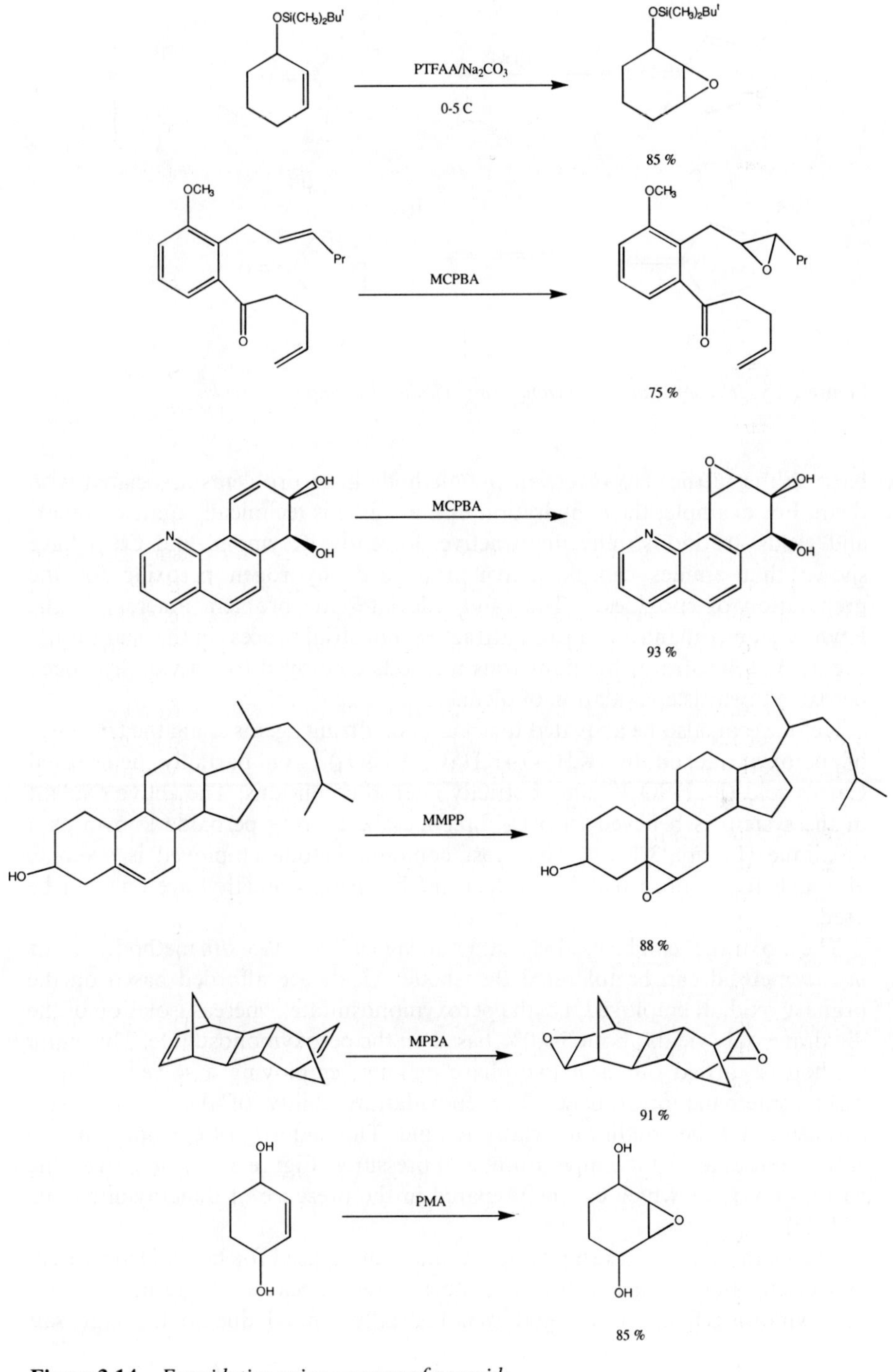

Figure 3.14 *Epoxidation using a range of peracids.*

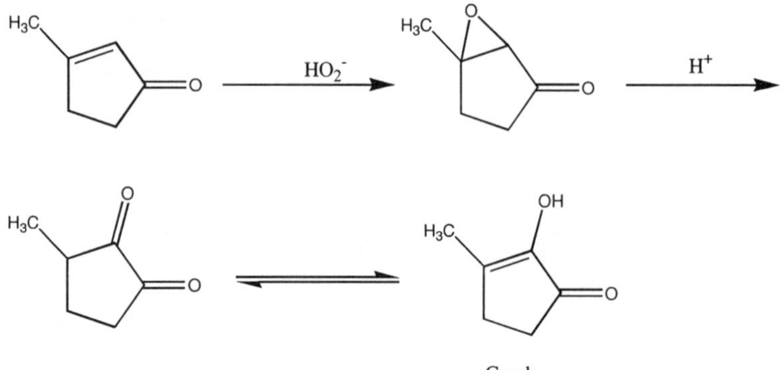

Corylone

Figure 3.15 *Preparation of corylone using alkaline hydrogen peroxide.*

back to the nitrile. However, all the methods have problems associated with them. For example, the dehydration of the amide is technically quite complex, and yields are industrially unattractive. Recently, Reymond and Chen have shown that amides can be activated towards hydrogen peroxide for the preparation of epoxides.[68] The route alleviates the problem inherent in the Payne system and may lead to an attractive industrial process in the near future. Figure 3.16 illustrates the numerous methods employed to activate hydrogen peroxide towards epoxidation of olefins.

Ketones can also be activated to act as epoxidizing agents using the triple salt of peroxymonosulfate $(KHSO_5 \cdot KHSO_4 \cdot K_2SO_4)$[69] or partially neutralized Caro's acid $(NaHSO_5)$[70] under strictly neutral conditions. The active oxidant in the system is believed to be a three-membered ring peroxide known as a dioxirane (Figure 3.17).[71] The most common ketone employed is acetone, although for a more reactive system, methyl trifluoromethyl acetone can be used.

The dioxiranes can be used via either an *in situ*[70] or an *ex situ* method.[71] If the *in situ* method can be tolerated then better yields are afforded based on the primary oxidant employed, *i.e.* the peroxymonosulfate, whereas isolation of the dioxirane only yields about 5–10% based on the peroxymonosulfate. The *in situ* method is carried out in a two-phase manner, employing a solvent such as dichloromethane or toluene. The epoxidation ability of the dioxiranes is excellent, and the conditions relatively mild. The majority of epoxidations are carried out at ambient temperatures and pressures. Figure 3.18 summarizes the various epoxides which can be prepared in the presence of dimethyldioxirane (DMD).

The method is remarkably versatile, and can be used for both electron-rich, and electron-deficient substrates. Unfortunately, because of the need to use peroxymonosulfate, the method is industrially limited due to the high salt loading.

A wide range of catalytic systems using peroxides has been developed for the epoxidation of olefins. Inorganic compounds, in particular molybdenum,

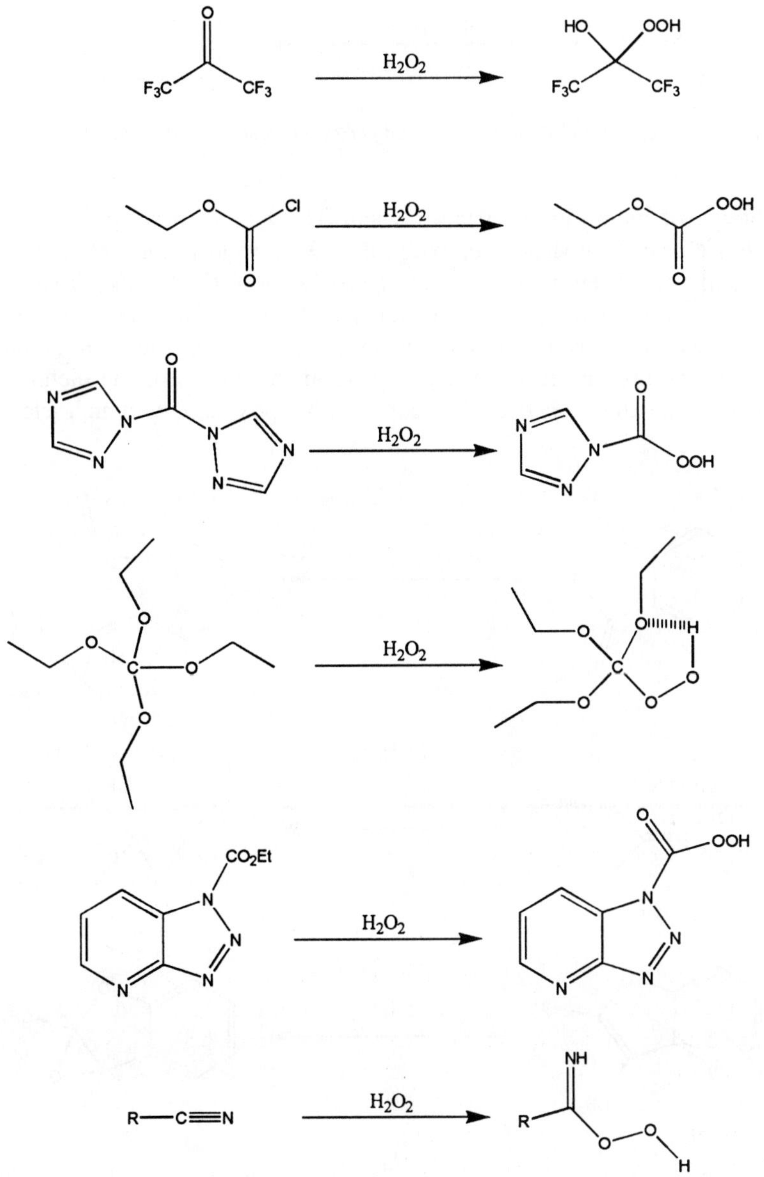

Figure 3.16 *Compounds employed for the activation of alkaline hydrogen peroxide.*

tungsten, selenium and boron will epoxidize olefins in the presence of hydrogen peroxide via the formation of metal hydroperoxides or metal peroxo species.[72] Olefins such as propylene and cyclohexene can be epoxidized in good yield with boron or arsenic compounds in the presence of hydrogen peroxide. The system works best under anhydrous conditions, and water is often removed azeotropically.[73] Good results have been obtained with a wide range of olefins using

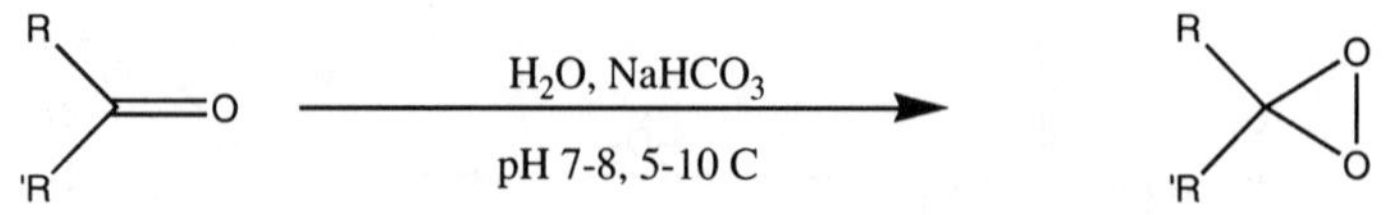

Figure 3.17 *Preparation of dioxiranes from peroxymonosulfate under neutral conditions.*

stoichiometric quantities of metaboric acid (HBO_2) due to *in situ* formation of orthoboric acid. Propylene is epoxidized in 99% yield using such a system.[74] Arsenic has also been used in a heterogeneous form, the arsenic species being bound to a cross-linked polystyrene matrix.[75] Selenium dioxide with hydrogen peroxide has also been found to be useful for the epoxidation of olefins. However, the product isolated depends upon the nature of the olefin.[76] For example, cyclooctene and cyclododecane can be epoxidized with high selectively

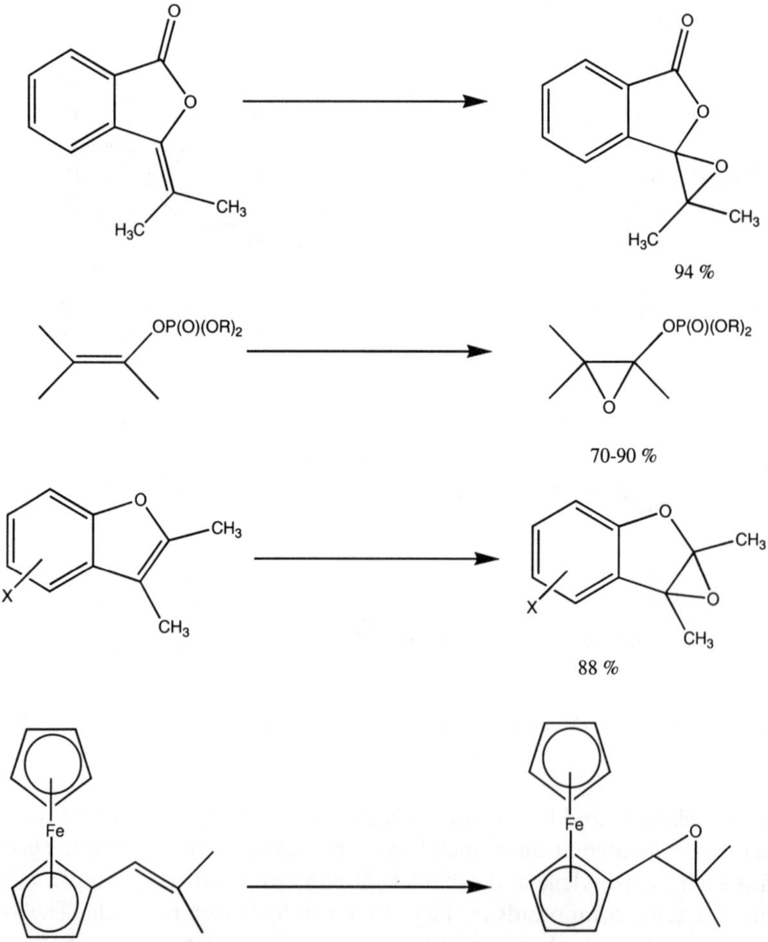

Figure 3.18 *Epoxidation of olefinic compounds with dimethyldioxirane.*

(97% and 96% respectively), whilst cyclohexene and norbornene under the same conditions give the corresponding diols. Selenium dioxide/hydrogen peroxide systems in the presence of aromatic nitrogen bases will epoxidize cyclohexene in 93% selectivity.[77] Inorganic rhenium compounds such as Re_2O_7 or ReO_3 have been known for a long time to exhibit modest catalytic activity for hydrogen peroxide based oxidations.[78] However, methylrhenium trioxide (MTO)/hydrogen peroxide systems are a relatively recent addition to the arsenal for the epoxidation of olefins.[79] The work of Hermann is seminal in this area of oxidation chemistry and has focussed on the preparation of organometallic rhenium compounds. However, the presence of water in such systems tends to be detrimental to the epoxide yield, due to acid catalysed side-reactions. Hermann's group only managed modest increases in selectivity.[80] The epoxide instability was somewhat alleviated by Sharpless and co-workers[81] when it was found that addition of an excess of a tertiary nitrogen base, *e.g.* pyridine, dramatically suppressed epoxide ring opening reactions (Figure 3.19).

The pyridine co-catalyst performs three important functions. Firstly, it accelerates the catalytic turnover, secondly, it prevents decomposition of the epoxide product and finally, in sufficient concentration, the pyridine increases the lifetime of the catalyst. Table 3.3 illustrates some of the substrates which have been epoxidized using the MTO/hydrogen peroxide/pyridine system.

Tungstate-catalysed epoxidation of olefins dates back to about 1949 with the work of Mugden and Young.[82] The influence of reaction conditions with tungstate/hydrogen peroxide systems has been extensively studied with maleic acids as substrates[83] and good yields of the epoxy acids were obtained.[84] The active catalyst was suggested by Mimoun to be a ligated metal-peroxo species.[85] The peroxo complexes can be used to carry out epoxidation stoichiometrically[86] but also have the capacity to be employed in a catalytic fashion, usually in a two-phase mixture for the epoxidation of water-immiscible compounds.[87] Selectivity towards olefin oxidation in bifunctional molecules is affected by

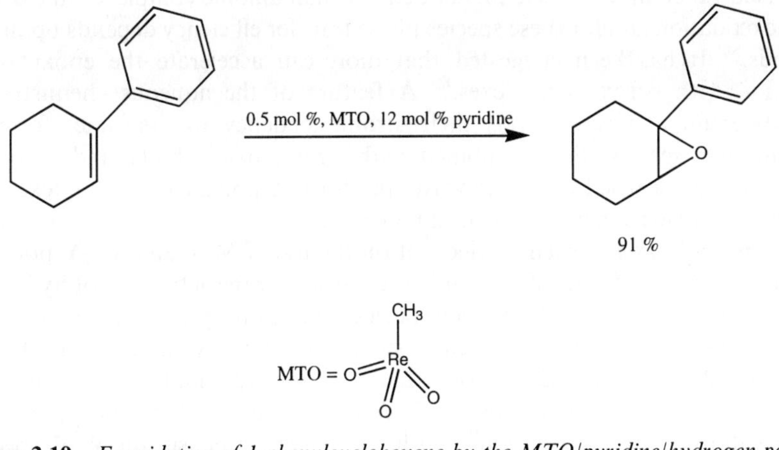

0.5 mol %, MTO, 12 mol % pyridine

91 %

$$MTO = O = \overset{\overset{\displaystyle CH_3}{|}}{\underset{\overset{\displaystyle ||}{O}}{Re}} \diagdown O$$

Figure 3.19 *Epoxidation of 1-phenylcyclohexene by the MTO/pyridine/hydrogen peroxide system.*

Table 3.3 *MTO/hydrogen peroxide catalysed epoxidation of olefins*[a]

Substrate	Product	Isolated yield (%)	Selectivity (%)
		82	98
		96	99
		86	98
		97	99
	1.2 : 1	95	99
	1 : 100	97	99

pH. Neutral complexes have higher activity than anionic complexes in electrophilic oxidation and for these species phase transfer efficiency depends upon the ligands.[88] It has been suggested that diols can accelerate the epoxidation reaction with peroxo complexes.[89] A feature of the aqueous chemistry of molybdenum and tungsten is their strong tendency to associate, forming polynuclear species; this, combined with ligand oxidizability and recovery problems, has caused a bias towards pre-formed polynuclear complexes for practical epoxidation with hydrogen peroxide.

Extensive work has been carried out on the use of M_4Y and $M_{12}Y$ polyoxo complexes (M = W, Mo; Y = P, As) as catalysts for the activation of hydrogen peroxide towards epoxidation. Such species are negatively charged and are used with a cationic phase transfer agent, usually a quaternary ammonium salt in a chlorinated solvent like 1,2-dichloroethane. Various epoxidations of olefins and allylic alcohols have been performed.[90] Venturello has developed phosphotungstate catalysts for epoxidation[91] with particular emphasis on W_4P species (Figure 3.20).[92]

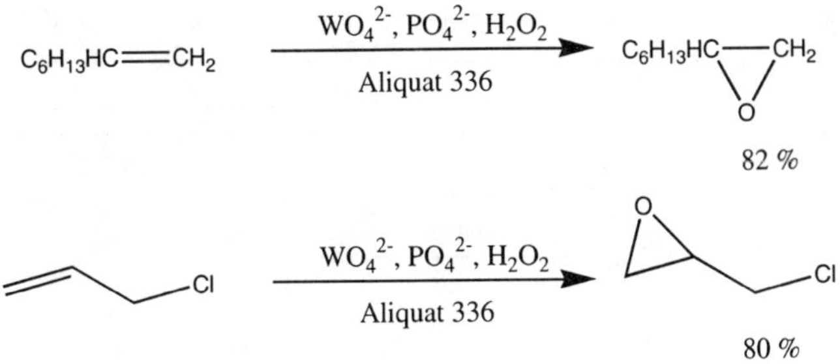

Figure 3.20 In situ *formation of phosphotungstate species for epoxidation of olefins.*

The Aliquat^{TM} 336 (methyltrioctylammonium chloride) is used as a phase transfer agent. The system does show signs of inactivity as the reaction proceeds which is believed to be due to poisoning by the epoxide product. However, new metal substituted heteropolyacid complexes have been developed which do not suffer from such deactivation. For example, $[Fe_4^{II}(BPW_4O_{34})_2]^{10-}$ is one such complex which is also stable to solvolysis by hydrogen peroxide.[93] Recently, a catalytic system consisting of sodium tungstate dihydrate, (aminoethyl)phosphonic acid and methyltrioctylammonium hydrogensulfate has been developed for the epoxidation of olefins using 30% m/m hydrogen peroxide.[94] The reaction proceeds in high yield without solvent, or alternatively with added toluene under entirely halide-free conditions. Lipophilic ammonium hydrogensulfate replaces the conventional chloride and an (α-aminoalkyl)phosphonic acid is crucial for high reactivity. The method is operationally simple, environmentally benign and much more economical than oxidation with MCPBA. The process was found to epoxidize a range of substrates, including terminal olefins such as 1,1- and 1,2-substituted olefins, and tri- and tetra-substituted olefins, as well as allylic alcohols, esters, α,β-unsaturated ketones and ethers (Figure 3.21).

The other main group elements which form peroxo complexes are d^6 and d^8 systems in group VIII including iridium, palladium and platinum. The μ-peroxo complexes do not generally catalyse the epoxidation of olefins with hydrogen peroxide,[95,96] but it has been found that trifluoromethyl-substituted Pd(II) and Pt(II) hydroperoxides will perform such a transformation.[97]

Metalloporphyrins are a versatile class of redox active species which occur widely in nature. Peroxidase enzymes usually based on iron porphyrins catalyse bio-oxidation with peroxides. Even more powerful are the cytochrome P450 mono-oxygenases which normally use molecular oxygen with a co-reductant, but can equally well be employed with peroxides. Work aimed at understanding these systems has led to the development of simple porphyrins which are efficient oxidation catalysts.[98] The best epoxidation catalyst based on a porphyrin system has been prepared by Mansuy's group,[99] which uses a manganese porphyrin together with an imidazole base as an axial ligand. The base also reduces side-reactions. Iron-based porphyrins have also been used with hydrogen peroxide as efficacious epoxidation catalysts.[100]

A number of heterogeneous metal peroxo systems have been developed for

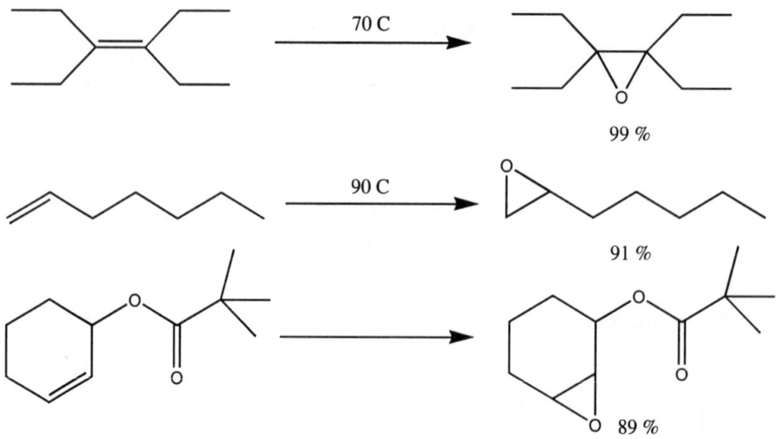

$Na_2WO_4/NH_2CH_2PO_3H_2/[CH_3(n-C_8H_{17})_3N]HSO_4/H_2O_2.$

Figure 3.21   *Epoxidation of olefins with 30% m/m hydrogen peroxide.*

use as epoxidation catalysts. Titanium(IV) oxide supported on silica has been claimed to be an effective catlyst for the epoxidation of allyl chloride with cumene hydroperoxide.[101] Titanium silicalites, which have titanium incorporated in the zeolite framework, have been developed for a wide range of oxidations including epoxidation with aqueous hydrogen peroxide.[102] The use of the system is limited by the size of the substrate, which all have to pass down zeolite channels of about 5.5 Å in diameter. Solvay Interox have developed a supported heteropolyacid catalyst system which makes catalyst recovery easier and does not significantly leach catalyst into the liquid phase.[103] Tellurinic acid immobilized on polystyrene has been reported to oxidize cyclohexene to the epoxide in the presence of hydrogen peroxide.[104] It should be noted that the system has no homogeneous analogue. The application of heterogeneous oxidation catalysts will be expanded upon in Chapter 4.

Finally, *t*-butyl hydroperoxide (TBHP) is an important epoxidizing agent, particularly when coupled with titanium(IV) and organic chiral templates such as D-(−)- or L-(+)-dialkyltartrate for asymmetric synthesis.[105] Enantioselective epoxidation has been attempted with several other peroxygen systems including (+)-2-heptahelicenonitrile with alkaline hydrogen peroxide,[106] quinium benzyl chloride with alkaline hydrogen peroxide,[107] polypeptides with alkaline hydrogen peroxide,[108] chiral molybdenum(VI) diperoxo complexes[109] and chiral dioxiranes (Figure 3.22).[110]

3   Hydroxylation of Olefins

The hydroxylation of olefins can be considered as the addition of hydrogen peroxide across a double bond to produce a vicinal diol. Where the epoxide is formed as an intermediate, the diol product is usually *trans* with respect to the olefin (Figure 3.23).

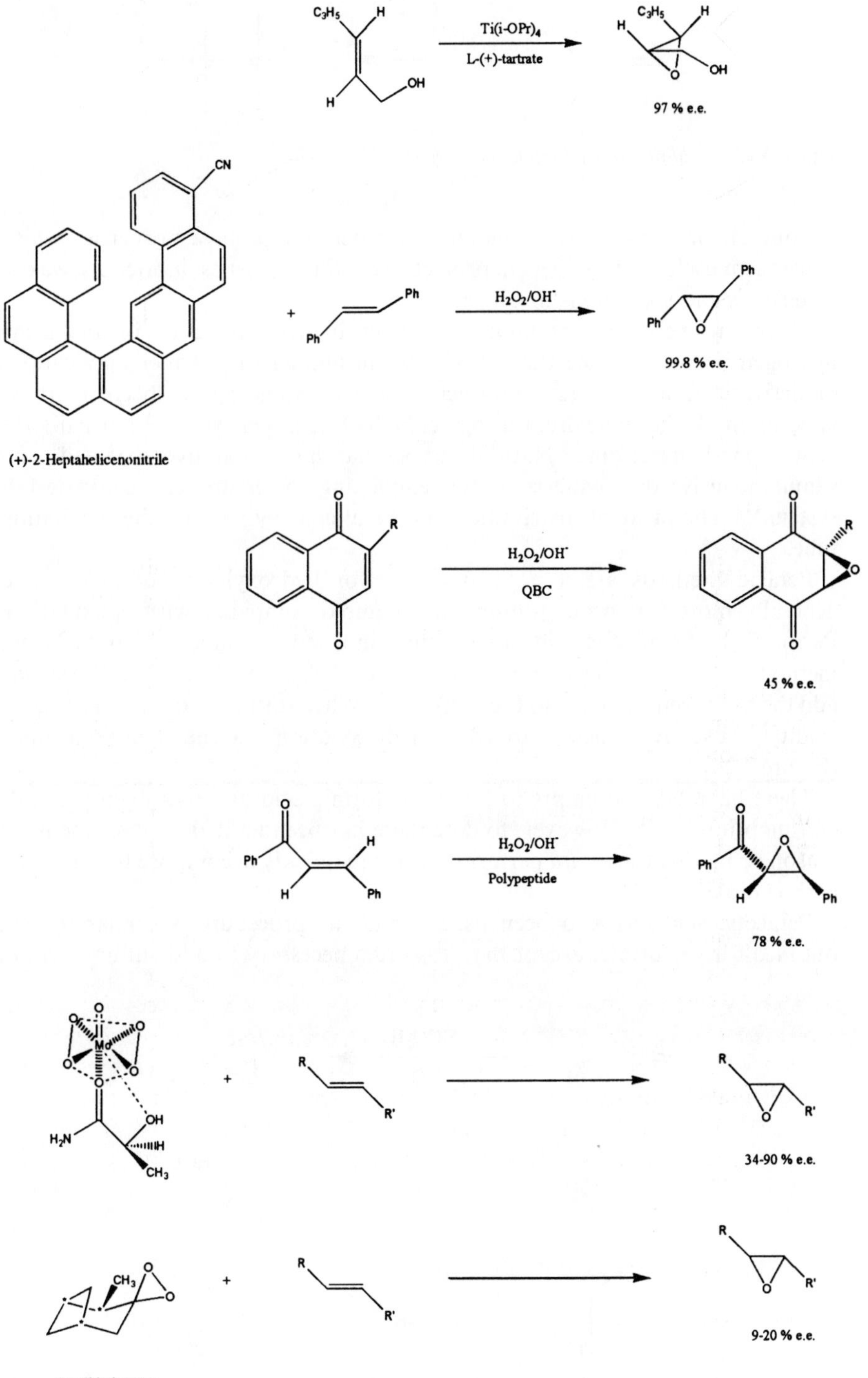

Figure 3.22 *Stereospecific epoxidation of olefins with hydrogen peroxide.*

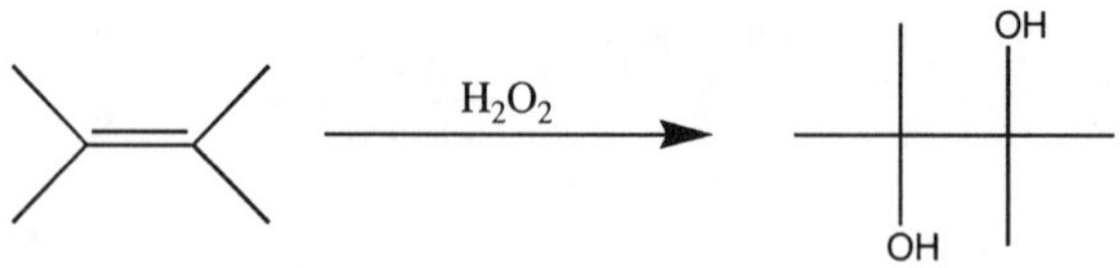

Figure 3.23   *Addition of hydrogen peroxide to olefins.*

Long chain aliphatic *vic*-diols and their derivatives have properties which make them useful as modifiers in protective coatings, plastics, lubricants, waxes, emulsifiers and textile finishing agents.

There are very few examples of direct hydroxylation of olefins using hydrogen peroxide, since these methods are limited to polymer applications or derivatization of natural products. Vinyl monomers have been hydroxylated in an alcoholic medium using acidic hydrogen peroxide;[111] normally the acid is methanesulfonic. Natural rubber has also been hydroxylated, and simultaneously depolymerized by employing a hydrogen peroxide/UV system.[112] The product distribution can be altered by varying the irradiation time.

Peracid methods are a popular choice for hydroxylation of olefins and generally more forcing conditions are required compared with epoxidation. Performic and peracetic acids are usually employed. A convenient and efficient method is to dissolve the olefin in an excess of the carboxylic acid and to slowly add the hydrogen peroxide to the mixture.[113] When formic acid is employed, the acidity is usually sufficient to effect hydroxylation via the hydroxyformate (Figure 3.24).

There have been attempts to lower the formic acid and hydrogen peroxide consumption.[114–116] However, to date there has been limited success, the main reason being that the *in situ* performic acid can quickly decompose to water and carbon dioxide.

Peracetic acid has also been used[117] and the procedure is similar to that outlined for performic, except that it is often necessary to add sulfuric acid in

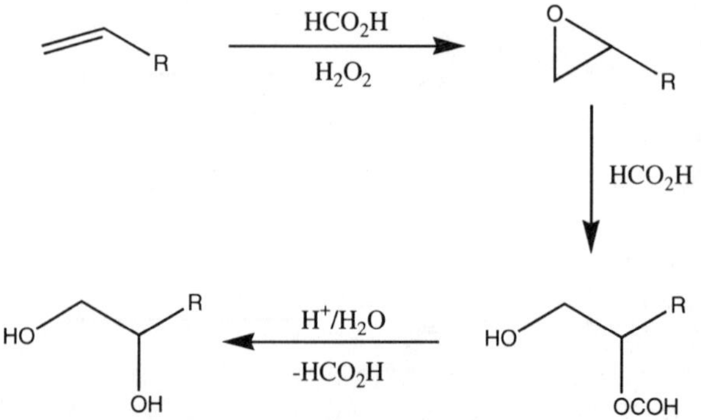

Figure 3.24   *Formation of 1,2-vic-diols from olefins in the presence of hydrogen peroxide/ formic acid.*

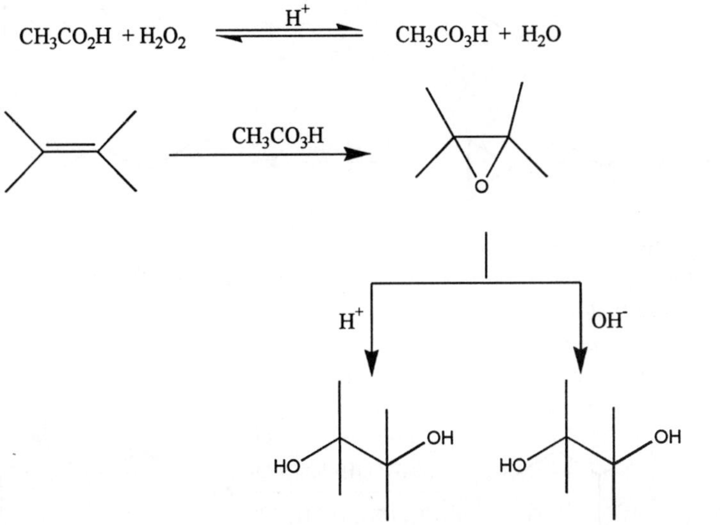

$$CH_3CO_2H + H_2O_2 \xrightleftharpoons{H^+} CH_3CO_3H + H_2O$$

Figure 3.25 *Hydroxylation of olefins to vic-diols using* in situ *formed peracetic acid.*

order to attain equilibrium to the peracid more quickly. The intermediate epoxide can either be isolated or directly acid-catalysed through to the diol. The isolated epoxide is usually treated with excess hydroxide to form the diol (Figure 3.25).

The peracid methods invariably open the epoxide with reversion of configuration, *i.e. trans*-diol formation. Aryl substituents, however, are converted to the *cis*-diols with retention of configuration.[118–120] Olefins which have been hydroxylated by means of *in situ* percarboxylic acid techniques include cyclohexene (65–73%),[121] dodecane (91%)[122] and oleic acid (99%).[123] Chlorestrol has been *trans*-hydroxylated with performic acid in high yield (91%).[124]

Pentane-1,2-diol is an important intermediate for the production of fungicides[125] and can be continuously prepared by reacting pent-1-ene with a solution of perpropionic acid in benzene, and subsequent hydrolysis of the epoxide.[126] The diol yield of 92% from the alkene is much higher than with performic acid (Figure 3.26).[116]

A number of catalytic systems have been devised based on the activation of hydrogen peroxide for olefin hydroxylation. Olefins have been hydroxylated using tungstic acid/hydrogen peroxide in unbuffered media.[127] Olefins which have been hydroxylated in this way are oleic acid, methyl oleate and maleic acid,[128] and if a suitable alcohol is employed (methanol, ethanol, isopropanol or tertiary butanol),[129] cyclohexene can be hydroxylated. The kinetics and mechanism of the system have been investigated.[130] A number of other catalyst systems based on heteropolyacids have been used to hydroxylate allyl alcohol to glycerol. The workers believed the rate-determining step was the breakdown of the phosphotungstate complex which dissociates under acidic conditions.[131] The sodium tungstate/hydrogen peroxide/phosphoric acid system in the

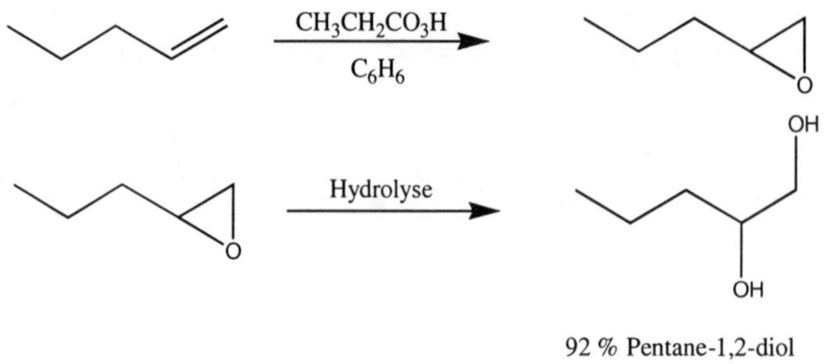

Figure 3.26   *Hydroxylation of pent-1-ene using perpropionic acid.*

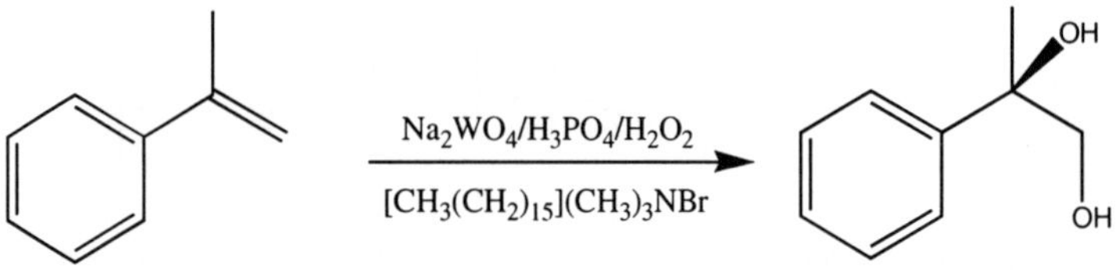

Figure 3.27   *Hydroxylation of α-methylstyrene using a phosphotungstate system.*

presence of a quaternary ammonium salt has been effectively employed to hydroxylate α-methylstyrene (Figure 3.27).[132]

Homogeneous oxidation of isolated isomers of methylcyclohexene into the corresponding glycols has been studied with a hydrogen peroxide/molybdenum oxobromide complex.[133] The yield of the glycols was optimized using statistically designed experiments. Oxidation of cyclic olefins over mixed Mo/W heteropolyacids has also been applied to hydroxylation with hydrogen peroxide. The workers found that the ratio Mo:W had a dramatic effect on the yield of the *vic*-diols.[134] Finishina and co-workers have reported the preparation of 2-methylfuran from penta-1,3-diene.[135] The oxidation goes through hydroxylated intermediates. The reaction is performed with palladium(II) salts and hydrogen peroxide. Varying the solution pH, penta-1,3-diene:oxidant ratio and temperature allows for an optimal yield of 2-methylfuran (*ca.* 88%) to be realized. The workers derived a mechanism based on ^{1}H NMR spectroscopic data and kinetic rate determination (Figure 3.28).

In acidic solutions, the reaction was found to proceed through pathway (a). Isomeric dihydroxypentanones (I and II) were formed in small yields, and led to the formation of side-products. The mechanism of the reaction is still unclear, however the specific oxidizing properties of hydrogen peroxide probably plays an important role.

Metal silicalite catalysts based on ZSM-5 zeolite type materials have been used for the hydroxylation of olefins. The silicalites need to contain a certain amount of AlO_2 units to be active.[136]

Oxidation systems based on osmium have been extensively researched for the hydroxylation of olefins. Hydrogen peroxide/OsO$_4$ (Milas reagent) will hydro-

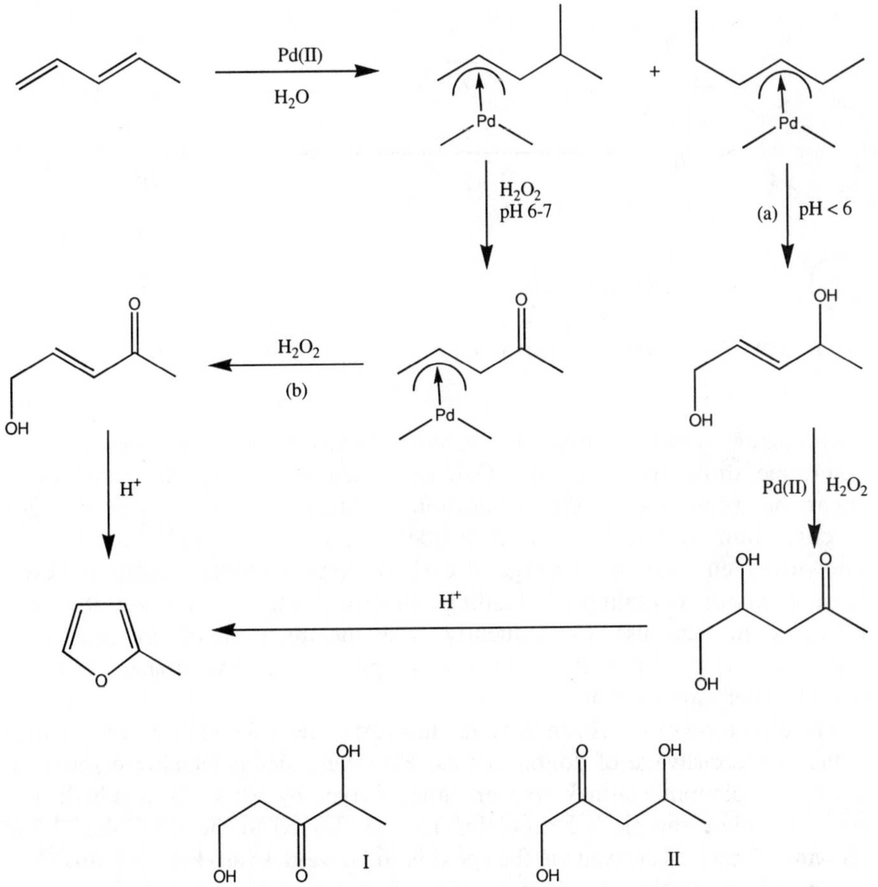

Figure 3.28 *Palladium(II)/hydrogen peroxide route to 2-methylfuran involving hydroxylation.*

xylate olefins stereospecifically to *cis*-glycols.[137] Osmium tetraoxide has also been immobilized on cross-linked resins bearing a tertiary amine group, and used in the presence of secondary oxidants (hydrogen peroxide, peracetic acid, trimethylamine oxide) to hydroxylate olefins. For example, styrene affords 80% of the diol product (Figure 3.29).[138]

An osmium catalyst with an alkyl hydroperoxide, halide promoter and an alkaline earth co-catalyst can be used to produce *vic*-diols in moderate to good yield (44–70%) with excellent selectivity up to 95%.[139]

4 Oxidative Cleavage of Olefins

Oxidative cleavage of carbon–carbon double bonds or *vic*-diols can lead to aldehydes, ketones or carboxylic acids depending upon the conditions employed. Oxidative cleavage of olefins is important from the viewpoint of

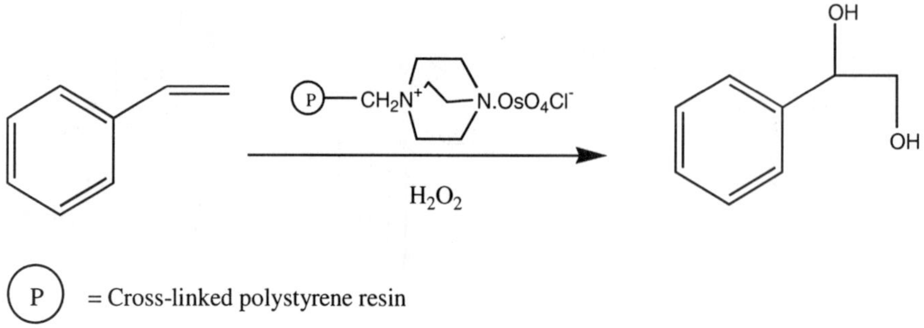

P = Cross-linked polystyrene resin

Figure 3.29 *Hydroxylation of styrene using a supported osmium catalyst.*

synthesizing systems containing carbonyl functions and has been used to determine molecular structure. Oxidative cleavage of *vic*-diol moieties in sugars has been used for the preparation of chiral synthons.[140] Oxidants such as chromium trioxide,[141] lead tetraoxide[142] and periodic acid[143] are all well known reagents for the cleavage of carbon–carbon double bonds. However, these oxidants contain high loadings of heavy metals, and are therefore restricted in their use. Consequently, it is the intention of this section to introduce the reader to the concept of employing peroxygen species for the oxidative cleavage of olefins.

The direct use of hydrogen peroxide has rarely been shown to be of practical value in the cleavage of double bonds. However, olefins which are activated towards nucleophilic attack are sometimes cleaved by excess alkaline hydrogen peroxide. For example, α-keto olefins can be cleaved to the aldehydes[144] and cinnamaldehyde is cleaved via the epoxide to benzaldehyde (Figure 3.30).[145]

Peracids themselves produce epoxides and diols from alkenes but are not powerful enough to oxidize these further by cleaving the carbon–carbon bond. However, in the presence of transition metals they will cleave alkenes and diols to give, usually, carboxylic acids. For example, peracids in combination with ruthenium compounds are well known in this capacity (Figure 3.31).[146,147] Warwel and co-workers have reported the cleavage of alkenes using peracetic acid and the ruthenium catalyst, Ru(acac)$_3$.[148]

The oxidation catalyst is believed to be ruthenium tetraoxide based on work by Engle,[149] who showed that alkenes could be cleaved with stoichiometric amounts of ruthenium tetraoxide. Suitable solvents for the Ru/peracid systems are water and hexane, the alkene (if liquid) and aromatic compounds. Complexing solvents like dimethylformamide, acetonitrile and ethers, and the addition of nitrogen-complexing agents decrease the catalytic system's activity. It has also been found that the system has to be carefully buffered otherwise the yield of the resulting carboxylic acid drops drastically.[150] The influence of various ruthenium compounds has also been studied, and generally most simple and complex ruthenium salts are active. The two exceptions are Ru-red and Ru-metal, which are both inferior to the others. Ruthenium to olefin molar ratios as low as 1/20 000 will afford excellent cleavage yields (> 70%). *vic*-Diols are also

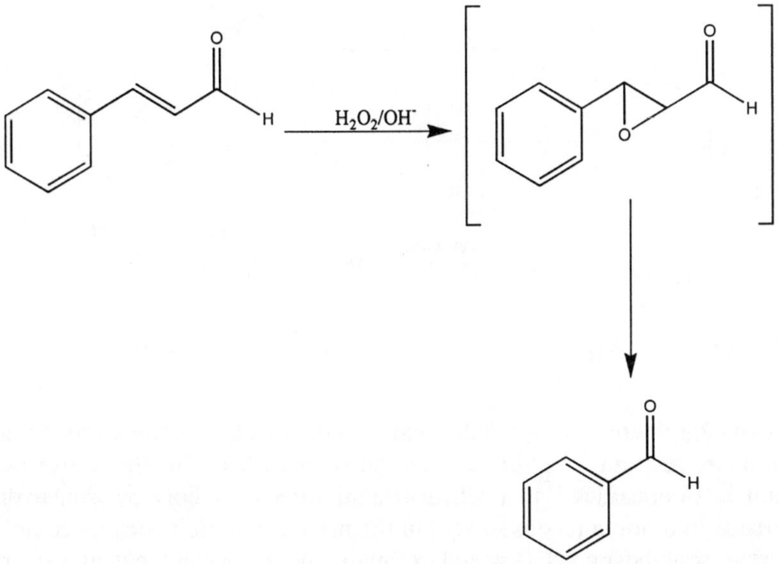

Figure 3.30 *Oxidative cleavage of cinnamaldehyde to benzaldehyde using alkaline hydrogen peroxide.*

converted to carboxylic acids in the presence of metal ions (Co, Mn, *etc.*) under essentially non-aqueous conditions.[151]

As with percarboxylic acids, hydrogen peroxide usually requires metal activation.[151–154] Peroxo complexes of d^0 transition metals can often be used to cleave olefins. Tungstic acid cleaves olefins to the carboxylic acid using 35% *m/m* hydrogen peroxide.[154] Sodium tungstate has also been reported to

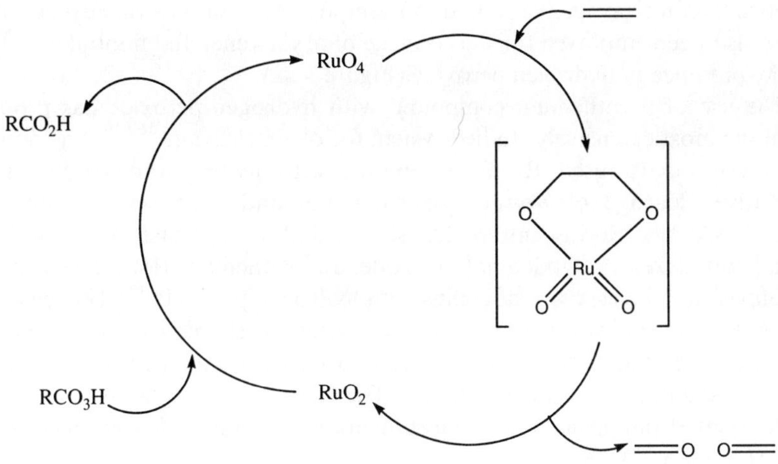

Figure 3.31 *Catalytic cycle of the ruthenium-catalysed cleavage of olefins in the presence of percarboxylic acids.*

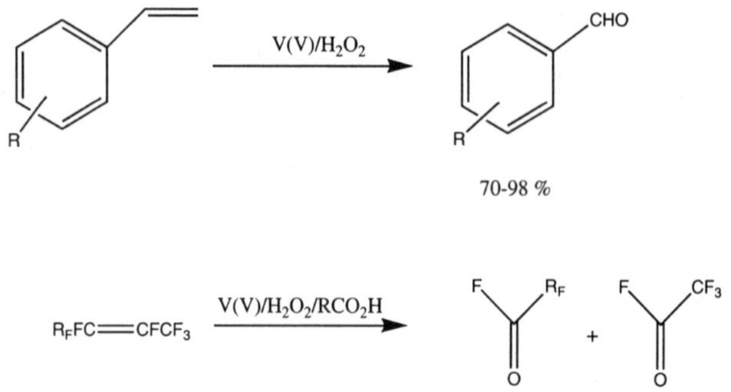

70-98 %

Figure 3.32 *Cleavage of olefins with hydrogen peroxide/vanadium(V) systems.*

cleave 1,2,3,6-tetrahydrophthalic acid to butane-1,2,3,4-tetracarboxylic acid, a useful sequestrant.[155] Processes are also described for the conversion of olefins[156] or epoxides[157] to aldehydes (in mixtures with diols) by using hydrogen peroxide in a non-aqueous solvent in the presence of molybdenum compounds such as molybdenum(III) acetylacetonate. A subsequent patent uses boron compounds such as boric anhydride or esters as additional reactants when epoxides are the starting materials.[158] From the fact that epoxide cleavage required no metal catalysts but only boron compounds,[159] it may be concluded that the role of the metal catalyst is in epoxide formation. The subsequent steps promoted by the boron compounds might well involve formation of a hydroxy–hydroperoxy intermediate and its fragmentation.[160] The formation of glutaraldehyde from cyclopentene is a typical example.

Mo(VI) peroxo complexes have also been used in the cleavage of olefins both stoichiometrically and catalytically.[161] A phase transfer agent can also be used with substrates insoluble in water. Cetylpyridinium chloride[162] and tetra-alkylphosphonium chlorides[163] have been used with metal peroxo compounds to effect olefin cleavage. Vanadium(V) compounds[164] such as vanadyl acetate[165] have also been employed for the cleavage of arylalkenes and fluorinated olefins in the presence of hydrogen peroxide (Figure 3.32).

The use of a ruthenium compound with hydrogen peroxide has probably been the most extensively studied system for olefin cleavage.[167,168] Sheldon and co-workers have used Ru–Bi complexes with hydrogen peroxide for the oxidative cleavage of *trans*-cyclohexane diol and methyl-α-D-glucopyrano-side.[168] Work at Solvay Interox has shown that a two-phase system with two metal complexes, molybdenum(VI) oxide, and ruthenium(III) chloride, can be employed for the cleavage of olefins with hydrogen peroxide.[169] The reason for using the mixed metal system was an attempt to reduce the amount of hydrogen peroxide which normally has to be used with ruthenium catalysts due to dismutation side-reactions. The molybdenum catalyst effectively hydroxylated the olefin, whilst the ruthenium compound cleaved the resulting diol (Figure 3.33).

Table 3.4 illustrates the effective cleavage of several olefins in the presence of the Ru(III)/Mo(VI)/hydrogen peroxide system.

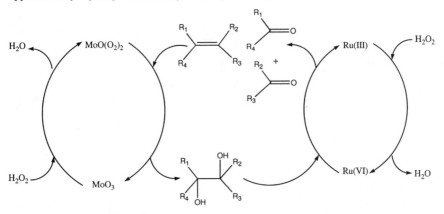

Figure 3.33 *Catalytic cleavage of olefins using a mixed ruthenium(III)/molybdenum(VI)/hydrogen peroxide system.*

Table 3.4 *Cleavage of olefins using a Ru(III)/Mo(VI)/hydrogen peroxide system*[a]

Substrate	Conversion (%)	Acid yield (%)	Aldehyde yield (%)
Oleic acid	100	Azelaic (100)	—
		Nonanoic (43)	
Styrene	100	42	49
Non-2-ene	100	62	—
Stilbene	100	—	65
4-Chlorostyrene	100	30	33
Castor oil	—	Heptanoic (42)	—

[a] t-Butanol (50 ml), MoO_3 (0.2 g), $RuCl_3$ (0.03 g), didecyldimethylammonium bromide (0.4 g), substrate/hydrogen peroxide ratio 1:8, 80 °C, 4 h.

Hydrogen peroxide with Milas reagent is also suitable for the cleavage of olefins. The system has been used to prepare aromatic aldehydes such as anisaldehyde from anethole, pipernal from isosafrole and vanilin from isoeugenol.[170,171] Hydrogen peroxide and chromium(III) can be used to cleave methyl methacrylate to methyl pyruvate.[172]

Recently, a number of *vic*-diols have been efficiently cleaved employing a heteropolyacid/hydrogen peroxide system under phase-transfer conditions.[173] The conversions of the substrates were 60–100% and selectivities to the carboxylic acids were 30–98%.

Heterogeneous systems have also been developed for the cleavage of olefins with hydrogen peroxide.[174,175] Titanium-containing zeolites can be used to cleave olefins.[176] Adam and co-workers have recently shown that acetophenone, an oxidation product from the Ti-zeolite catalysed oxidation of α-methylstyrene, derives from 2-hydroxyperoxy-2-phenylpropan-1-ol as an intermediate (which they detected and isolated) (Figure 3.34).[177]

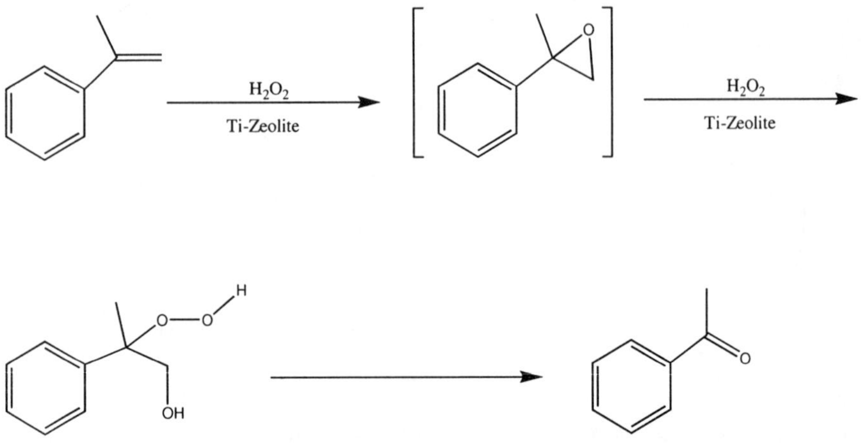

Figure 3.34   *Cleavage of α-methylstyrene with a hydrogen peroxide/titanium zeolite system involving the production of 2-hydroxy-2-phenylpropan-1-ol.*

On an industrial scale, the traditional method for cleavage of carbon–carbon double bonds is ozonolysis, used for the manufacture of azelaic acid and nonanoic acids from oleic acid, and of butane tetracarboxylic acid from tetrahydrophthalic anhydride. The process is effectively a quantitative and mild process.[178] However, it is capital and energy intensive. The intermediate ozonide is worked up either reductively or oxidatively to produce the aldehyde, ketone or carboxylic acid. Hydrogen peroxide is the common oxidizing agent used in the second step.[179–181] Oxygen can also be used either alone[182] or in combination with zeolites.[183] Reviews on ozonolysis are available and the reader is directed to reference 184 for further information.

5   Oxidation of Alcohols

The oxidation of alchols can be readily carried out using peroxygen reagents. The extensive range of methods available makes it possible to achieve the oxidation of a wide variety of both aliphatic and aromatic primary and secondary alcohols, often in a highly selective manner.

Direct activation methods are not normally successful for the oxidation of alcohols using hydrogen peroxide. An advantage of this is that alcohols can often be used as solvents for other oxidations, which do not use direct activation methods. However, hydroxy ketals can be cleaved with aqueous hydrogen peroxide, and the ene–diol of L-ascorbic acid can be oxidized with sodium perborate.[185] Under the conditions employed, it is likely that the active oxidant is free hydrogen peroxide rather than any boron species.

Peroxides may be used to oxidize alcohols, but additional activation is usually necessary for successful reaction to take place. An early example of this was the use of catalytic amounts of 2,2,6,6-tetramethylpiperidine hydrochloride

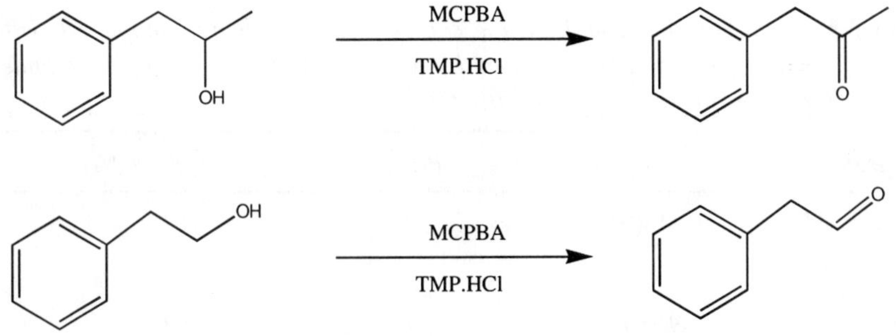

Figure 3.35 *Activation of MCPBA with TMP·HCl for the oxidation of alcohols.*

(TMP·HCl) in conjunction with 3-chloroperbenzoic acid (MCPBA) to oxidize alcohols to aldehydes or ketones in good yield (Figure 3.35).[186]

A number of variations on the system, including the use of the nitroxyl radical of TMP·HCl[187] and of perbenzoic acid,[188] have been described. Unsaturated substrates are converted to epoxyketones.[187] By adding excess peracid, further conversion of ketones to esters via Baeyer–Villiger re-arrangement is possible. MCPBA may be used to oxidize sterically unhindered, acid-stable alcohols in the presence of hydrochloric acid using dimethylformamide or tetrahydrofuran solvents.[189]

Simpler, aliphatic peracids can also be used for alcohol oxidation under certain conditions. Morimoto and co-workers have reported the smooth oxidation of secondary alcohols and benzylic alcohols to the corresponding carbonyl compounds using peracetic acid (PAA) in the presence of 0.5–1.5 equivalents of sodium bromide in acetic acid solution (Figure 3.36).[190]

Primary aliphatic alcohols were oxidized slowly using the system but gave only poor yields ($<10\%$).

Metal species, including chromium,[191] iron,[192] cobalt,[193] cerium[194] and ruthenium[195] compounds, have also been used to catalyse alcohol oxidation with peracetic acid.

A wide range of catalytic systems, with hydrogen peroxide as the primary

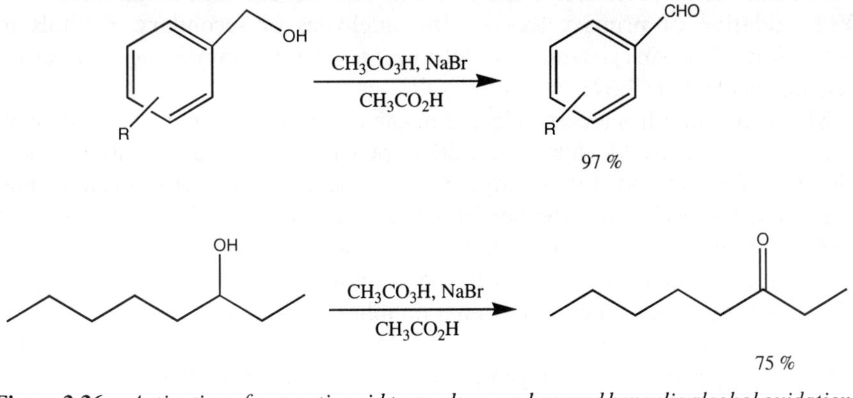

Figure 3.36 *Activation of peracetic acid towards secondary and benzylic alcohol oxidation.*

Table 3.5 *Oxidation of alcohols to the corresponding carbonyl compounds with hydrogen peroxide catalysed by Mo(VI) or W(VI) complexes under phase-transfer conditions at 70 °C^a*

Substrate	Catalyst	Product	Yield (%)
(cyclohexanol, OH)	Mo(VI)	(cyclohexanone, O)	88
(cyclohexanol, OH)	W(VI)	(cyclohexanone, O)	97
(benzyl alcohol, CH₂OH)	Mo(VI)	(benzaldehyde, CHO)	85
(menthol, OH)	W(VI)	(menthone, O)	93
(menthol, OH)	Mo(VI)	(menthone, O)	97

^{a} Yields based on alcohol; phase-transfer agent is Aliquat$^®$ 336.

oxidant, are available for the selective oxidation of alcohols. Tungsten and molybdenum species used under phase-transfer conditions (Table 3.5) have been found to be particularly effective and versatile for such oxidations.[196,197] The oxidation of primary alcohols to aldehydes or secondary alcohols to ketones do not, in general, over-oxidize to the carboxylic acid or ester, respectively (Figure 3.37).

Muzart's group has recently described the use of molybdenum catalysts with the hydrogen peroxide adduct, sodium percarbonate and a phase-transfer agent.[198] The molybdenum catalyst used in the study was $MoO_2(acac)_2$, and the solvents screened were dichloroethane and acetonitrile. The active species is a Mo peroxo complex and in common with other methods based on Mo and W catalysts, secondary, allylic and benzylic alcohols react quickly, and give higher yields of carbonyl product than primary aliphatic alcohols.

Mechanistic studies on the behaviour of Mo(VI) peroxo complexes reveal that the anionic peroxospecies of general formula $[Mo(O_2)_n(OH)(H_2O)_{m-1}]^-$ are particularly efficacious for the oxidation of secondary alcohols.[199] Tungsten(VI)

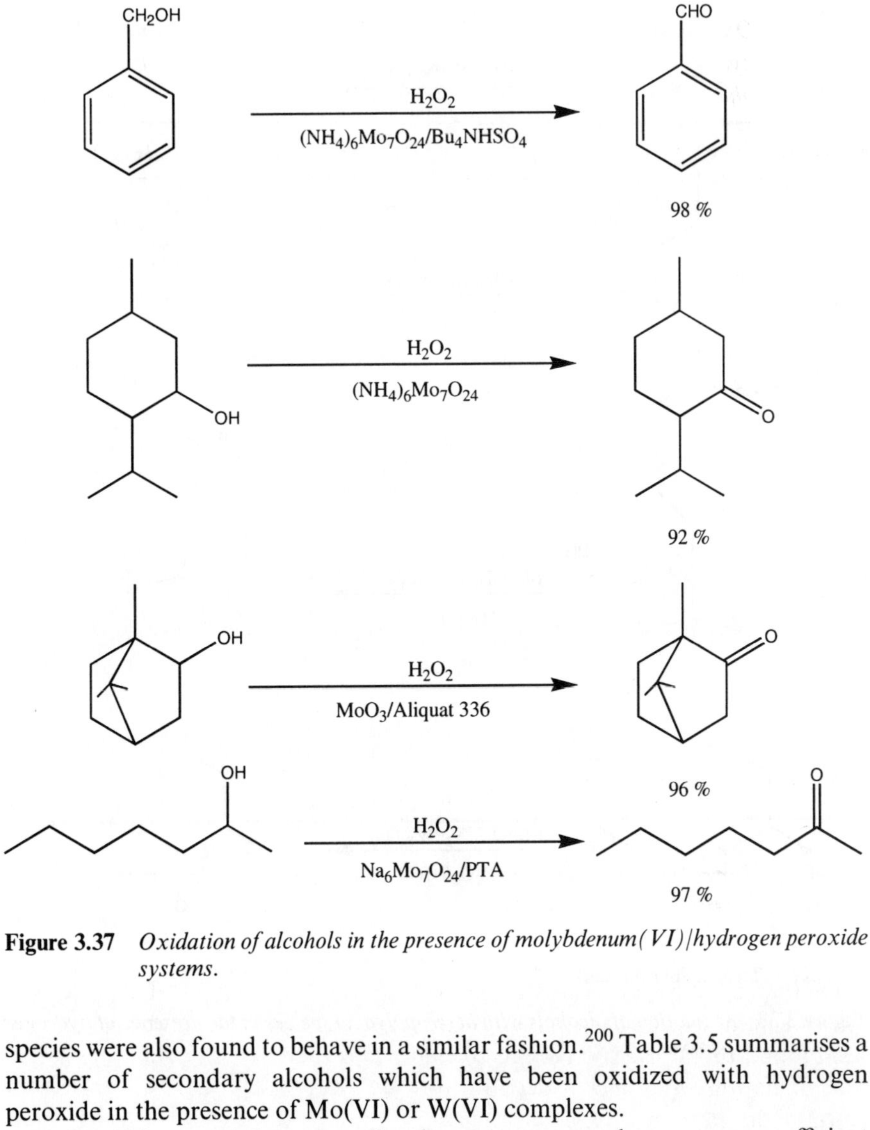

Figure 3.37 *Oxidation of alcohols in the presence of molybdenum(VI)/hydrogen peroxide systems.*

species were also found to behave in a similar fashion.[200] Table 3.5 summarises a number of secondary alcohols which have been oxidized with hydrogen peroxide in the presence of Mo(VI) or W(VI) complexes.

It should be noted that generally, the tungsten complexes are more efficient catalysts than their analogous molybdenum complexes. One reason for the difference may be due to the slight differences in the μ-oxo (M–O) bond lengths in the two complex types.

Heteropolyacid species of tungsten and molybdenum in the presence of hydrogen peroxide have also been employed for alcohol oxidation, again under phase-transfer conditions (Figure 3.38).[201]

Griffith has recently reported the synthesis and use of lanthanide hetero-polyacid complexes with aqueous hydrogen peroxide for the oxidation of secondary alcohols.[202] The complexes were $[LnW_{10}O_{36}]^{9-}$ and $[Ln(PW_{11}O_{39})_2]^{11-}$ (Ln = La, Pr, Sm, Tb). However, the active oxidant was

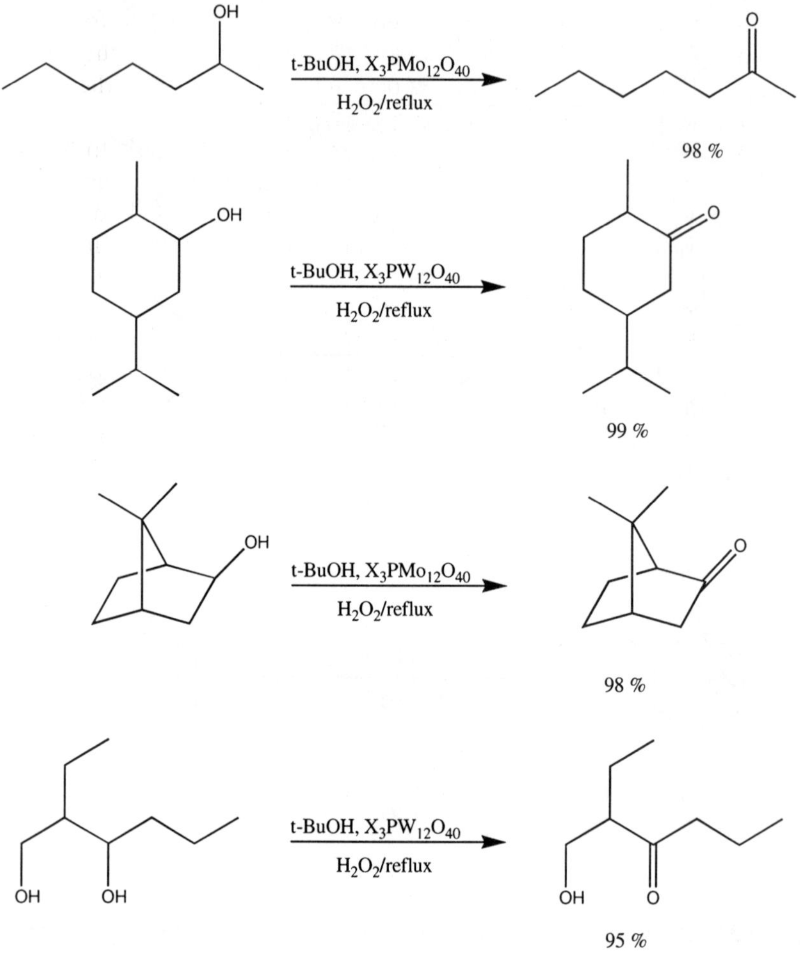

X_3 = Tris(cetylpyridinium).

Figure 3.38   *Oxidation of alcohols with heteropolyacid species in the presence of hydrogen peroxide.*

found, via Raman studies, to be $[W_2O_3(O_2)_4(H_2O)_2]_2$,[203] *i.e.* the lanthanide was effectively dormant.

Ruthenium(III) chloride has been shown to be particularly effective with hydrogen peroxide for the oxidation of alcohols under phase-transfer conditions.[204] Primary alcohols are converted to acids, allylic and secondary alcohols to ketones, and benzyl alcohols to either benzaldehydes or benzoic acids.

The problem of heavy metal contamination of effluent streams which can result from the use of homogeneous catalysts, whilst not insurmountable, can be eliminated by the use of heterogeneous catalysts such as titanium silicalite[205] and phosphotungstic acid on alumina[206] with hydrogen peroxide as the primary oxidant.

A number of other peroxygen based methods are known for alcohol oxidations. These include both organic peroxides, such as *t*-butyl hydroperoxide, and inorganic species such as perdisulfates. In all cases, the presence of metal species to catalyse the reaction is required.

The ability of hydrogen peroxide to oxidize hydrogen bromide to bromine, and of bromine to oxidize alcohols, has also been exploited widely. Such reactions may proceed smoothly giving high yields of the desired products. Any excess hydrogen bromide can be re-cycled efficiently, thereby minimizing any problems associated with effluent. Photolytic conditions are, however, necessary in some instances (Figure 3.39).[207]

However, benzyl alcohols can be converted to benzaldehydes even in the dark.[208] It has been reported that catalytic amounts of HBr may be used for this reaction.[209] Primary aliphatic alcohols give rise to carboxylic esters. Catalytic amounts of bromine itself have been used, instead of HBr in the oxidation of lactate esters to pyruvates.[210]

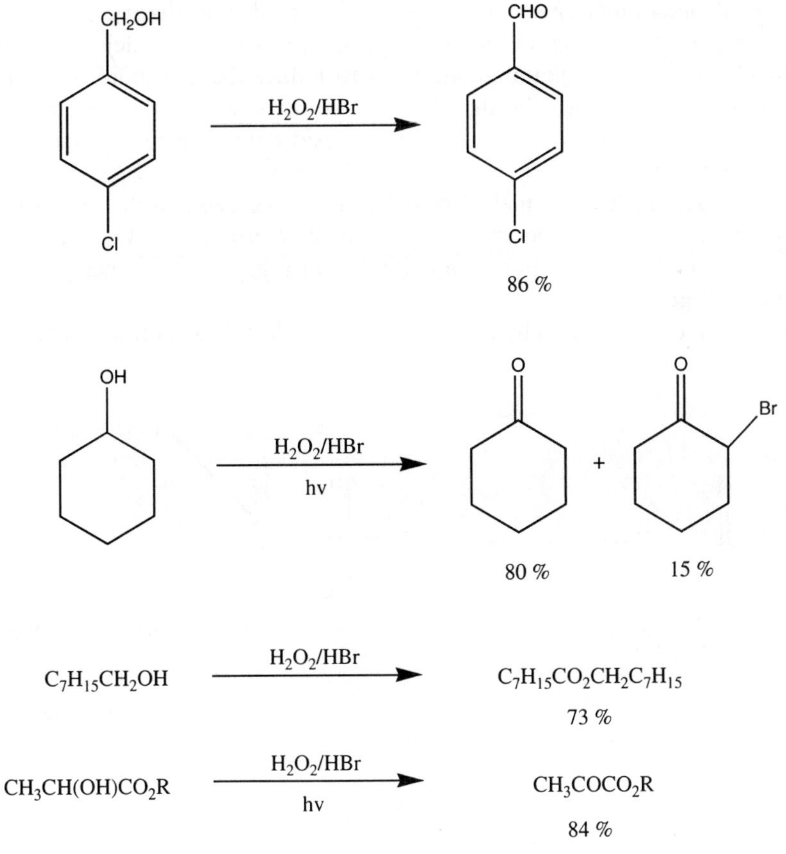

Figure 3.39 *Oxidation of alcohols using a hydrogen peroxide/hydrogen bromide oxidation system.*

6 Oxidation of Carbonyl Compounds

6.1 Oxidation of Aldehydes

Aldehydes are readily oxidized by peroxygen species. Two reaction pathways generally operate (Figure 3.40): oxidation to carboxylic acids or, less commonly, rearrangement to formate esters (*cf.* Baeyer–Villiger oxidation in 6.2) which may hydrolyse to a hydroxy compound containing one less carbon atom compared with the starting material. Both pathways involve a common hydroxyperoxy intermediate, rearrangement only occurring when the attached group has a high migratory tendency.

Whilst the reactivity of the aldehyde function offers a number of synthetic possibilities, care must be taken in oxidizing aldehydes with peroxygens, especially in neutral, anhydrous conditions. The hydroxyperoxy intermediate may be stable enough for dangerous concentrations to accumulate in the reaction mixture, or it may react further, including self-condensation, to form highly explosive organic peroxygen species.[211,212] Safe procedures which avoid the formation of such intermediates are highlighted below. Further information on the safe handling of peroxygen reagents is provided in Chapter 1.

Hydrogen peroxide is an effective oxidant for aromatic aldehydes.[213] It is normally used under alkaline conditions to reduce the possibility of forming explosive intermediates. In many cases the products of the reaction are carboxylic acids. Piperonal, for example, is oxidized to the corresponding acid (Figure 3.41).[214]

Aldehydes which contain electron-releasing substituents in the *ortho* or *para* position give phenol products via a formate intermediate. This reaction is known as the Dakin reaction, the oxidation of salicylaldehyde being a classic example (Figure 3.42).[215]

Although dilute sodium hydroxide is often employed to maintain the alkaline

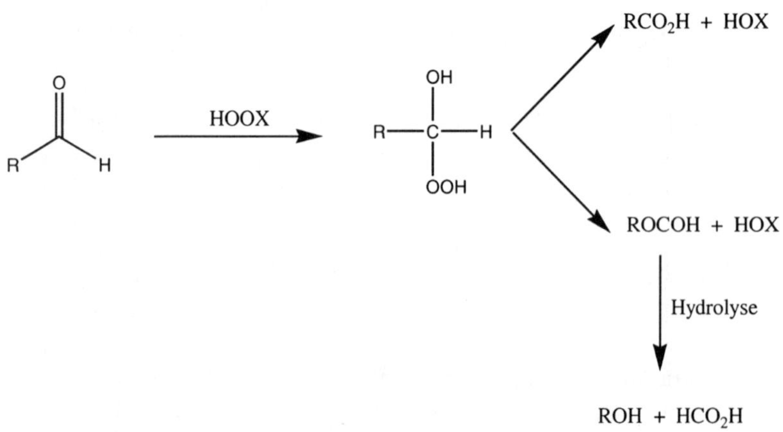

Figure 3.40 *Two competing pathways operating during the oxidation of aldehydes with peroxygen species.*

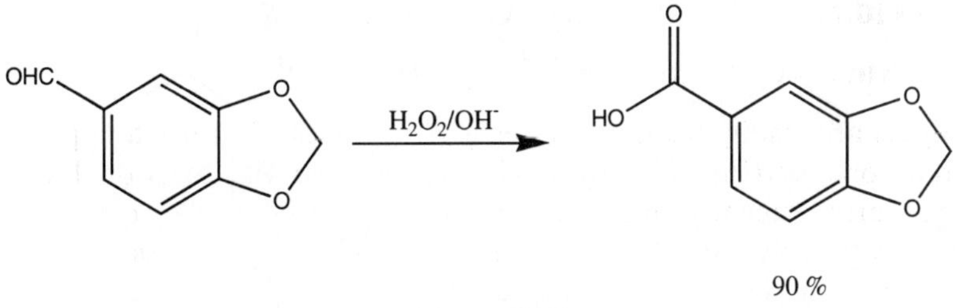

90 %

Figure 3.41 *Oxidation of piperonal with alkaline hydrogen peroxide.*

69-73 %

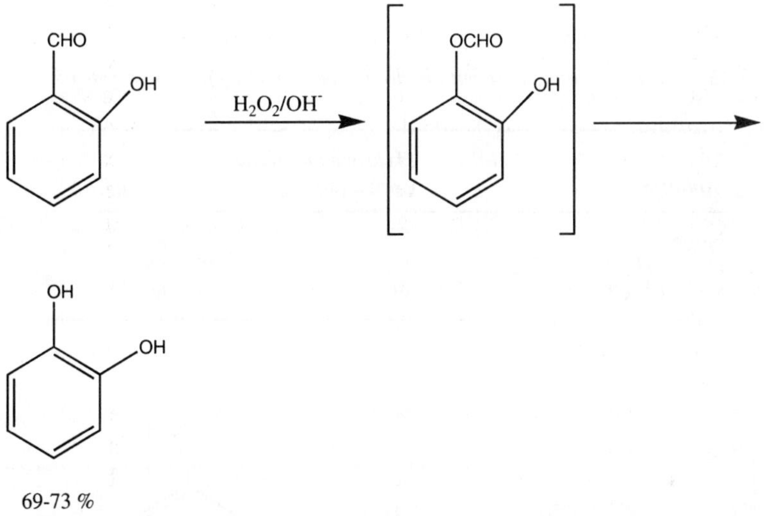

Figure 3.42 *Dakin reaction of aromatic aldehydes to phenols in the presence of alkaline hydrogen peroxide.*

conditions of these reactions, other bases may be used. pH control is often critical to minimize formation of alcohol by-product via the Cannizzaro reaction. Amines are particularly effective in the oxidation of heterocyclic aldehydes, including thiophene-2-carboxaldehyde and furfural (Figure 3.43).[216]

Hydrogen peroxide is prone to decomposition under alkaline conditions. To counteract this problem, stabilizers, which may be used under strongly alkaline conditions, can be employed. The oxidation of 4-isopropylbenzaldehyde to 4-isopropylbenzoic acid requires significantly less hydrogen peroxide when the appropriate stabilizer is used (Figure 3.44).

Although alkaline conditions are normally employed, for reasons of safety it is occasionally advantageous to operate under neutral conditions where, for example, oxidation of benzaldehydes to benzoic acids may be desired rather than the Dakin reaction.[210] Conversely alkyl benzaldehydes, which under most conditions would be expected to yield the corresponding benzoic acids, have been reported to undergo the Dakin-type reaction yielding alkyl phenols in the presence of strong acids.[220]

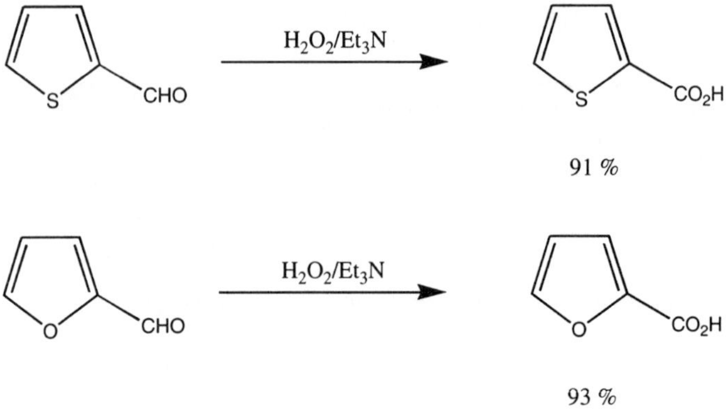

91 %

93 %

Figure 3.43   *Use of an amine base with hydrogen peroxide to effect aldehyde oxidation.*

Stabilizer	Hydrogen peroxide (mol equivalent)	Yield (%)
None	5.1	85
Dequest® 2060 (ref. 217)	4.4	88
Mykon® CIX (ref. 218)	3.9	90

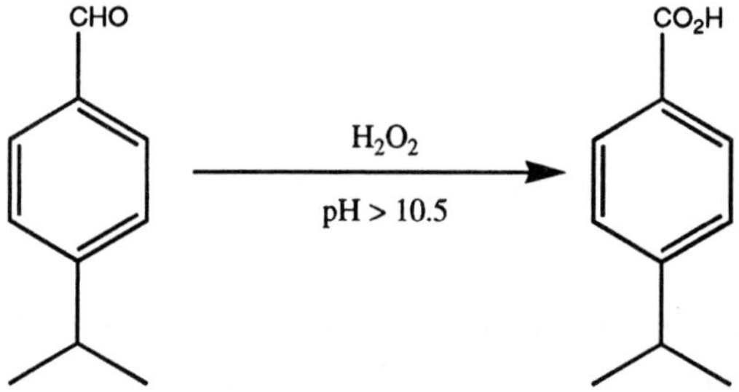

Figure 3.44   *The effect of stabilizer on the oxidation of 4-isopropylbenzaldehyde.*

Percarboxylic acids readily oxidize both aliphatic and aromatic aldehydes to carboxylic acids.[221] Peracetic acid has been the most widely applied. Both pre-formed peracetic acid and *in situ* species may be used (Figure 3.45).[211]

It should be noted that unsaturated aldehydes undergo carbonyl oxidation rather than epoxidation or hydroxylation.[222] Aromatic aldehydes containing electron-releasing groups in the *ortho* or *para* positions give products arising from migration in a manner analogous to the Dakin reaction.

More complex pre-formed peracids, such as 3-chloroperbenzoic acid[223] or the magnesium salt of monoperoxyphthalic acid,[224] may also be used to oxidize aromatic aldehydes, either to carboxylic acids or phenols.

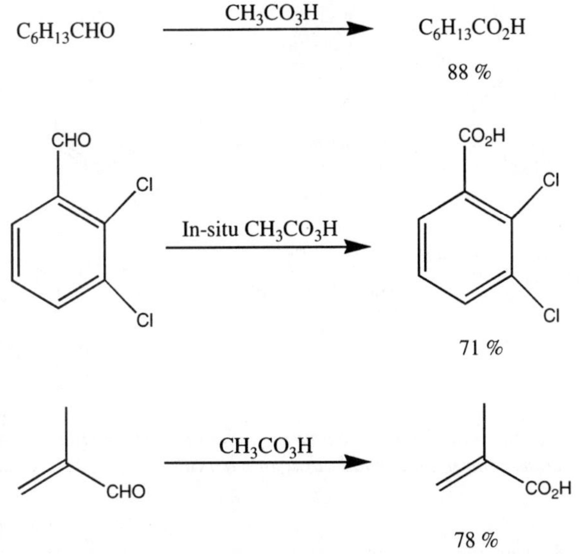

$$C_6H_{13}CHO \xrightarrow{\text{CH}_3\text{CO}_3\text{H}} C_6H_{13}CO_2H$$

88 %

In-situ CH$_3$CO$_3$H

71 %

CH$_3$CO$_3$H

78 %

Figure 3.45 *Use of peracetic acid to effect the oxidation of aldehydes to carboxylic acids.*

Peroxymonosulfuric acid (Caro's acid, H_2SO_5) and its salts may be used to oxidize aldehydes. Although early results were poor compared with the use of organic peracids, good yields of esters have been obtained when reactions are carried out in the presence of alcohols.[225] Unsaturated and aromatic aldehydes undergo analogous reactions. It is believed that hemiacetal formation occurs in these reactions, and that it is this species which is oxidized, rather than the aldehyde.

The ready oxidation of aldehydes using direct activation and peracid methods have meant that there has been little need for the development of catalytic methods for these oxidations. A number of catalytic methods involving selenium species have, however, been examined.[226,227] Such methods may be useful for reactions which are difficult using normal reagents. For example, aromatic aldehydes having polycondensed ring systems or electron-donating substituents may be converted to phenols via formates by the action of hydrogen peroxide in the presence of a range of organoselenium compounds. Materials which would not normally be considered to be sufficiently activated to undergo Dakin type reactions may react in this way (Figure 3.46).[228]

A few examples of the use of transition metal catalysis in conjunction with hydrogen peroxide for aldehyde oxidations have also been reported.[228,229]

A number of other peroxygen reagents have been used for aldehyde oxidation. In particular, solid adducts of hydrogen peroxide such as urea–hydrogen peroxide (UHP) can generate peracids conveniently *in situ*.[230] This can be applied to the oxidation of benzaldehydes (Figure 3.47). The products are those expected to form typical peracid reactions and are obtained under mild conditions in good yield.

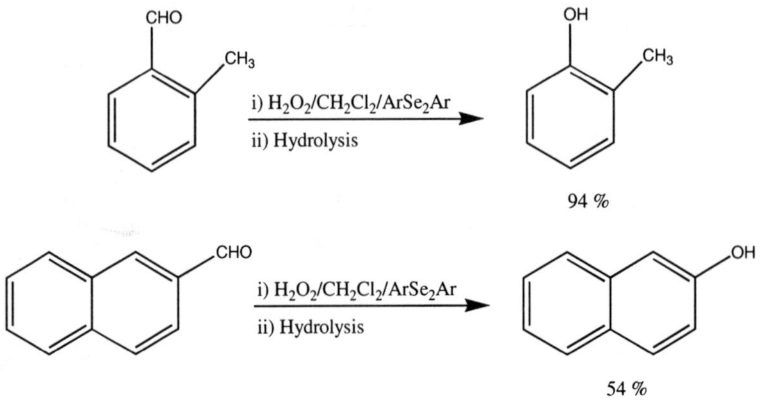

Figure 3.46 *Catalytic oxidation of aromatic aldehydes to phenols in the presence of selenium compounds and hydrogen peroxide.*

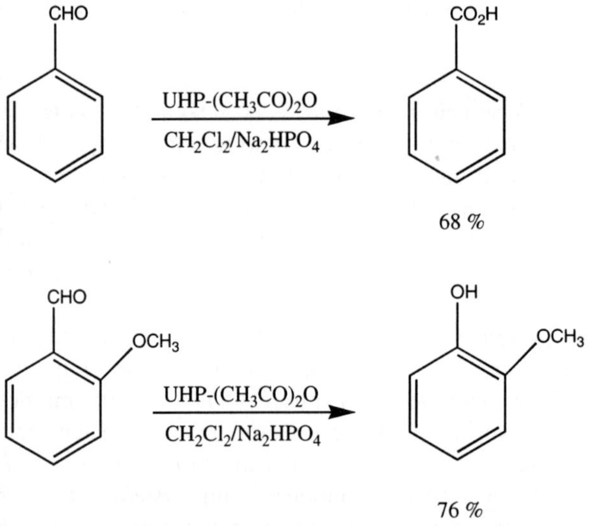

Figure 3.47 *UHP/acetic anhydride systems used for the oxidation of aromatic aldehydes.*

Sodium perborate tetrahydrate (PBS-4) has been shown to oxidize a wide range of aromatic aldehydes to acids in excellent yield under mild conditions in acetic acid solvent (Figure 3.48).[231]

Organic peroxygen compounds other than peracids have not found wide use in aldehyde oxidation, although some examples of the use of organic hydroperoxides, including *t*-butyl peroxides (TBHP) and 2-hydroperoxyhexafluoropropan-2-ol have been reported.[213]

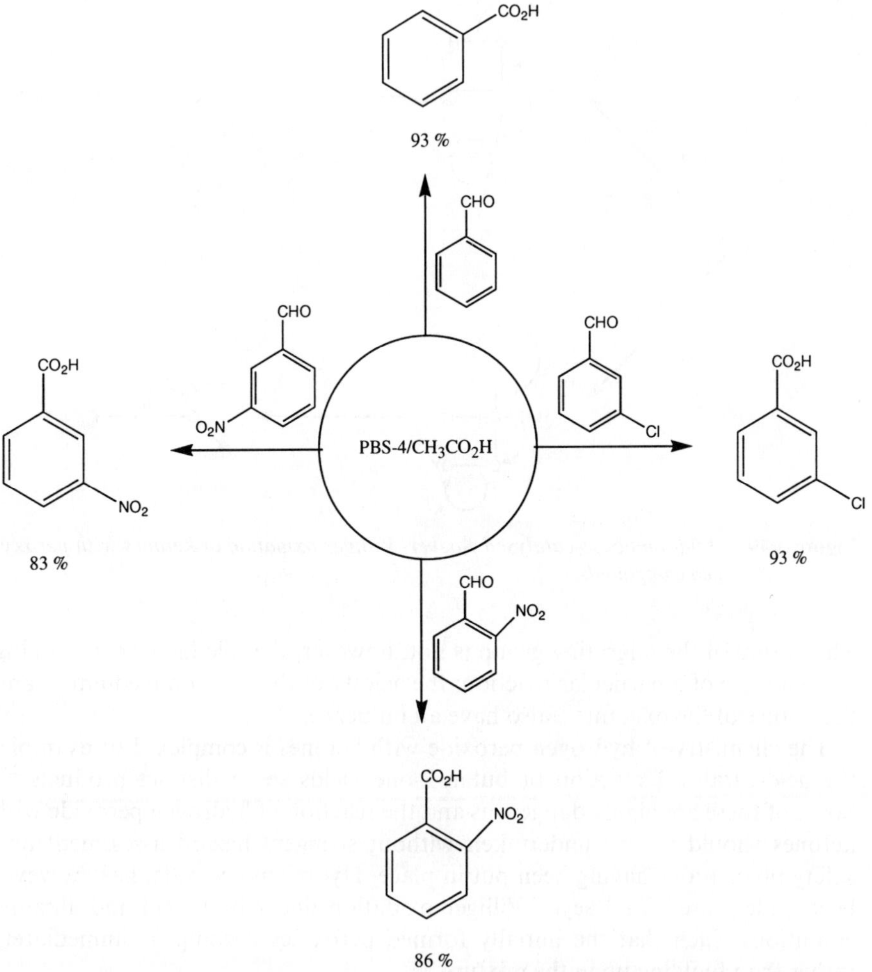

Figure 3.48 *Use of PBS-4/acetic acid systems for the oxidation of aromatic aldehydes to benzoic acids.*

6.2 Oxidation of Ketones

Ketones may be oxidized by peroxygen species under a range of conditions. The most common reaction involves the formation of esters, or in the case of cyclic ketones, lactones. This is known as the Baeyer–Villiger reaction.[232–235] This reaction can occur under acidic or alkaline conditions (Figure 3.49), first outlined by Criegee.[236]

The ability of the R or R′ groups to migrate determines the ease of reaction and which products dominate in the case of unsymmetrical ketones. The following sequence is observed for the migratory tendency of different groups:

tertiary alkyl > cyclohexyl > secondary alkyl > benzyl > phenyl > primary alkyl > methyl > haloalkyl.

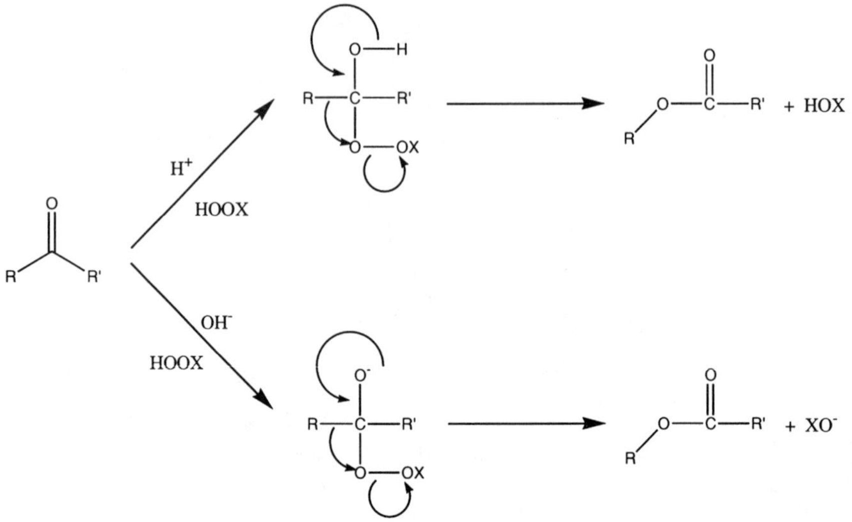

Figure 3.49 *Acid- and base-catalysed Baeyer–Villiger oxidation of ketones with peroxygen compounds.*

The nature of the migrating group is not, however, the sole factor determining the outcome of a particular reaction; the acidity of the reaction medium[237] and the nature of the oxidant[238] also have an influence.

The chemistry of hydrogen peroxide with ketones is complex. For example, the acid-catalysed reaction of butan-2-one yields seven distinct products.[239] Some of these are highly dangerous and the reaction of hydrogen peroxide with ketones should not be undertaken without stringent hazard assessment and safety precautions having been put in place. Hydrogen peroxide, has, however been widely used for Baeyer–Villiger oxidation under both acid and alkaline conditions, such that the initially formed peroxides decompose immediately rather than building up in the mixture.

Relatively weak acids such as acetic acid are sufficient to catalyse the reaction of rigid polycyclic ketones to the corresponding lactones, as in the case of adamantanone.[239] Stronger protic acids have been used with substrates which are more resistant to oxidation, *e.g.* octan-2-one (Figure 3.50).[240]

Under these conditions, however, stable peroxides may form and polymerization of lactones can occur. Lewis acids, *e.g.* BF_3, SbF_5, have been used to activate hydrogen peroxide in similar Baeyer–Villiger oxidations.[241]

Treatment of cyclic ketones (Figure 3.51) under acid conditions can give rise to dimeric cycloalkylidene peroxides, which on thermolysis give lactones and CO_2 (the Story synthesis).[242]

Whilst yields are generally low, this is a useful method for preparing large ring compounds, including a number of important musk compounds used in perfumes.

Baeyer–Villiger reactions also occur under alkaline conditions, although poorer results are usually obtained compared with the use of peracids (see

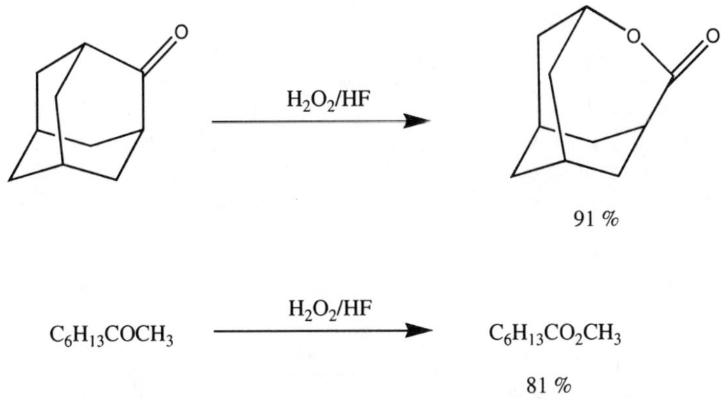

91 %

$C_6H_{13}COCH_3$ $\xrightarrow{\text{H}_2\text{O}_2/\text{HF}}$ $C_6H_{13}CO_2CH_3$

81 %

Figure 3.50 *Acid-catalysed Baeyer–Villiger oxidation of ketones with hydrogen peroxide.*

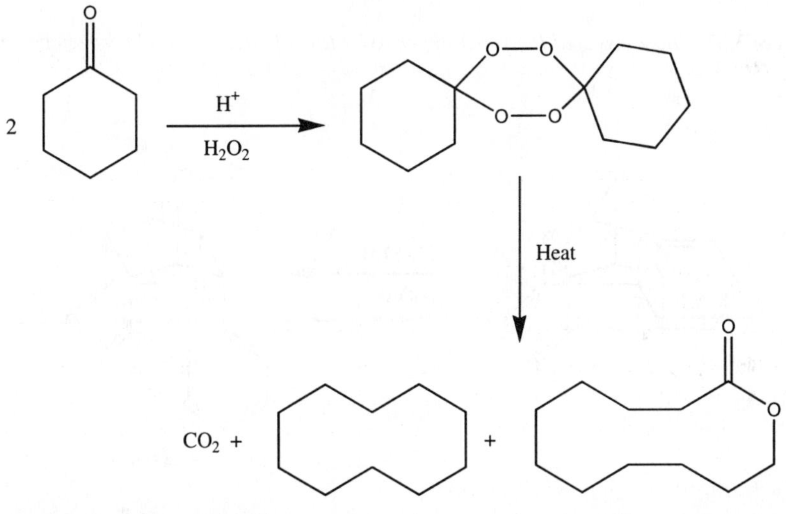

Figure 3.51 *Story synthesis of lactones.*

below). A notable exception is the conversion of cyclobutanones to butyrolactones.[243,244] Alkaline hydrogen peroxide may also be used to cleave hydroxyketones to carboxylic acid products,[245] *e.g.* benzoin to benzoic acid (Figure 3.52).

Percarboxylic acids have been used widely in Baeyer–Villiger oxidation. Peracetic acid is one of the most commonly used peracids; distilled peracetic acid is employed commercially by Solvay Interox for the production of ε-caprolactone.[246] The use of distilled peracetic acid is essential, as it contains no strong protic acids which can catalyse polymerization of the resulting lactone and cause other side reactions. Figure 3.53 illustrates the use of pre-formed peracetic acid for the Baeyer–Villiger oxidation of ketones.[247,248]

Peracids formed *in situ* (*e.g.* performic acid) may also be used for the Baeyer–

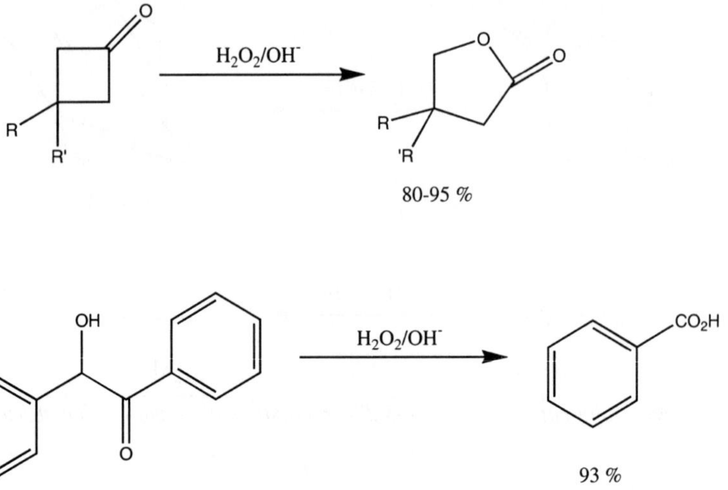

80-95 %

93 %

Figure 3.52   *Base-assisted Baeyer–Villiger oxidation of ketones with hydrogen peroxide.*

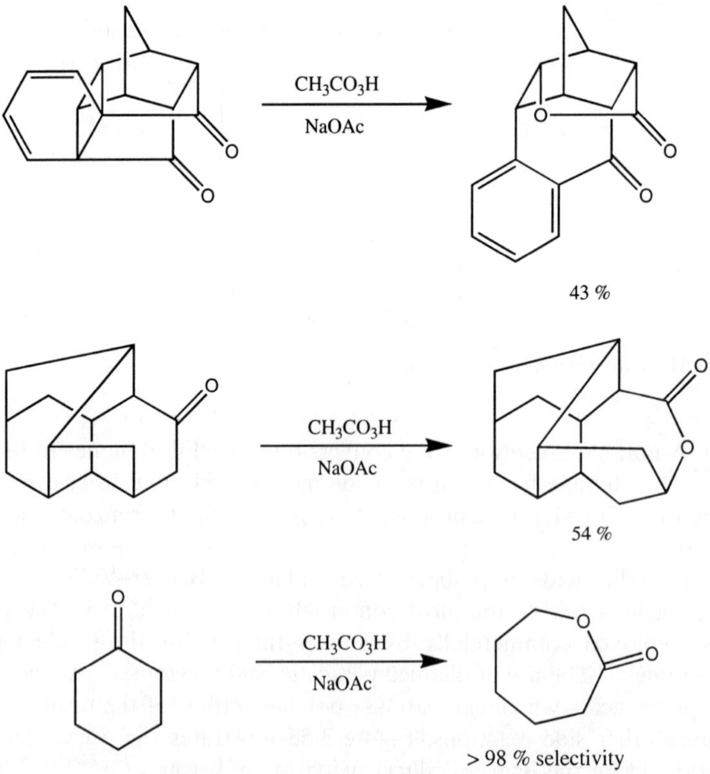

43 %

54 %

> 98 % selectivity

Figure 3.53   *Baeyer–Villiger oxidation of ketones using pre-formed peracetic acid.*

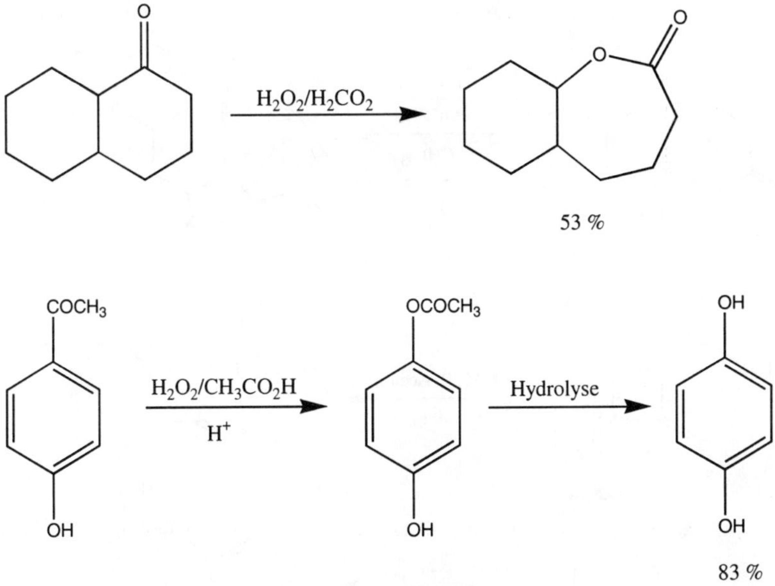

53 %

83 %

Figure 3.54   *Oxidation of ketones using* in situ *formed peracids.*

Villiger reactions.[249] Acetophenones may be converted readily into phenyl acetates,[250] which may be hydrolysed to phenols in their own right, *e.g.* hydroquinone monoacetate (HQMA) (Figure 3.54).

For reactions where the presence of strong acids may give undesirable side reactions, such as lactone ring-opening, pre-formed peracids such as MCPBA,[251] permaleic acid,[252] monoperphthalic[253] or MMPP[254] may be used as alternatives to distilled peracetic acid (Figure 3.55).

Some ketones are resistant to oxidation with simple peracids in which case trifluoroperacetic acid is the reagent of choice (Figure 3.56).[255]

Whilst catalysts are not normally required for Baeyer–Villiger oxidation using peracids, boron trifluoride etherate has been used to activate peracetic acid in the preparation of macrocyclic ketones (Figure 3.57).[256]

Cyclic ketones may be converted to cyclic ethers via the oxidation of the corresponding diethyl ketal using MCPBA.[257] This reaction is formally a double Baeyer–Villiger type oxidation followed by elimination of diethylcarbonate (Figure 3.58).

Peroxymonosulfuric acid (Caro's acid H_2SO_5) was the reagent employed originally by Baeyer and Villiger in 1899 for the oxidation of alicyclic ketones.[258] Since then it has only been used to a limited extent, probably because under aqueous conditions it favours formation of peroxides, and in organic solvents the use of organic peracids is more convenient. In some cases, however, peroxymonosulfuric acid, generated *in situ* from hydrogen peroxide and concentrated sulfuric acid, may give good yields where percarboxylic acids fail or are only successful under hazardous conditions. One example is the oxidation of 4,4'-difluorobenzophenone.[258]

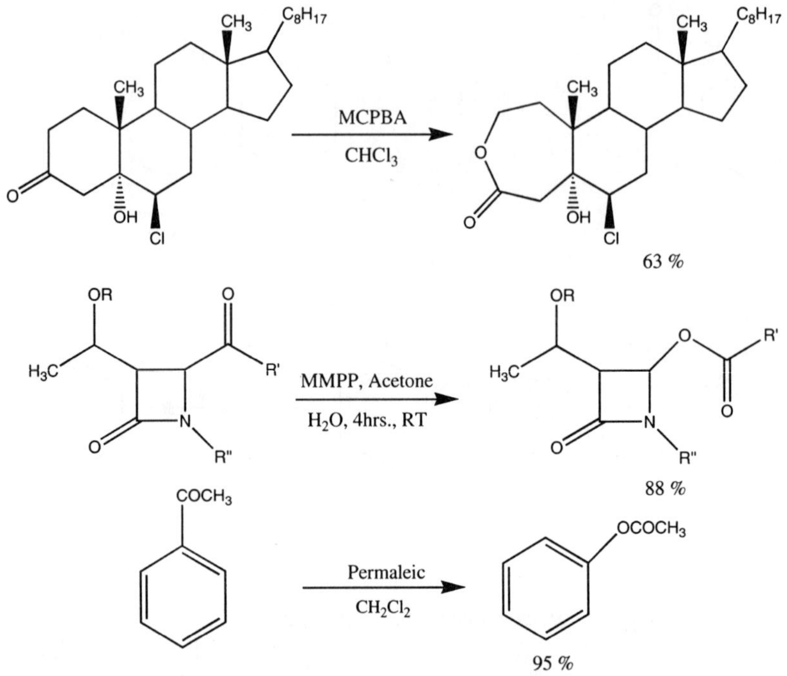

Figure 3.55 *Oxidation of ketones with non-peracetic acid oxidants.*

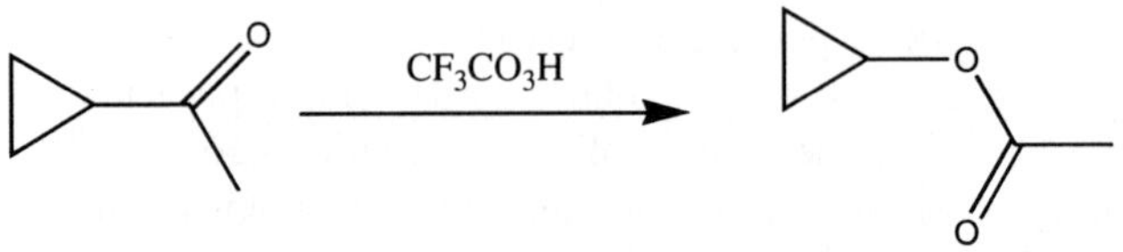

Figure 3.56 *Baeyer–Villiger oxidation with trifluoroperacetic acid.*

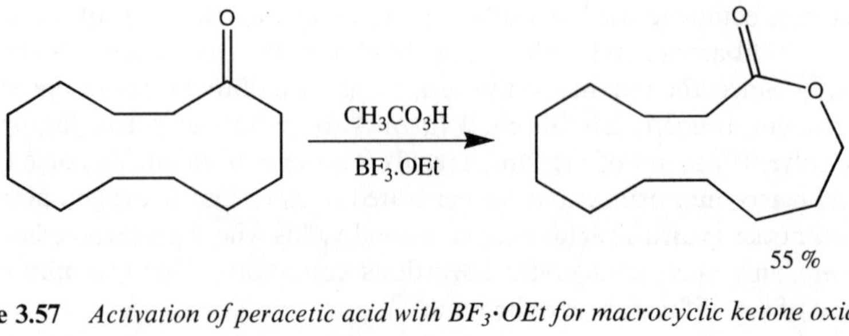

Figure 3.57 *Activation of peracetic acid with BF$_3$·OEt for macrocyclic ketone oxidation.*

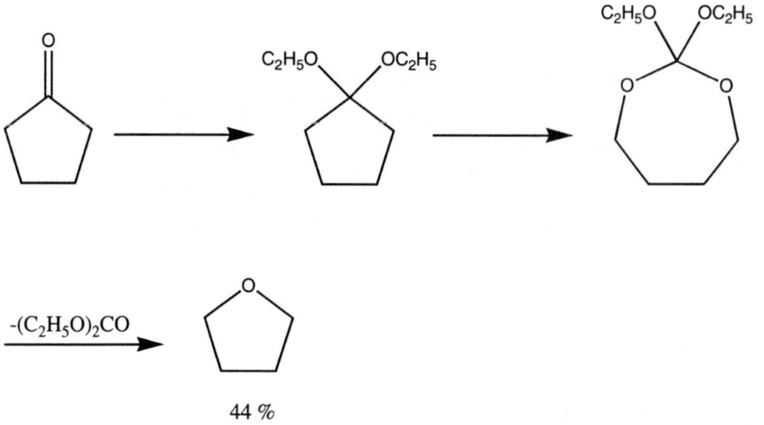

44 %

Figure 3.58 *Formation of ethers via the double Baeyer–Villiger reaction using MCPBA as the oxidant.*

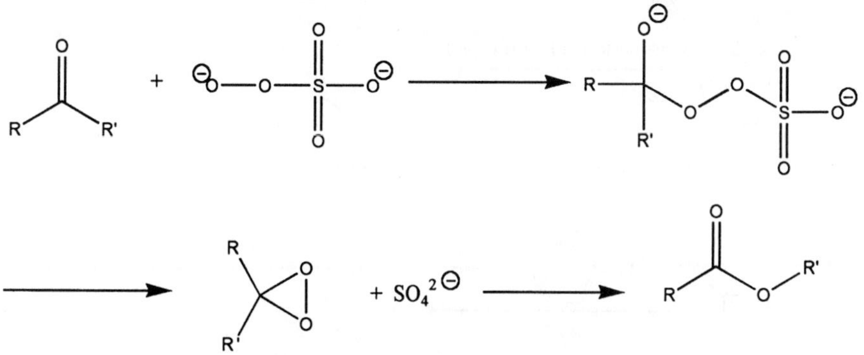

Figure 3.59 *Use of dioxiranes for the Baeyer–Villiger oxidation of ketones.*

Under neutral conditions, peroxymonosulfate can be used to convert some ketones to stable dioxiranes.[259] The dioxiranes then re-arrange to esters or lactones (Figure 3.59). Dioxirane formation may involve the same intermediate as proposed by Criegee for the Baeyer–Villiger reaction. The combination of an oxidant, which provides a good leaving group, and a ketone-bearing group with a high migratory tendency should be used for Baeyer–Villiger transformations.

In addition to simple acids and bases, other materials have been used as catalysts for Baeyer–Villiger oxidations using hydrogen peroxide. Selenium compounds have received most attention,[260–263] one example being the use of polymer-bound benzene selenic acid for the oxidation of small-ring ketones.[264] Under acidic or neutral conditions, selenium dioxide catalyses oxidative ring contraction with hydrogen peroxide as the primary oxidant (Figure 3.60).[265]

This reaction probably involves a selenium(VI) species rather than the selenium(IV) species believed to catalyse the Baeyer–Villiger reaction.[258]

Other species (Figure 3.61) which have been examined as catalysts for

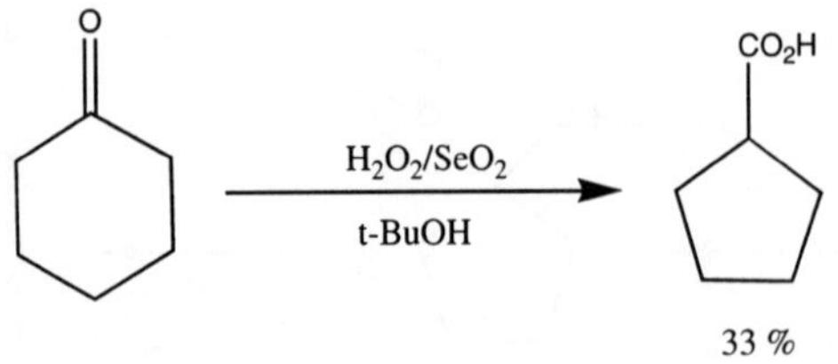

Figure 3.60 *Ring contraction of cyclic ketones with selenium dioxide in the presence of hydrogen peroxide.*

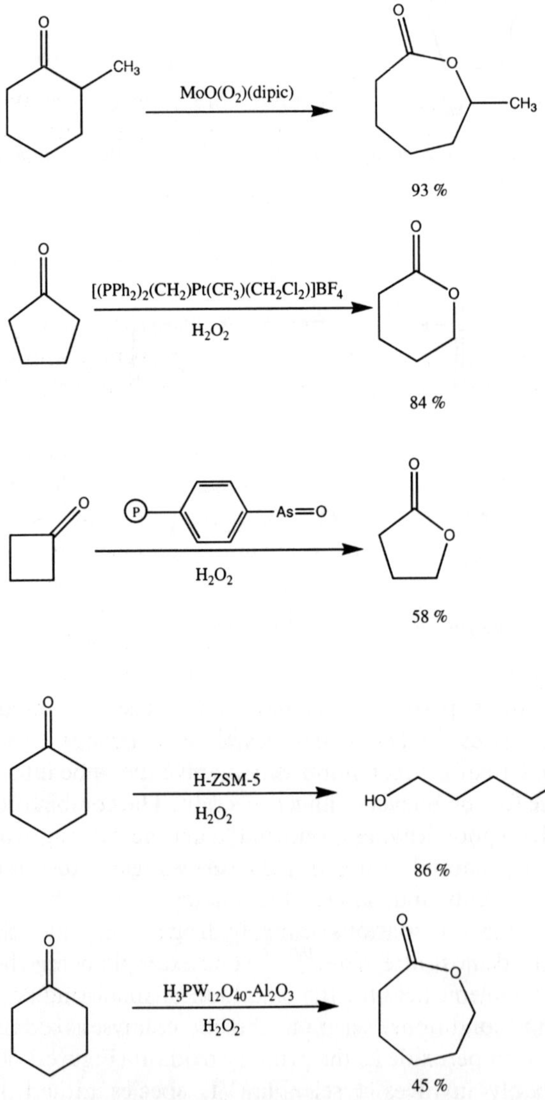

Figure 3.61 *Catalytic activation of hydrogen peroxide for the Baeyer–Villiger oxidation of cyclic ketones.*

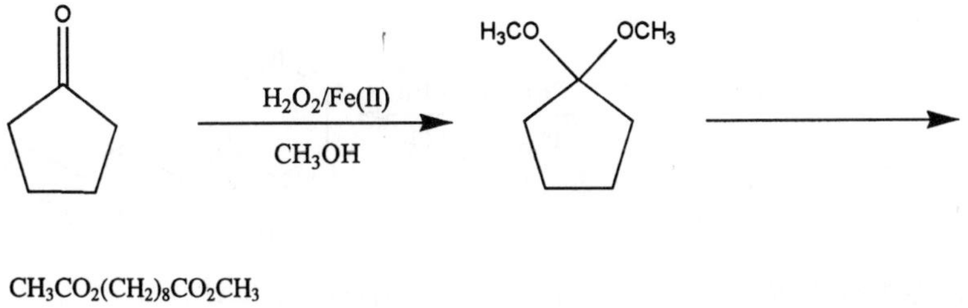

$CH_3CO_2(CH_2)_8CO_2CH_3$

Figure 3.62 *Iron(II) activation of hydrogen peroxide to produce ring-opened methyl esters from cyclic ketones.*

Baeyer–Villiger oxidation include molybdenum and tungsten compounds,[266] platinum complexes[267] and arsenic derivatives impregnated onto polymeric supports.[268] Zeolites have also been used as heterogeneous catalysts in the conversion of cyclic ketones to lactones or the corresponding hydroxy carboxylic acids.[269] Alumina-supported phosphotungstic acid has been found to be useful for catalysing the conversion of cyclohexanone to ε-caprolactone in acetic acid solvent.[270]

In the presence of iron(II) catalysts, cyclic peroxides formed from hydrogen peroxide and the corresponding ketones undergo ring-opening and dimerization to give diacid derivatives, *e.g.* the dimethyl ester of sebacic acid (Figure 3.62).[271]

Baeyer–Villiger oxidation of ketones using persalts and acetic anhydride has been studied by Tao and co-workers. Sodium percarbonate is said to perform well for cyclic ketones with yields of lactone being typically 80%, but aromatic ketones only give good yields if the ring is activated. Reactions are assisted by ultrasound.[272] Sodium perborate similarly gives good yields of lactones from cyclic ketones with acetic anhydride.[273] Sodium perborate in acetic acid gives good results for Baeyer–Villiger oxidation of aromatic ketones having at least one group which migrates readily.[274] Replacement of some or all of the acetic acid by trifluoroacetic acid increases the potency of the reaction mixture, and extends the scope of the reaction. Sodium perborate in acetic acid has also been used to cleave aromatic 1,2-diketones and hydroxy-ketals to carboxylic acids.[275]

Peroxymonophosphoric acid has also been used for Baeyer–Villiger oxidation. Acetophenones are reported to be converted to aryl acetates at higher rates and in better yields when this reagent is used in preference to peracetic or perbenzoic acid.[276]

Organic peroxides are not generally useful reagents for Baeyer–Villiger type oxidation under basic conditions, however, aryl ketones may be oxidized to Baeyer–Villiger type products in low yields.[277] Under these conditions, alkyl groups migrate in preference to aryl groups, and the intermediate ester is not isolated.

Boric acid derivatives are reported to catalyse lactone formation from cyclic ketones using hydroperoxide oxidants.[278]

Bis(trimethylsilyl)peroxide may be used for the oxidation of molecules

Figure 3.63 *Preparation of jasmine lactone with bis(trimethylsilyl)peroxide.*

containing acid sensitive groups.[279] Jasmine lactone has been produced from Z-2-(pent-2-enyl)cyclopentanone without the need to protect the double bond (Figure 3.63).

7 Oxidation of Aromatic Side-chains

The oxidation of an aromatic side chain is an industrially important process. However, the transformation of toluenes to benzaldehydes and benzoic acids is relatively difficult to perform. The most demanding stage is the introduction of the initial oxygen, as subsequent conversion between oxidation levels is easier (Figure 3.64).

The products of such oxidations are useful intermediates in areas such as agrochemical, perfumery, pharmaceutical and polymer chemicals. Traditionally, transition metal oxidants such as manganese and chromium compounds have been used in the chemical industry over many years.[280] Some side-chain oxidation is carried out on a very large scale using air as the oxidant.[281] Such oxidation involves transformation of methyl functions to the corresponding carboxylic acid and isopropyl groups to the phenol and acetone. The former category includes liquid-phase oxidation with cobalt and/or a manganese catalyst with an inorganic bromide promoter[281] for p-xylene to terephthalic acid, m-xylene to isophthalic acid and 1,2,4-trimethylbenzene to trimellitic anhydride, and gas-phase oxidation of toluene to benzoic acid.

Peroxygen systems have three distinct advantages in industrial side chain oxidation compared to the employment of air. Firstly, the reactions can be conducted at atmospheric pressure, and moderate temperature, which makes this an option for small- to medium-volume production in general purpose (low cost) plant. Secondly, the selectivity to produce alcohols or aldehydes is easier to

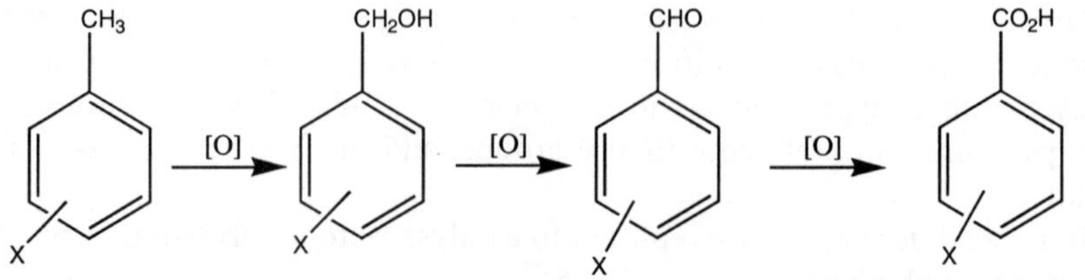

Figure 3.64 *Side-chain oxidation of toluenes.*

control. Finally, highly deactivated substrates can be successfully converted by one of the systems described in this section.

Although both acid-activated and hydroxyl radical systems are capable of oxidizing aromatic side-chains, they also attack the aromatic nucleus and this lack of selectivity limits their specific use for side-chain oxidation. However, a metal-based hydroxyl radical system, which has been extensively studied, is based on Fenton chemistry [hydrogen peroxide in the presence of iron(II) salts]. The selectivity and yield to any particular compound are usually low.[282–284] However, a two-phase process employing iron(III) activation of hydrogen peroxide for the oxidation of toluene to benzaldehyde does provide a system with better yield and selectivity to the benzaldehyde product.[285] The workers claim that selectivity is best when the substrate is in excess and the hydrogen peroxide concentration is kept at a low throughput during the course of the reaction.

Alkaline hydrogen peroxide does not attack alkyl side chains directly, but it can be very useful for the conversion of the aromatic aldehyde to the carboxylic acid in the side chain oxidation sequence. The chemistry is covered in Section 6.1. Peracids usually do not attack alkyl side chains, but can oxidize benzylic alcohols and aldehydes to the carboxylic acids and benzylic ketones to phenyl ester. These reactions are covered in sections on alcohol (5), aldehyde (6.1) and ketone (6.2) oxidation.

Toluenes can be oxidized by Caro's acid (H_2SO_5) activated by iron(II) salts.[286]

Catalytic systems are by far the most studied methods for oxidizing alkyl side chains. Cobalt(II) acetate and cerium(III) acetate in the presence of a bromide ion activator in acetic acid with hydrogen peroxide are used for the transformation of toluenes to benzaldehydes, carboxylic acids and benzyl bromides (Figure 3.65).

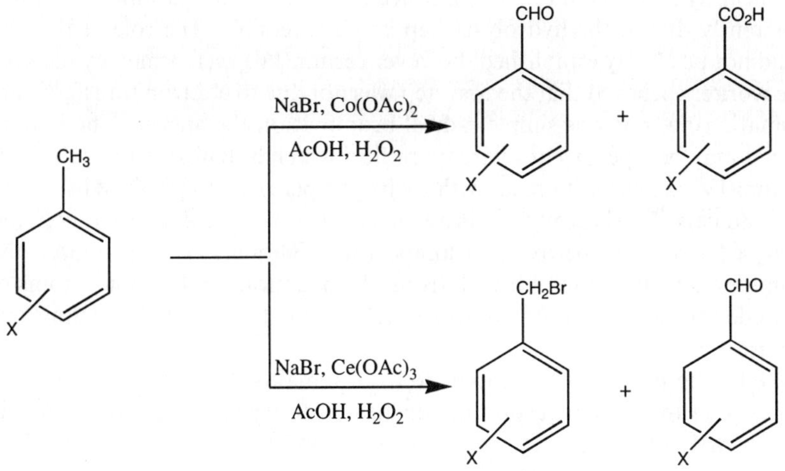

Figure 3.65   *Metal catalysed side-chain oxidation of toluenes.*

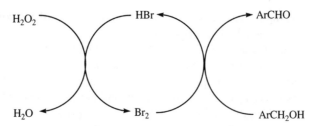

Figure 3.66 *Possible role of in situ bromine formation during the side-chain oxidation of toluenes.*

The cobalt(II) acetate/acetic acid/sodium bromide (CAB) combination is well known for the autoxidation of alkyl benzenes.[287] It is normally employed for the production of terephthalic acid from 1,4-dimethylbenzene.[281] However, the use of hydrogen peroxide, as mentioned earlier, means the systems are generally more selective, and can be operated at ambient pressure and relatively low temperatures.[287]

The CAB/H_2O_2[281] and the analogous CeAB/H_2O_2[288] processes should be compared. The CAB/H_2O_2 system affords typically benzaldehydes and benzoic acid mixtures. However, the CeAB/H_2O_2 system gives benzaldehydes and benzyl bromide compounds. Benzyl alcohol is formed in only small amounts if the CeAB/H_2O_2 route is used and not at all if the CAB/H_2O_2 system is used, owing to its rapid oxidation under the reaction conditions. The oxidation of the alcohol may be via the *in situ* formation of bromine (Figure 3.66).

The oxidation of 4-*t*-butyltoluene with the CeAB/H_2O_2 system has been studied by Thomas.[289] The work has shown that conversion of the substrate proceeds via benzylic bromination, followed by the rapid oxidation of the alcohol to the aldehyde by *in situ* generated bromine (see Figure 3.66). Reaction was found to be ineffective in the absence of bromide anion, but was also significantly inhibited if too high a concentration of bromide anion was present, apparently due to the hydrolysis step being reversible. The role of the cerium could not be clearly established, however cerium(IV) was formed by the system. The workers believed that the first step was not due to electron transfer from the aromatic ring, nor was simple radical bromination the answer. The important species may be a peroxy or hydroxyl radical which initiates radical substitution. Cerium(IV) is alleged to react with hydrogen peroxide to produce hydroxyperoxy radicals.[290] Hydroxyl radicals may also be present from cerium(III) catalysed hydrogen peroxide decomposition. Thomas *et al.* proposed that a bromine radical anion, formed from the reaction of bromine atoms with bromide ions at a diffusion controlled rate, is the hydrogen abstracting species (Figure 3.67).[291]

Whilst the nature of the initial step still remains unclear, Figure 3.68 is the likely route for the conversion of substituted toluenes to benzaldehydes using the CeAB/H_2O_2 system. However, the CAB/H_2O_2 system may not follow the same pathway (Figure 3.69).[292]

Both systems will only function in the presence of a carboxylic acid.[293,294] The

$$Ce(III) + H_2O_2 \longrightarrow Ce(IV) + HO^{\bullet} + HO^{\bullet}$$

$$Ce(IV) + H_2O_2 \longrightarrow Ce(III) + HO_2^{\bullet} + H^{\oplus}$$

$$Br^{\bullet} + Br^{\ominus} \longrightarrow 2Br^{-\bullet}$$

Figure 3.67 *Possible initiating species present in the CeAB/hydrogen peroxide system for the oxidation of toluenes.*

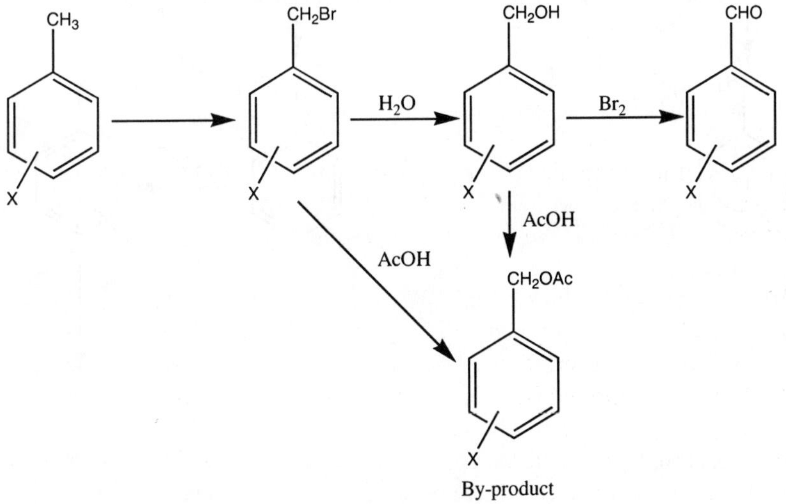

Figure 3.68 *Formation of benzaldehydes from toluenes in the CeAB/hydrogen peroxide system.*

reason for this has been explained in terms of the stability of the higher oxidation state metal in the carboxylic acid (see Figure 3.70). The complex can then break down to an alkyl radical and CO$_2$. The alkyl radical can then abstract a hydrogen atom, and the resulting side chain radical proceeds as outlined in Figure 3.69.

The oxidation of 4-methoxytoluene and 4-t-butyltoluene using the CAB/ H$_2$O$_2$ process affords the resulting aldehydes which are of commercial value as flavour and fragrance intermediates (Figure 3.71). The CeAB/H$_2$O$_2$ system will produce 4-t-butylbenzaldehyde from the parent toluene, however the ring brominated substrate is formed when 4-methoxytoluene is used as the substrate (Figure 3.71).

The ease of oxidation in the presence of the CAB system with hydrogen peroxide is generally benzyl alcohols > aldehydes > toluenes. This series accounts for the reactivity and subsequent product distributions when 1,4-dimethybenzene is oxidized with the CAB system. Here, 4-methylbenzoic acid forms preferentially to attack at the second methyl group. Similarly, oxidation of 1,3,5-trimethylbenzene yields 3,5-dimethylbenzoic acid. This is complementary to the reactivity observed with the HBr/hydrogen peroxide/hν method (see

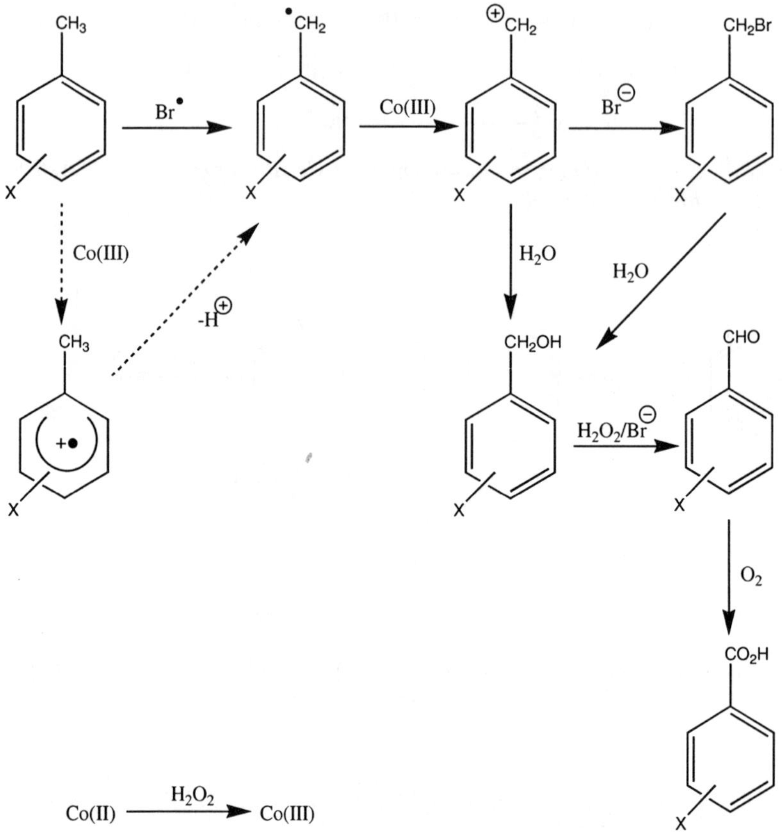

Figure 3.69   *Possible mechanisms operating during the side-chain oxidation of substituted toluenes employing the CAB/hydrogen peroxide system.*

below). In the latter case, the second methyl group on 1,4-dimethylbenzene will be oxidized in preference to the aldehyde group (Figure 3.72).

Another important contrast between the CAB/H_2O_2 and the halogenation based systems is seen in the reactivity towards substituted toluenes. The halogenation method works best for deactivated molecules, whereas the CAB/H_2O_2 system is better suited to activated arene centres. If electron-withdrawing groups are present on the ring, low conversions are observed and this point is emphasized in Figure 3.73 for CAB/H_2O_2.

The oxidation of ethylbenzene proceeds smoothly when the CAB/H_2O_2 process is employed, typically 93% conversion and 93% selectivity to acetophenone. Diphenylmethane is also readily oxidized to benzophenone using the CAB/H_2O_2 system, affording 97% conversion and 71% selectivity to the aromatic ketone.

The oxidation of substituted toluenes to benaldehydes employing the CAB system with solid peroxide sources has also been studied by Jones and co-workers.[292] The group screened a range of solid peroxygens in place of aqueous

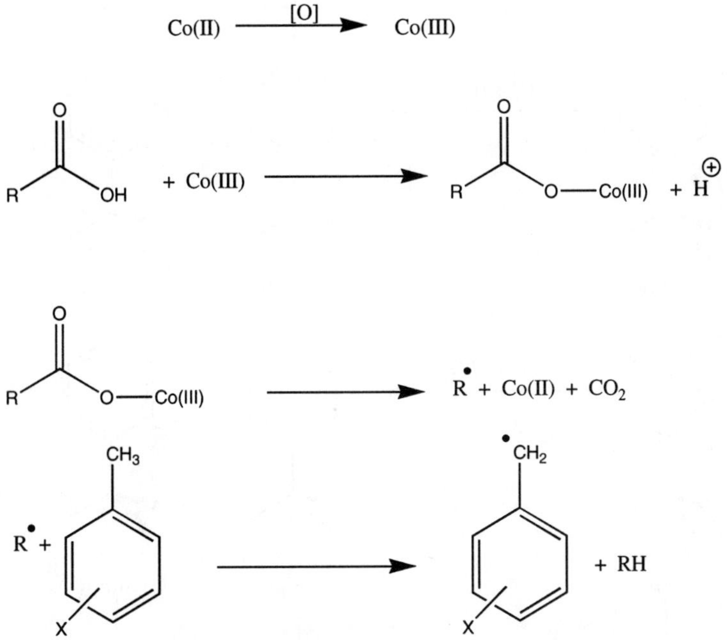

Figure 3.70 *Possible role of carboxylic acids during the oxidation of substituted toluenes using the CAB/hydrogen peroxide system.*

hydrogen peroxide. Sodium perborate monohydrate was found to be the most selective primary oxidant screened for adehyde production (Table 3.6).

The analogous system which employs cerium(III) acetate and hydrogen peroxide offers three principle advantages over the CAB/H$_2$O$_2$ process. Firstly, only low levels of cerium(III) acetate are required for effective catalysis, typically about five times less on a molar basis compared to the CAB/H$_2$O$_2$

Table 3.6 *Side-chain oxidation of substituted toluenes using PBS-1 in the CAB system[a]*

Substrate	Oxidant	Aldehyde selectivity (%)
4-*t*-Butyltoluene	Hydrogen peroxide	40
4-*t*-Butyltoluene	PBS-1	76
4-Methoxytoluene	Hydrogen peroxide	45
4-Methoxytoluene	PBS-1	72
Diphenylmethane	Hydrogen peroxide	79
Diphenylmethane	PBS-1	81
1,4-Dimethylbenzene	Hydrogen peroxide	53
1,4-Dimethylbenzene	PBS-1	76

[a] Acetic acid (50 g) at 45 °C over 3 h, substrate (33 mmol), cobalt(II) acetate (2 mmol), PBS-1 (0.05 mol) or hydrogen peroxide (35% *m/m*, 0.05 mol), sodium bromide (5 mmol).

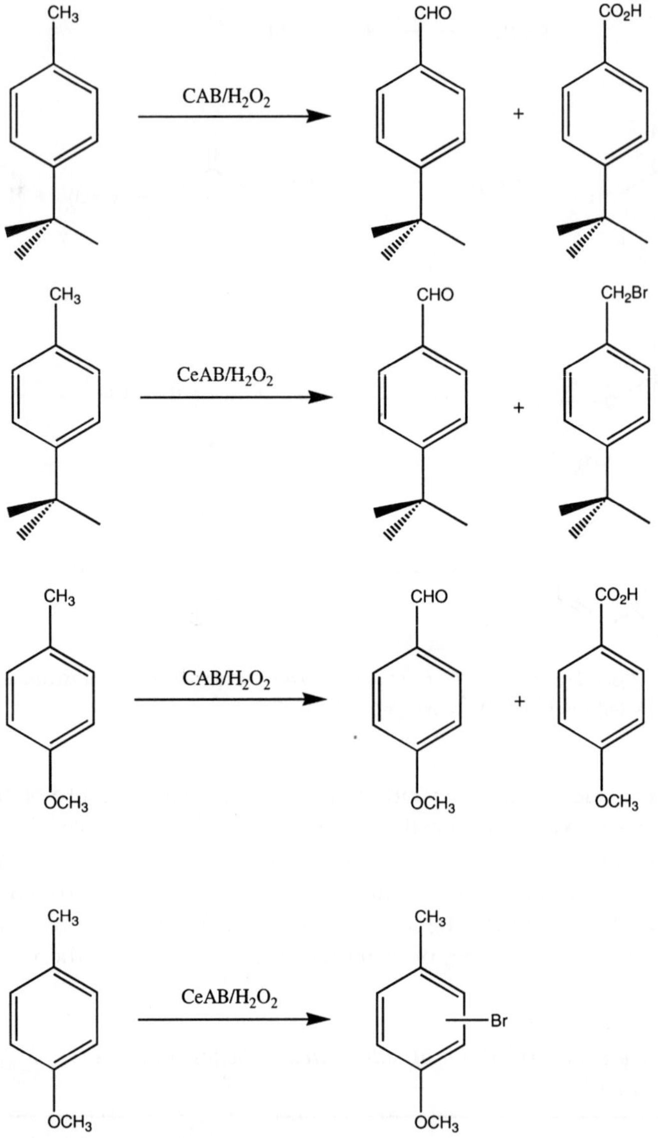

Figure 3.71 *Oxidation of 4-t-butyltoluene and 4-methoxytoluene in the presence of the CAB and CeAB/hydrogen peroxide systems.*

system. Secondly, the cerium system is not deactivated by relatively large amounts of water, and thirdly, no over-oxidization to the carboxylic acid is observed. A range of substrates has been oxidized using the CeAB/H_2O_2 system (Figure 3.74).

The *in situ* generation of bromine from H_2O_2 and HBr provides an extremely versatile indirect method for oxidizing toluenes and higher alkylbenzenes. It is

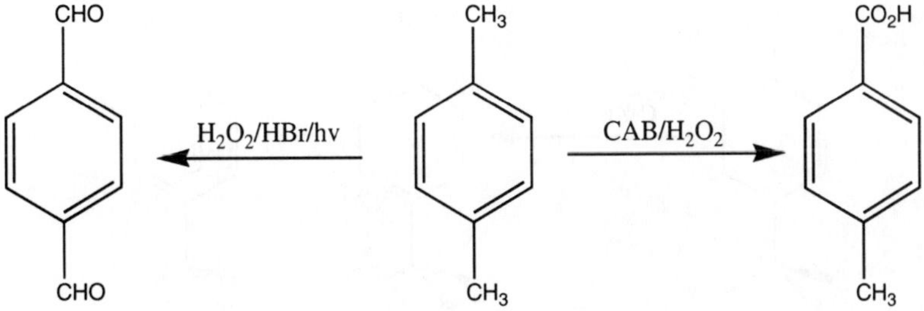

Figure 3.72 *Reactivity differences between the CAB/hydrogen peroxide and HBr/hv/ hydrogen peroxide systems.*

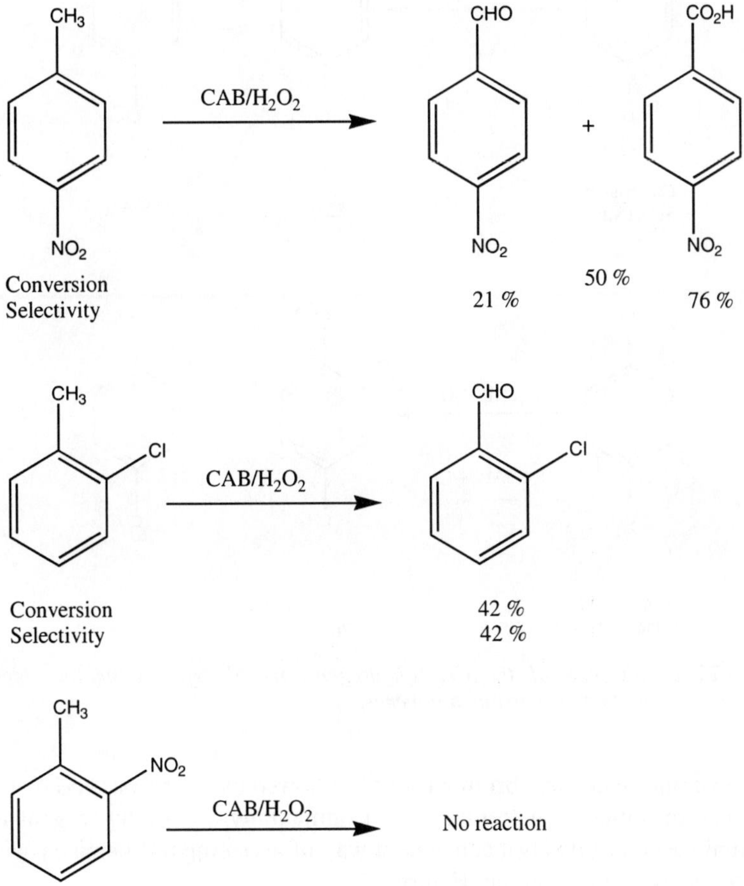

Figure 3.73 *Poor performance of the CAB/hydrogen peroxide system in the presence of deactivated substrates.*

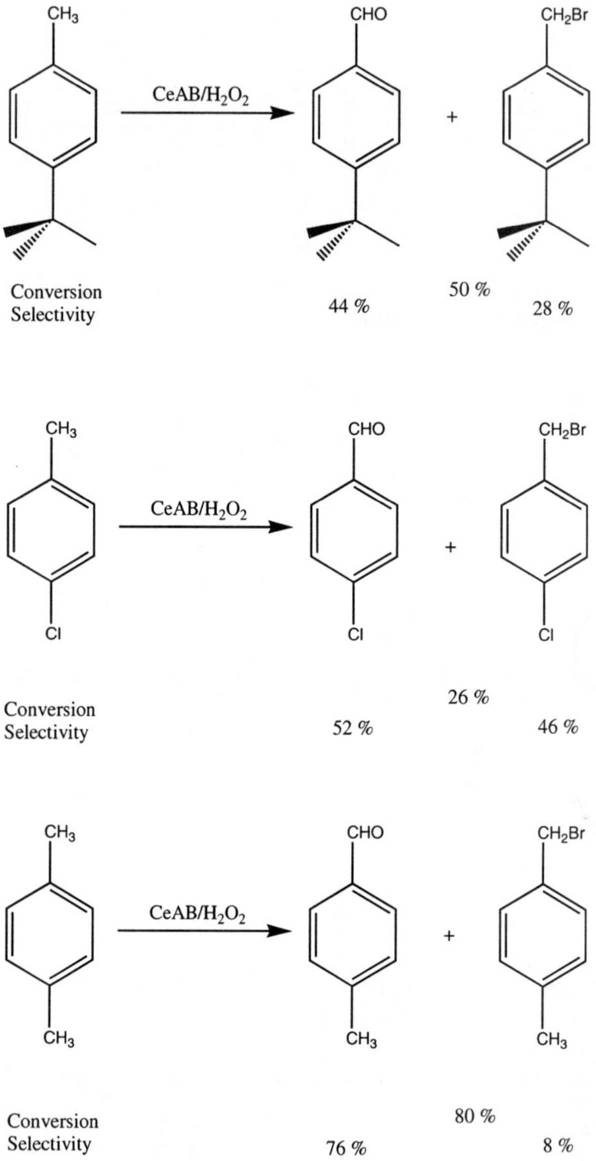

Figure 3.74 *Application of the CeAB/hydrogen peroxide system for the side-chain oxidation of substituted toluenes.*

based on a photochemical bromination[295] followed by hydrolysis and oxidation steps. The bromine is a potentially environmentally unfriendly reagent in the elemental form and this is a convenient way of accessing its chemistry without employing elemental bromine (Figure 3.75).

The technology works best with deactivated substrates, and has been adapted to give a wide range of products (Figure 3.76).

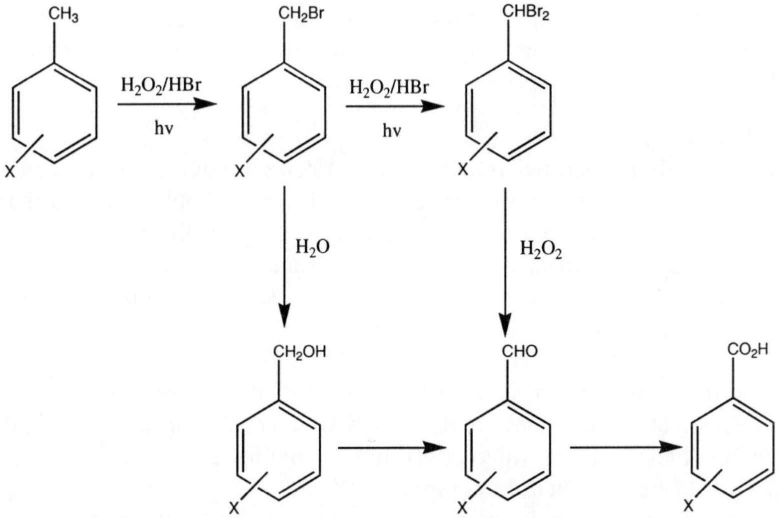

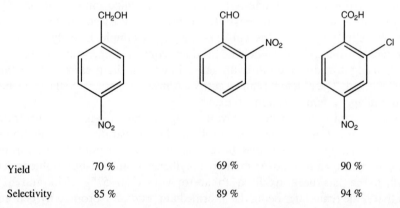

Figure 3.75 *Indirect oxidation of substituted toluenes using the hydrogen peroxide/hv/ HBr system.*

	CH$_2$OH NO$_2$	CHO NO$_2$	CO$_2$H Cl NO$_2$
Yield	70 %	69 %	90 %
Selectivity	85 %	89 %	94 %

Figure 3.76 *Side-chain oxidation of deactivated toluenes using the hydrogen peroxide/hv/ HBr system.*

The advantages of using hydrogen peroxide and hydrogen bromide can be summarized as follows:

- The system is easier to handle due to the *in situ* generation of elemental bromine.
- The bromine generated is used more efficiently. Using elemental bromine alone means that only half of the molecule is employed, whereas using hydrogen peroxide, the resulting HBr generated from the organic molecule can be oxidized.

- Bromide released in the hydrolysis step can be recycled.
- Bromine concentration can be controlled in the reactor by altering the flow of hydrogen peroxide.

The reaction can be operated as a two-phase system; in this case the process can be employed at the reflux temperature of the solvent. It should be noted that in some cases, no solvent need be used[296] and this is especially efficacious where mono-brominated products are the target. An example of the system's versatility in solvent-free mode can be seen in Table 3.7. Here we see that at a molar ratio of 1:1.1:1 (hydrogen peroxide/hydrogen bromide/2-nitrotoluene) over 90% selectivity to the benzyl bromide is observed at a conversion of 76%. However, increasing the oxidant ratio to 3:3.2:1 (hydrogen peroxide/hydrogen bromide/2-nitrotoluene), a selectivity of 98% to the benzal bromide is afforded. Diphenylmethane can be fully converted to oxidized products with 96% selectivity to the benzophenone compound.[297]

The benzyl and benzal bromides can be hydrolysed to the corresponding alcohol or aldehyde using a variety of methods, depending on the substrate. The simplest method allowing ready re-cycle of hydrogen bromide is steam hydrolysis. The agrochemical intermediate, 4-fluorobenzoic acid, can be prepared via the bromination route in a one-pot process (*i.e.* bromination/hydrolysis).[298] A simple hydrolysis will not be possible when de-activated substrates are used. Here, more forcing conditions will be needed. For example, hydrolysis of the 2-nitrobenzal bromide to the aldehyde product requires temperatures in excess of 100 °C using aqueous sodium hydrogencarbonate in the presence of a phase-transfer agent. The aldehyde can be easily oxidized to the carboxylic acid which is a useful agrochemical intermediate.

In the circumstances where activated ring systems are used as substrates, nuclear bromination is sometimes a problem. Two substrates worth mentioning which are mildly activated and have been used in the photolytic hydrogen peroxide/hydrogen bromide system are ethylbenzene and 4-*t*-butyltoluene. The ethylbenzene has been oxidized to acetophenone in 57% yield and 61% selectivity, the remainder being the intermediate product 1-bromoethylbenzene. The preparation of 4-*t*-butylbenzaldehyde from the toluene affords a yield of 58% (Figure 3.77).

Table 3.7 *Control of product distribution in the hydrogen peroxide/hydrogen bromide/hv system*

H_2O_2/HBr/substrate	Products	Substrate conversion (%)	Product selectivity (%)
1:1.1:1	2-Nitrobenzal bromide	76	2
	2-Nitrobenzyl bromide		92
3:3.2:1	2-Nitrobenzal bromide	100	98
	2-Nitrobenzyl bromide		2

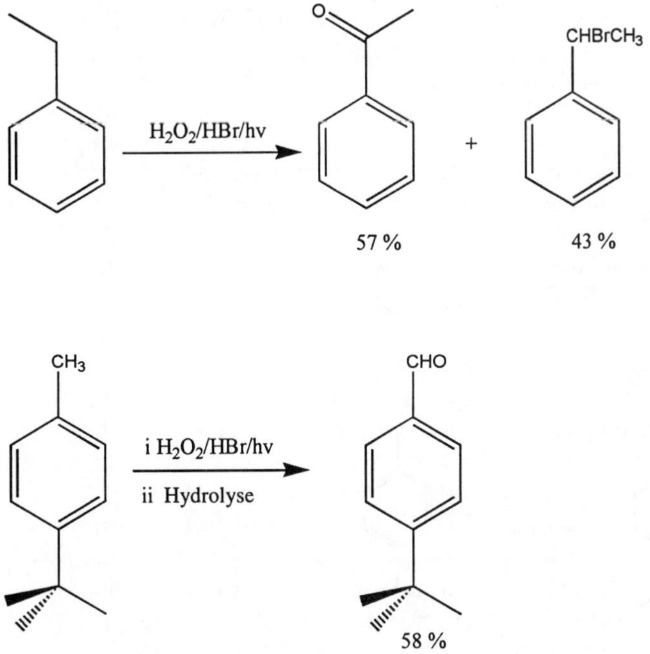

Figure 3.77 *Use of the photolytic hydrogen peroxide/HBr system for the side-chain oxidation of mildly activated alkylbenzenes.*

Toluene has been oxidized by the silver ion catalysed reaction with peroxydisulfate. The reaction produces a mixture of bibenzyl, benzaldehyde and benzoic acids.[299] Russian workers have described the conversion of 4-methoxytoluene to the benzaldehyde by oxidation with peroxydisulfate in the presence of silver or copper ions and oxalic acid.[300] The presence of copper salts in iron or copper catalysed peroxydisulfate oxidation is believed to suppress side-reactions.[301] Phillips have patented a palladium(II)/tin(IV)/persulfate system for the oxidation of toluene derivatives.[302] The reactions are carried out in carboxylic acid solvents (Figure 3.78).

8 Oxidation of Organo-nitrogen Compounds

Industrial nitrogen oxidation is of immense importance. This section will briefly review the various methods employed for the transformation of nitrogen compounds using hydrogen peroxide and its derivatives. It should be noted that most amines produce alkaline solutions, and as such it is common to add stabilizers when using hydrogen peroxide for organo-nitrogen oxidation. Tertiary amines are usually oxidized by peroxygens to *N*-oxides, which can be reacted further to give either a di-*N*-substituted hydroxylamine, and an olefin (Cope elimination) or an *N,N,O*-trialkyl hydroxylamine (the Meisenheimer rearrangement) (Figure 3.79).

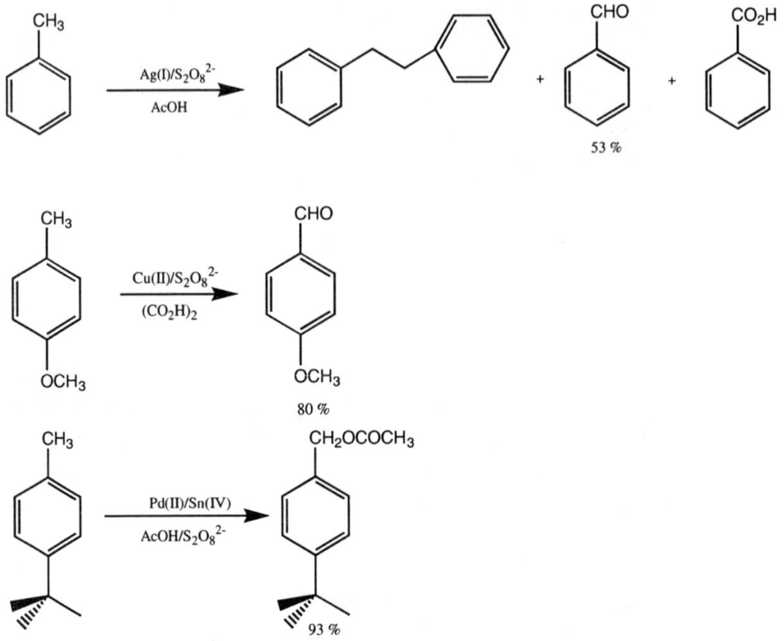

Figure 3.78 *Metal catalysed peroxydisulfate oxidation of toluenes.*

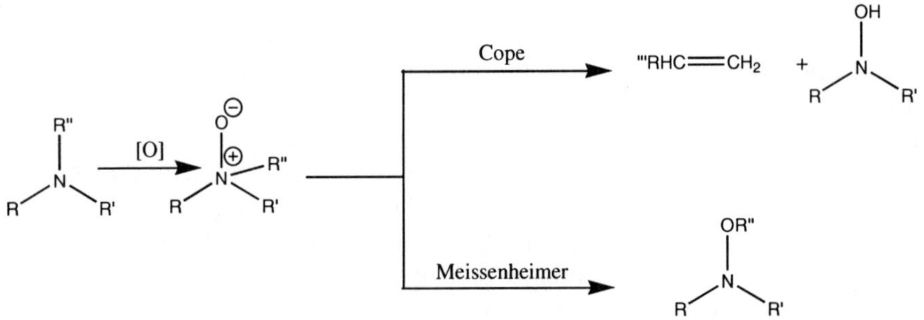

Figure 3.79 *Rearrangement products from N-oxides.*

Heterocyclic amines require an activated peroxygen source. The resulting *N*-oxides can sometimes be used indirectly to functionalize the ring, *e.g.* the Katada reaction (Figure 3.80).

Secondary amines can be oxidized at the N–H bond to hydroxylamines and nitroxides, and via nitrones via C–N oxidation. Nitrones are valuable intermediates in the production of isoxazolines. Initial C–N oxidation of secondary amines gives imines which can react further to oxaziridines. The latter can be converted to nitrones, and both to amides. Primary amines are oxidized at the N–H bond to mono-substituted hydroxylamines, which are readily converted further to nitroso and nitro compounds by the more activated peroxygen

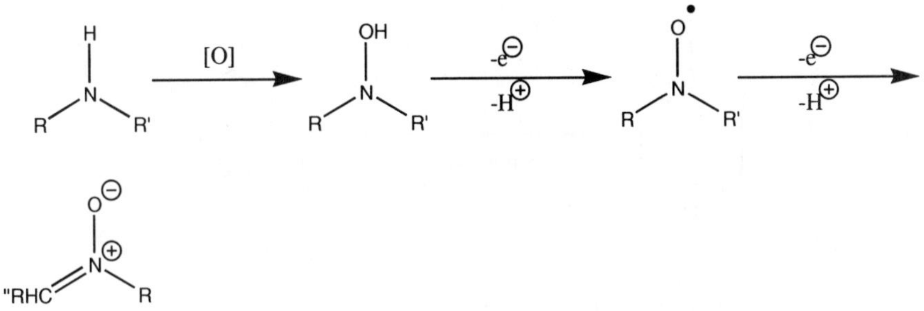

Figure 3.80 *Katada reaction.*

species. Diazo compounds are alternative products formed when oxidation is slow or low oxidant excess is employed. Diazo compounds result from condensation of the nitroso compound with the initial amine (Figure 3.81). Once formed, diazo compounds can be oxidized to the azoxy derivative which is comparatively inert to further oxidation.

Some catalysts with hydrogen peroxide can oxidize ammonia to hydroxylamine. There is also a process for the oxidation of ammonia to hydrazine. Other special *N*-oxidation with hydrogen peroxide includes hydrazo to diazo, which are valuable free radical initiators and blowing agents. The catalysed N–C cleavage, used in the production of the systemic herbicide, glyphosate and oxidative cleavage of hydrazones to form ketones, a deprotonation reaction in multi-stage synthesis, are both important applications of hydrogen peroxide.

Aliphatic tertiary amines can be oxidized to *N*-oxides using 35% *m/m* aqueous hydrogen peroxide alone,[303] typically in water at 1.1:1 molar ratio of oxidant to amine for 4–5 h at 60–65 °C. This method is employed worldwide for the production of 25 kilo-tons per annum of fatty amine oxides which are invariably used as surfactants[304] in personal care products, and as thickener compounds in hypochlorite household bleaches. *N*-Methylmorpholine oxide

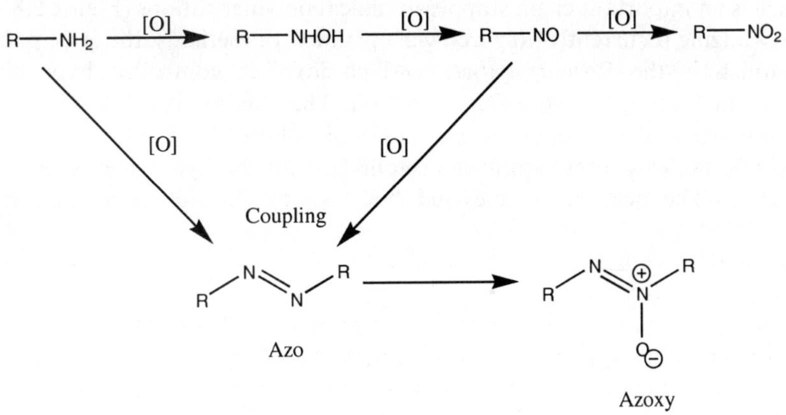

Figure 3.81 *Oxidation of primary amines.*

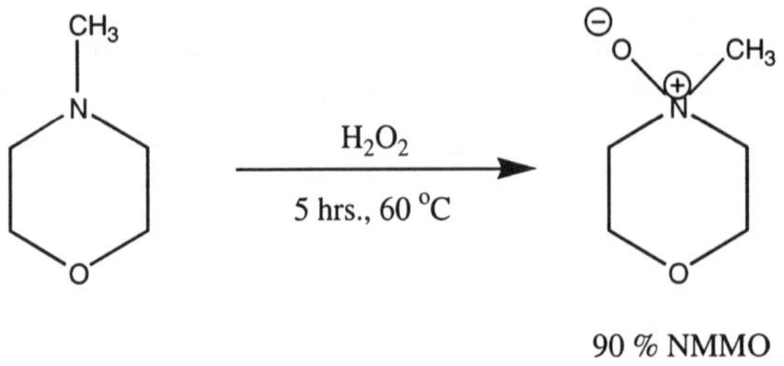

Figure 3.82 *Preparation of NMMO using aqueous hydrogen peroxide.*

(NMMO) has a growing use as a solvent in the manufacture of a new cellulose fibre, Tensil® (Figure 3.82).[305] Substituted oxazolidine *N*-oxides are relatively new user-friendly biocides.[306]

Although the above oxidations are relatively simple, concern has been raised that many nitrogen oxidation reactions, not just those with peroxygens, can lead to traces of nitrosamines in the product, which must be reduced from ppm to ppb levels in any product destined for human contact. Nitrosamine formation is exacerbated by impurities in the amine (*e.g.* secondary amines) and in the peroxide (*e.g.* metal ions), leading to radical chemistry in the slightly alkaline mixture. The biggest process improvements so far are the use of carbon dioxide,[307] and addition of stabilizers to the hydrogen peroxide such as metal sequestrants and free radical inhibitors.[308] Both these approaches together can achieve extremely low nitrosamine levels.

Aliphatic secondary amines can be oxidized to hydroxylamines using hydrogen peroxide without a catalyst in aqueous or alcoholic solution, depending on the solubility of the amine.[309,310] Addition of stabilizers improves both selectivity and efficiency of hydrogen peroxide usage. Among the effective stabilizers are magnesium and zinc salts.[311] These salts form colloidal suspensions of their hydroxides which occlude redox transition metals, and prevent them from decomposing the peroxygen. Diethylamine is oxidized to the hydroxylamine, which is an important chain stopper in radical polymerizations (Figure 3.83).

Hydrazine is currently prepared via a plethora of methods, the most popular of which is the Raschig process, which involves controlled hypochlorite oxidation leading to hydrazine formation. The Raschig process suffers from the formation of a stoichiometric quantity of sodium chloride by-product. The Bayer Ketazine process employs acetone to trap the hydrazine as dimethylketazine. The acetone is re-cycled but sodium chloride is again formed.

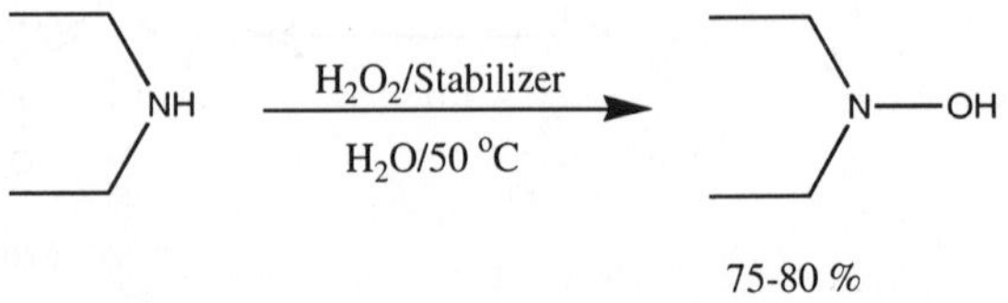

75-80 %

Figure 3.83 *Preparation of diethylhydroxylamine using aqueous hydrogen peroxide.*

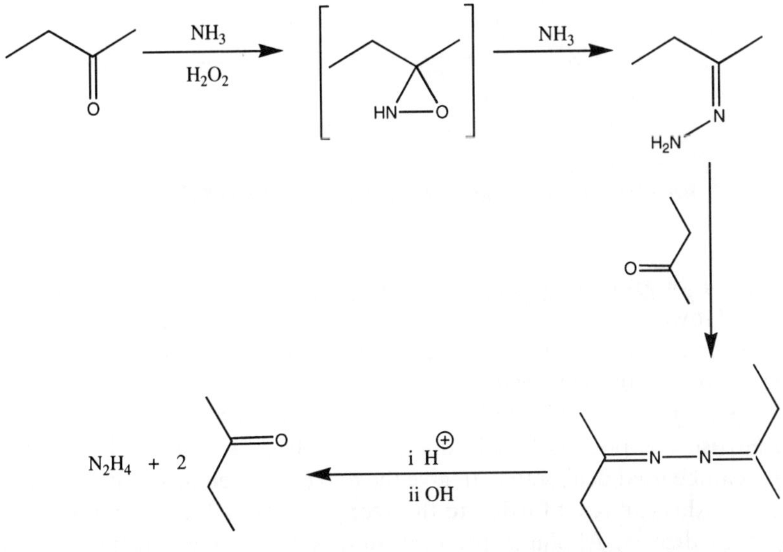

Figure 3.84 *Preparation of hydrazine using aqueous hydrogen peroxide.*

However, Atochem uses hydrogen peroxide as the oxidant with methyl ethyl ketone to form the ketazine intermediate.[313] The ketazine forms as a separate phase in the reaction, and is then purified before hydrolysis to hydrazine and ketone re-cycle. No salt or other aqueous effluent is produced, making the process relatively environmentally benign. It is operated at a 10 kilo-tonne per year scale and now represents nearly 25% of the world hydrazine production (Figure 3.84).

The oxidation of symmetric azo compounds from primary amines usually requires more active peroxygen systems than just hydrogen peroxide. Of industrial importance is the oxidation of hydrazo compounds to azo species using hydrogen peroxide. Azobis-isobutyronitrile (AIBN) is widely used as a radical initiator for polymerizations and as a blowing agent for foam and plastics. It is made from acetone cyanohydrin and hydrazine followed by oxidation in acid for which stoichiometric chlorine[314] or hydrogen peroxide with catalytic addition of hydrogen bromide[315] can be used, the latter leading to less effluent. Analogous chemistry relates to azodicarbonamide, a blowing agent for foams and baking, made from urea and hydrazine followed by oxidation.

Traditionally, percarboxylic acid systems such as extracted peracetic acid,[316] trifluoromethylperacetic acid[317] or magnesium monoperphthalate[318] have been used for the oxidation of heterocyclic tertiary amines (azines). Pyrimidine N-oxide derivatives such as minoxidil (2,4-diamino-6-piperidinopyridine-3-oxide) are important anti-hypertensive agents. Up to now, these have usually been prepared using peracid methods.[319] Neutralized Caro's acid ($NaHSO_5$) oxidizes 2-chloropyridine in high yield to the N-oxide.[320] The Payne system [$R–C(NH)OOH$] is also effective for cyano-pyridines.[321]

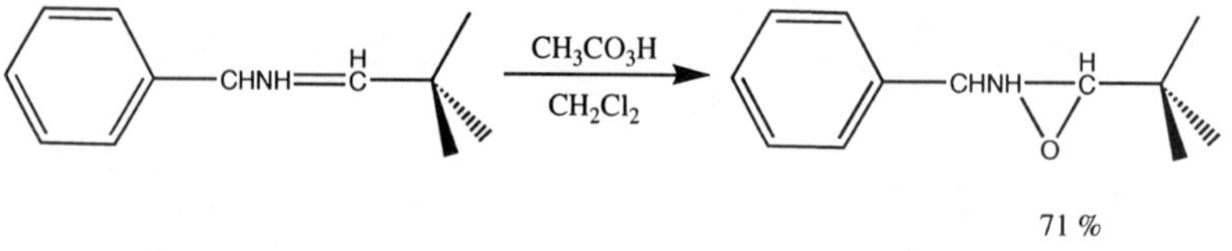

71 %

Figure 3.85 *Preparation of oxaziridines in the presence of peracetic acid.*

Oxidation of aromatic primary amines to nitro compounds is not generally required. However, in the case of deactivated molecules or in order to obtain specific substitution patterns, this transformation may be useful. The reaction can be performed in the presence of peracids.[322]

Less forcing conditions with organic peracids or Caro's acid can be used to make nitroso compounds.[323] Although, as mentioned earlier, a low excess of oxidant can be used deliberately to give the diazo-coupled material as the major product,[324] this can react further to the azoxy compound, but the latter is then hard to oxidize.[325] Aliphatic primary amines are more difficult to oxidize compared to the aromatics, but use of peracetic acid in a solvent will lead to the formation of nitro compounds.[322]

The —N=C— bond in imines, especially Schiff bases formed from aromatic aldehydes and amines, can be 'epoxidized' by peracids to form oxaziridines [N(O)C].[326] Unlike epoxides, oxaziridines will oxygen-transfer.[327] Chiral oxaziridines have been used to carry out enantioselective epoxidations,[328] although these compounds are often prepared by non-peroxygen routes.[329] Oxaziridines can also be rearranged to oximes or nitrones (Figure 3.85).[330]

Hydrazones containing the —C=N—N— linkage are often used in multi-step syntheses to protect carbonyl groups during other transformations. The carbonyl compound is most conveniently regenerated by oxidative deprotection, which amounts to —C=N— cleavage. A number of other peroxygen reagents perform well for this purpose, including MMPP.[331]

The highly acidic conditions required to generate peracids *in situ* from carboxylic acids and hydrogen peroxide are usually not conducive to efficient amine oxidation. *In situ* generation from anhydrides in neutral solution or other acylated agents in alkali can be employed. *In situ* perphthalic or permaleic have been reported for pyrimidine oxidation.[332]

Oxidation at the C–N bond gives imine as the first product but such species are highly labile where the nitrogen is unsubstituted, and readily undergo further oxidation to oximes. For example, benzylamine can be oxidized to benzaldoxime using alkaline peracetic acid generated *in situ*.[333]

It is possible to speed up aliphatic tertiary amine oxidation by adding tungstate or molybdate catalysts.[334] However, for oxidation of aromatic and particularly heterocyclic tertiary nitrogen, a stronger system than hydrogen peroxide alone is required. *N*-Oxidation of heterocycles is of pivotal importance in industrial chemical synthesis.[335] Catalysed systems have been applied and these are dominated by metal peroxo systems based on molybdenum or tungsten. For example, quinoxaline and pyrazine may be oxidized to mono- or

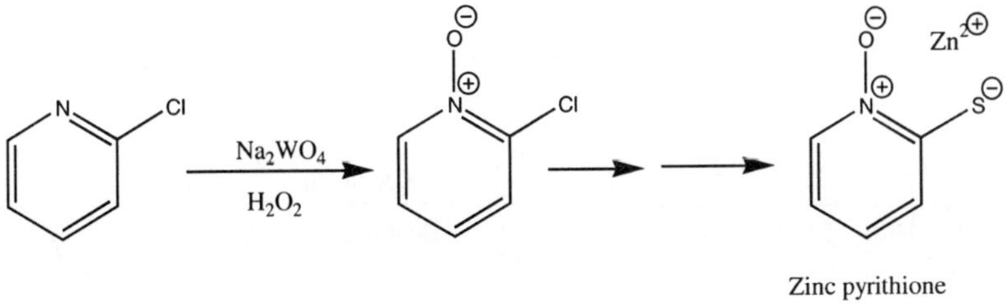

Figure 3.86 *Preparation of zinc pyrithione from 2-chloropyridine.*

di-*N*-oxides with hydrogen peroxide/molybdic acid.[336] Dipicolinic acid can be oxidized to the *N*-oxide with hydrogen peroxide/sodium tungstate and 2-carboxypyrazine-4-oxide, and the amide can be prepared in a similar manner.[337] 2-Chloropyridine-*N*-oxide is an important intermediate in the manufacture of zinc pyrithione, the leading anti-dandruff agent in shampoos (Figure 3.86). Deactivated chloropyridines can be oxidized with hydrogen peroxide/tungstate.[338]

When activated peroxygen systems are applied to secondary amines, oxidation often proceeds to the nitrone, provided the amine has an α-hydrogen. For example, hydrogen peroxide catalysed by selenium dioxide[339] or tungstate[340] readily oxidizes acyclic secondary amines to nitrones. If the oxidation is carried out in the presence of olefins *in situ*, cycloaddition occurs to give isoxazolidines.[341] Where α-hydrogen is absent, the oxidation step stops at the nitroxide,[342] a comparatively stable product which itself can be used catalytically in alcohol oxidation. Tetrahydroisoquinolines give nitrones, but tetrahydroquinones can be oxidized to cyclic hydroxamic acids.[343]

Hydroxylamine is currently manufactured using mainly the salt producing processes alluded to earlier.[344] The Raschig process involves reduction of nitrite by sulfur dioxide leading to co-production of hydroxylammonium sulfate and ammonium sulfate. In a newer process, nitric acid is catalytically hydrogenated in sulfuric acid, again making hydroxylammonium sulfate. Over 95% of the hydroxylamine production is employed to manufacture ε-caprolactam, the monomer for nylon-6. In this process, it is condensed with cyclohexanone to produce the oxime followed by a Beckmann rearrangement in oleum to form the product. It is known that cyclohexanone oxime can be formed from the ketone, ammonia, hydrogen peroxide, and tungstate.[345] However, it has been found that titanium silicalite (TS-1) also catalyses the reaction (Figure 3.87).[346]

An important current industrial process based on N–C cleavage using hydrogen peroxide is the manufacture of *N*-phosphonomethylglycine (systemic herbicide, glyphosate) from *N*-phosphonomethyl imido diacetic acid via the *N*-oxide (Figure 3.88).[347] *N*-Oxidation is catalysed by molybdate or tungstate and the intermediate formed can be decomposed with a second catalyst such as iron(II) to form the product.[348]

Dimethyldioxirane oxidizes primary amines to nitro compounds either preformed[349] in acetone or *in situ*.[350]

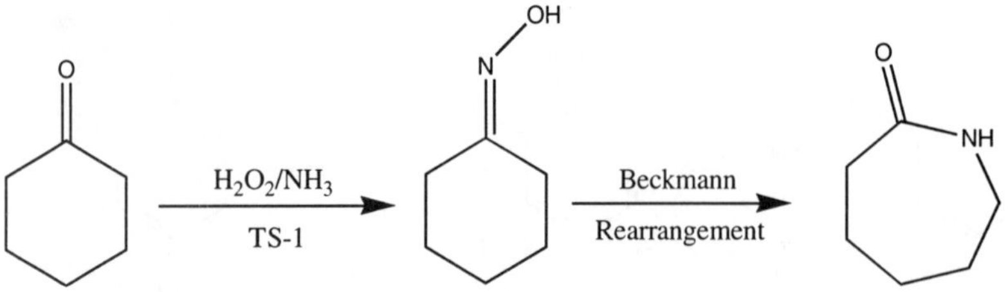

Figure 3.87 *Use of a TS-1/hydrogen peroxide system for the production of cyclohexanone hydroxylamine.*

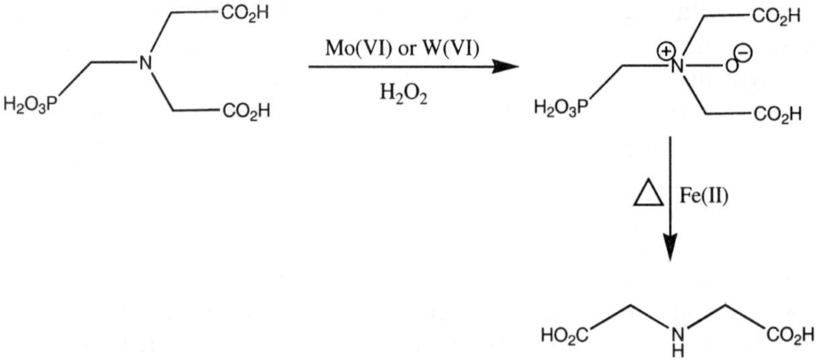

Figure 3.88 *Preparation of the systemic herbicide glyphosate.*

Many heterocyclic bases can be oxidized to *N*-oxides with sodium perborate and acetic acid in the absence of metal catalysts.[352] Use of a smaller excess of oxidant leads to diazo compounds.[353] Aliphatic amines can be converted to nitroso products. Sodium perborate/acetic acid systems can also cleave hydrazones, regenerating carbonyl compounds which have been protected by hydrazine formation.[354]

9   Oxidation of Organo-sulfur Compounds

The oxidation of organic sulfur compounds is a particularly broad subject, and chlorine or peroxygen based reagents are most often employed. The peroxygen routes are often to be preferred because they lead to lower salt effluent, and product purity can be superior. Many sulfur compounds are strongly reducing which means they can often be oxidized selectively in substrates containing other oxidizable functions by adjusting the reaction conditions.

Oxidation of a sulfide affords a sulfoxide as the first product, which can be further oxidized to a sulfone (Figure 3.89).

Control at the sulfoxide stage can be achieved by limiting conditions of temperature, molar ratio and addition rate of the peroxygen. Another means of control arises from an inherent feature of sulfur chemistry, namely oxidation

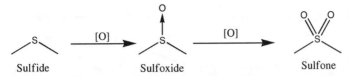

Figure 3.89 *Oxidation of sulfides to sulfoxides and sulfones.*

of sulfur(II) to sulfur(IV) requires electrophilic reagents, whereas oxidation of sulfur(IV) to sulfur(VI) is normally a nucleophilic process. Hence, sulfide or sulfoxide can often be selectively oxidized in the presence of another group by choosing an electrophilic or nucleophilic peroxygen source.

The oxidation is more complex (Figure 3.90) since the first stage is invariably sulfide coupling to form a disulfide, leading to a family of intermediate oxidation products, until the final stage of organic oxidation, *i.e.* the sulfonic acid is reached. Oxidation to the disulfide is remarkably facile, and often carried out in alkaline media, where the thiol is soluble but the sulfide precipitates. Further oxidation requires more forcing conditions, often using a peracid or with the addition of a metal catalyst to effect disulfide cleavage.

Hydrogen peroxide used over a wide pH range has been employed to oxidize sulfur compounds. The scope of the oxidation using hydrogen peroxide includes coupling of thiols to disulfides[355] and thiols coupled with amines,[356] which are used to produce industrially important compounds. For example, sulfenamides are accelerators for rubber vulcanization (Figure 3.91).[357]

The coupling of thiols with amines is an excellent example of the competitiveness between chlorine and peroxygen oxidants. Commercially, sodium hypochlorite or chlorine are employed for the preparation of sulfenamides. However, in addition to their inherent environmentally aggressive nature, chlorine-based oxidants have a number of other disadvantages: there is chlorinated by-product formation (normally chloramines), the process has a high salt loading, and the product requires several washing stages to reduce the salt content. Hydrogen peroxide circumvents all these problems, and in addition affords better quality products. The main disadvantage of the peroxide route is the lower chemical yield to the sulfenamide. In the past this has meant that the chlorine process was significantly cheaper, but increasing effluent treatment costs and environmental legislation is altering the balance towards hydrogen peroxide. As a compromise, some producers run mixed oxidant systems (Table 3.8); here, the yield is improved over using hydrogen peroxide alone, whilst the environmental disadvantages of sodium hypochlorite are somewhat alleviated.

The preparation of disulfides can be achieved by reacting the thiol or its salts with aqueous hydrogen peroxide[358] under alkaline conditions. However, the pH needs to be closely monitored, since if too high, the disulfide selectivity will drop.[359] Dialkyl dithiocarbamates can be oxidized to tetraalkylthiuram disulfides (Figure 3.92).[360]

Again, excess alkalinity must be avoided. The reaction is therefore carried out in aqueous solution using sulfuric acid to regulate the pH.[361] Monoalkyldithio-

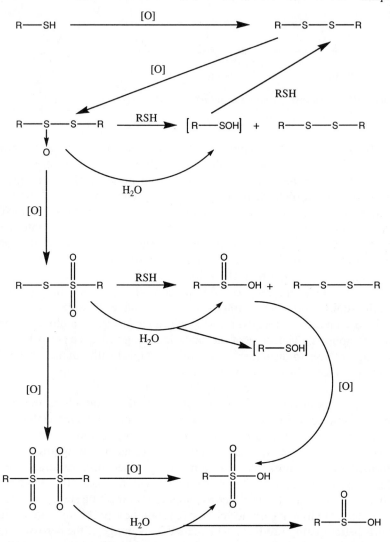

Figure 3.90   *Oxidation of thiols.*

Table 3.8 *Preparation of NCBS in the presence of a mixed oxidant system*[a]

NaOCl/H_2O_2	Yield (%)
1:0	95
1:3	82
0:1	77

[a] 20% excess of oxidant, 50 °C, pH 10–10.5.

$$2RSH \xrightarrow{H_2O_2} RSSR$$

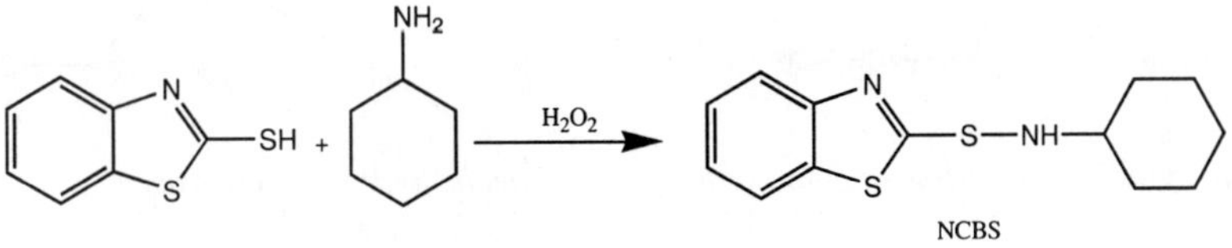

Figure 3.91 *Oxidative coupling of thiols with amines.*

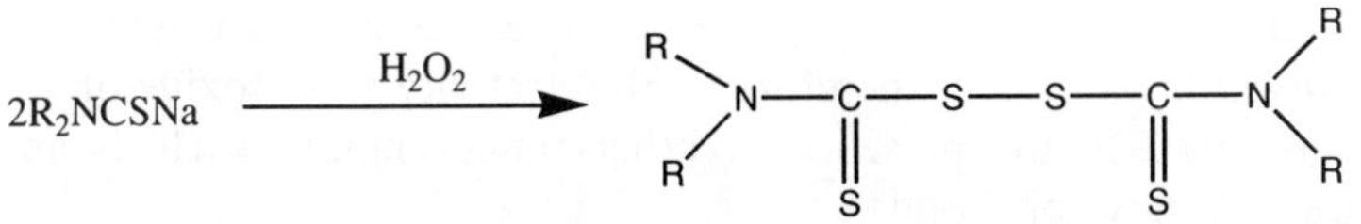

Figure 3.92 *Preparation of tetraalkylthiuram disulfides.*

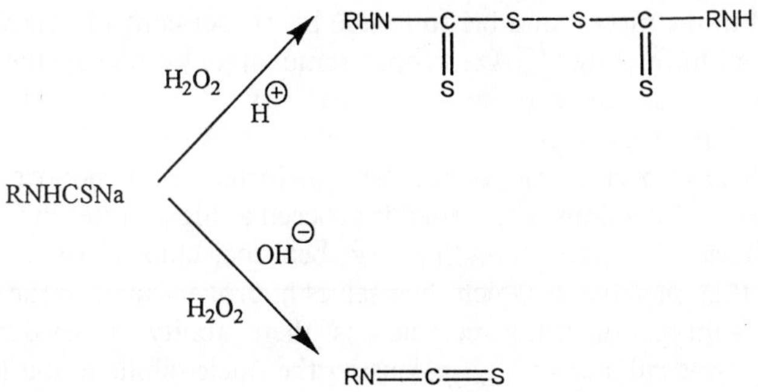

Figure 3.93 *Effect of pH on the product distribution when monoalkyldithiocarbamates are reacted with hydrogen peroxide.*

carbamates produce different products depending upon the pH of the system. For example, at acidic pH the thiuram disulfide is produced,[362] whereas at alkaline pH, the isothiocyanate is the main species (Figure 3.93).[363]

This dependence of product distribution on reaction conditions is also evident when 2-mercaptobenzothiazole is oxidized under acidic conditions. The material will form the disulfide provided a stoichiometric amount of hydrogen peroxide is employed.[364] However, when excess hydrogen peroxide is used, the sulfinic acid will be formed, which upon acidification liberates sulfur dioxide to give the benzothiazole.[365] Thus 2-mercaptothiazoles, under certain conditions, will undergo desulfurization, which is an important step in the preparation of chlormethiazol,[366] a vitamin B$_1$ intermediate (Figure 3.94).

Aqueous hydrogen peroxide is the oxidant of choice for the preparation of sulfoxides.[367] Acetic acid is often used as the solvent due to its specific solvation rather than due to peracetic acid formation.[368] Alcohols and ketones have been

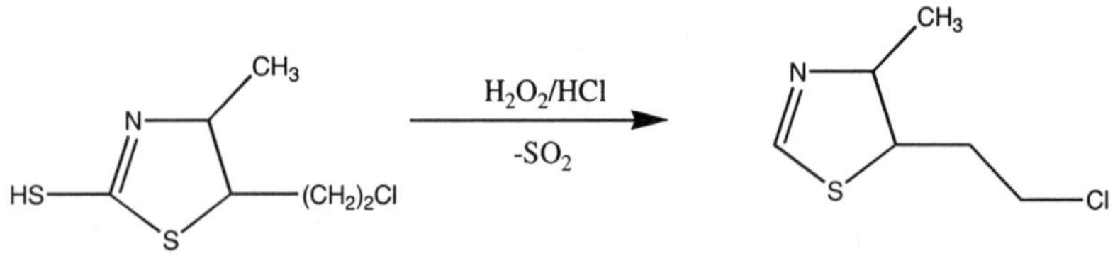

Figure 3.94   *Preparation of chlormethiazol using acidic hydrogen peroxide.*

used by some workers, however; because of the possible formation of hazardous peroxides, ketones are not recommended as solvents.

The sulfur atom generally tends to be more reactive than other functional groups towards oxidation; therefore, high selectivity to sulfoxide or sulfone[369] can be achieved in the presence of other functionalities such as amines,[370] amidines[371] or multiple bonds.[372]

It will often be necessary to avoid sulfone formation, and this can be achieved by carrying out the reaction below 50 °C, and using only stoichiometric amounts of hydrogen peroxide.[373] For example, 1,4-dithiane can be selectively converted to the mono- and bis-sulfoxide by proper control over the hydrogen peroxide addition rate.[374] Azeotropic removal of water by the addition of suitable solvents can also increase selectivity.[375] However, if sulfones are required, then elevated temperatures should be used.[376]

Peracids have been employed for the transformation of sulfides to sulfoxides or sulfones.[377] Reactions with peracids proceed at higher rates in solvents which favour internal hydrogen bonding, *e.g.* benzene, chloroform or tetrachloromethane, than in solvents which themselves hydrogen-bond to the peracid.

The advantage of using peracids is their ability to oxidize sulfoxides selectively over sulfides at high pH using the nucleophilic anion (Figure 3.95) since sulfoxide oxidation is nucleophilic, whereas sulfide oxidation is electrophilic. This behaviour is due to the formation of peroxidic intermediates by the addition of peroxy anion to the sulfoxide.[378]

Sulfoxides have been prepared using a covalently bound peracid cross-linked polystyrene resin.[379] The yields are dependent upon the sulfide structure. The common peracids have also been used for the preparation of sulfoxides,[380] but generally no advantage is gained over hydrogen peroxide. Penicillin sulfoxides,[381] which are intermediates in the synthesis of cephalosporins, are a special case, and are produced more efficiently using peracid oxidants.[382] The oxidation has also been successful for steroid thioethers[383] and thiuranes.[384] The conversion of thiols and disulfides into sulfonic acids is relatively easy to achieve using

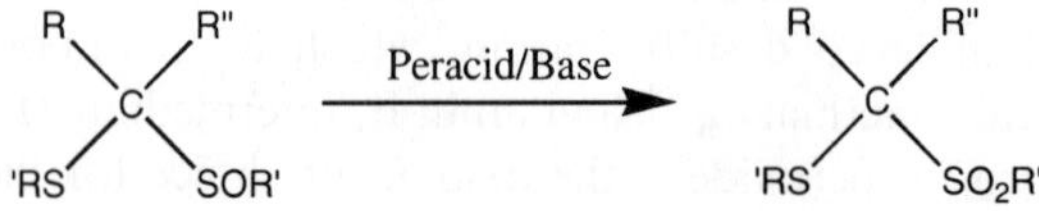

Figure 3.95   *Selective oxidation of sulfoxide over sulfide in the presence of peroxy acid anion.*

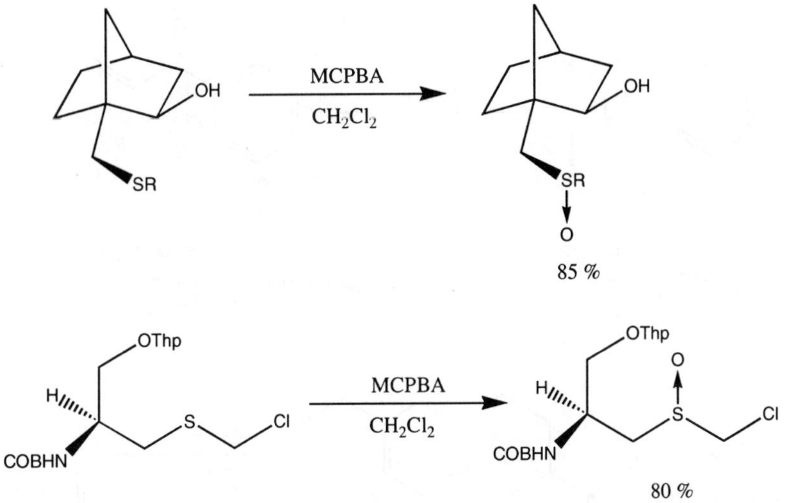

Figure 3.96 *Sulfide oxidation in the presence of MCPBA.*

peracids.[385] Thioacetic esters obtainable by free radical addition of thioacetic acid to terminal olefins yield sulfonic acids (free from isomers) upon oxidation with peracids.[386]

3-Chloroperbenzoic acid (MCPBA) has been frequently applied to sulfide oxidation (Figure 3.96). MCPBA has been the oxidant of choice for the diastereoselective oxidation of 2-*exo*-hydroxynorbornyl systems containing a sulfide group in the 10 position and a diastereomeric excess (d.e.) of 80–90% is obtained. However, if the hydroxide function is protected as the ether, the d.e. drops to 30%.[387,388]

A wide range of sulfides has been oxidized to the sulfoxide or sulfone in high yield using magnesium monoperoxyphthalate (MMPP) (Figure 3.97).[389-392]

The vinyl sulfoxides have a variety of synthetic uses, including their ability to serve as Michael acceptors[393] towards enolate anions and organometallics, and as moderately active dienophilies in Diels–Alder reactions.[394]

A range of metal catalysts can be employed with peroxygen species for the effective oxidation of sulfur compounds. For example, branched-chain high molecular weight mercaptans are difficult to oxidize with hydrogen peroxide. However, this difficulty is overcome if the reaction is conducted with hydrogen peroxide in the presence of a copper(II) salt.[395] The formation of a copper(I) mercaptide followed by its oxidation are believed to be the key steps.

The oxidation of sulfur nucleophiles by peroxo-metal complexes is believed to take place via attack of the substrate to the peroxygen atom in a simple bi-molecular reaction.[396] As a consequence of this, the order of reactivity found for typical substrates parallels their nucleophilicity ($k_{thioether} > k_{sulfoxide} > k_{alkenes}$). Table 3.9 illustrates the oxidation of thioethers to the sulfoxide with sodium molybdate or tungstate with hydrogen peroxide. The results show the high degree of selectivity, the sulfones often being present only in trace

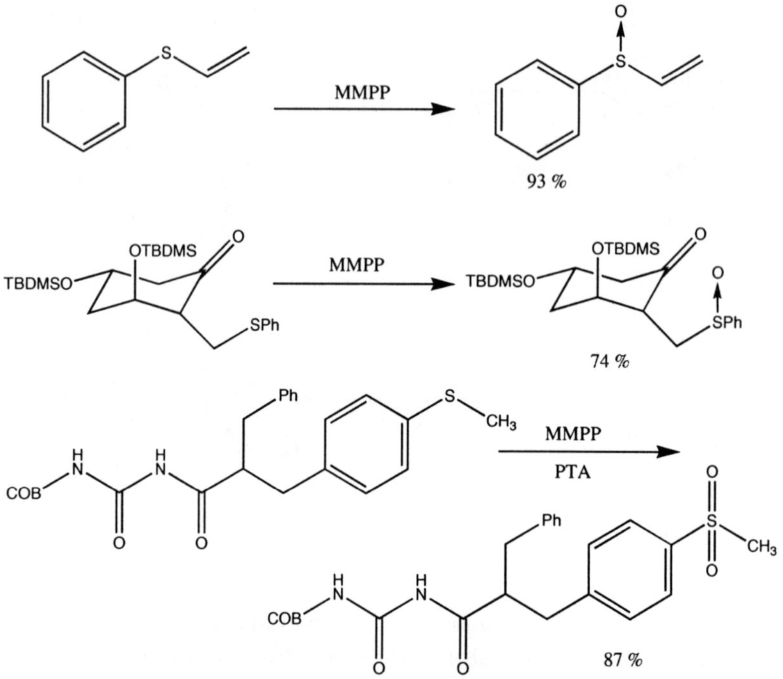

Figure 3.97   Sulfide oxidation in the presence of MMPP.

Table 3.9 Selective oxidation of thioethers to sulfoxides with hydrogen peroxide in the presence of molybdate or tungstate[a]

Substrate	Catalyst	Time (min)	Selectivity (%)
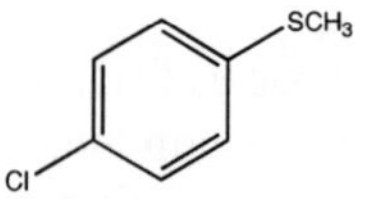	Mo(VI)	10	94
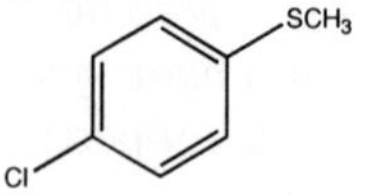	W(VI)	10	88
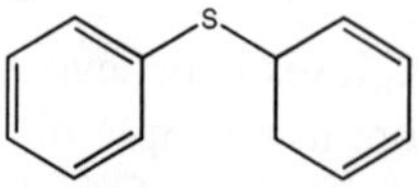	Mo(VI)	60	75
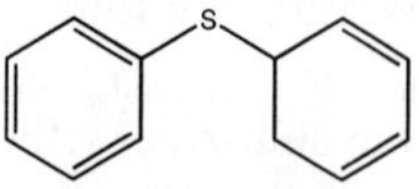	W(VI)	30	73

[a] DCE, hydrogen peroxide, Aliquat® 336, 40 °C.

amounts.[397] Generally, molybdenum(VI) compounds are more selective than tungsten(VI).

Cysteic acid is obtained in nearly quantitative yield from cysteine with aqucous hydrogen peroxide in the presence of iron(II) ions.[397] Molybdates and tungstates have also been used as effective catalysts for similar transformations.[398] An excellent route for the oxidation of 2-thioethanol to isothionic acid has been developed.[399] Heteropolyoxometallates supported on alumina[400] can also be used to oxidize a range of organo-sulfur compounds. For example, alkyl monosulfides to sulfoxides and sulfones, and thiols to sulfonic acids are a few possibilities (Figure 3.98).

Disulfides have been oxidized to sulfonic acids with hydrogen peroxide in the presence of tungstic acid.[401] Whilst sulfonic acids are the normal oxidation product, intermediates of the structure RSS(O)R have been isolated.[402]

Sulfoxides can also be transformed to sulfones using catalysed hydrogen peroxide systems. The catalysts can be molybdenum,[403] zirconium,[404] vanadium[405] and selenic[406] salts. The use of $MoO_2(acac)_2$ with t-butyl hydroperoxide (TBHP) as a diastereoselective oxidation system has been reported, but yields and selectivity are at best moderate (Figure 3.99).

$$HOCH_2CH_2SH \xrightarrow{H_2O_2/Na_2WO_4} HOCH_2CH_2SO_3H$$

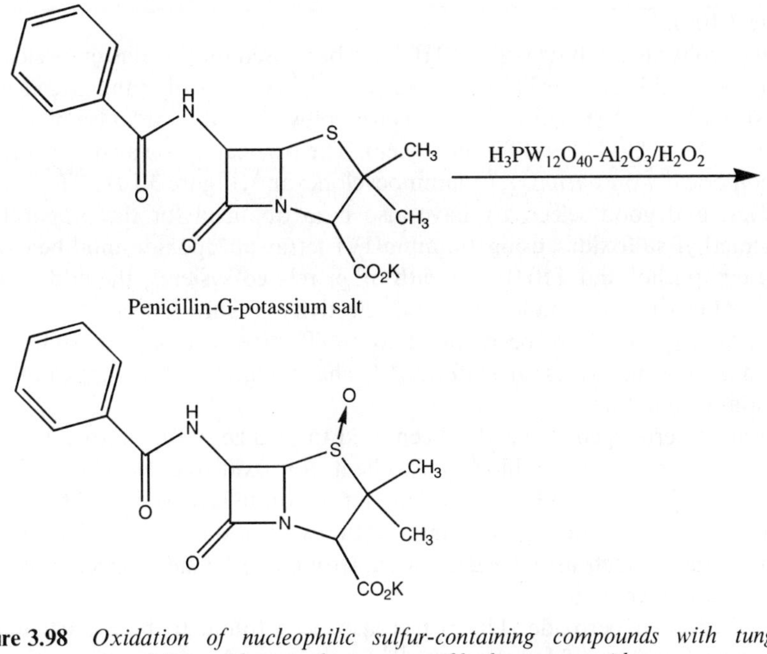

Penicillin-G-potassium salt

$$\xrightarrow{H_3PW_{12}O_{40}\text{-}Al_2O_3/H_2O_2}$$

Figure 3.98 *Oxidation of nucleophilic sulfur-containing compounds with tungsten-containing catalysts in the presence of hydrogen peroxide.*

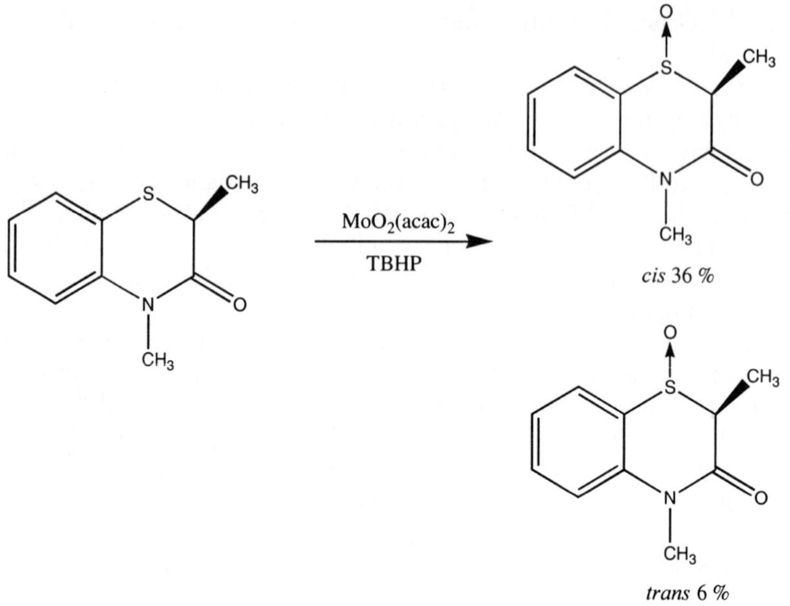

Figure 3.99 *Diastereoselective oxidation of sulfides using a MoO₂(acac)₂/TBHP system.*

The oxidation system VO(acac)₂/TBHP has been used for a series of β- and γ-hydroxy sulfides to sulfoxides with moderate to good diastereoselectivity (Figure 3.100).[407]

Iron porphyrin catalysts with TBHP have been used for the diastereoselective oxidation of sulfides affording up to 46% d.e.[408] A series of manganese(salen) catalysts with hydrogen peroxide has been employed for the oxidation of aralkyl sulfides in 34–70% d.e. and 80–90% yield. The best catalyst was derived from enantiomerically pure *trans*-1,2-diaminocyclohexane (Figure 3.101).[409]

Modest and good selectivity have also been obtained for the preparation of arylmethyl sulfoxides using titanium(IV) tetraisopropoxide modified with R-(+)-binaphthol and TBHP. As with other related systems, the addition of water is of crucial importance for good selectivity (Figure 3.102).[410]

Dioxirane species have been found to be effective oxidants for sulfides to sulfoxides and sulfoxides to sulfones.[411] The oxidations both appear to be electrophilic in nature.

Inorganic peroxygens have also been used to oxidize sulfur centres, particularly sodium perborate, which is excellent for oxidizing electron-deficient sulfides to sulfones.[412] Sodium perborate tetrahydrate has been used to stereospecifically oxidize α-methylbenzylamine thiols in acetic acid.[413] The use of organic peracids such as MCPBA was not found to be stereospecific for the above mentioned oxidation.

Urea hydrogen peroxide (UHP) has also been found to be a particularly effective oxidant for sulfide substrates.[414]

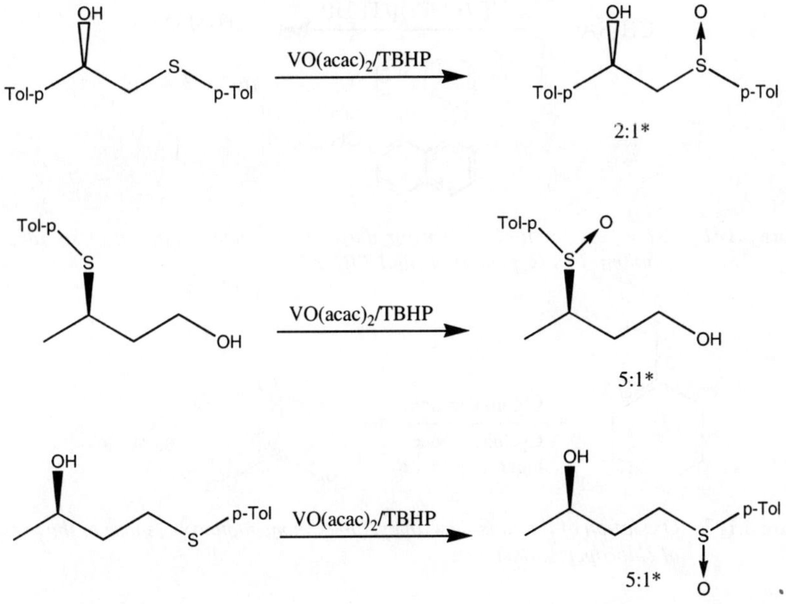

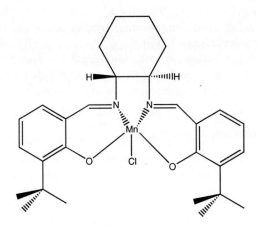

* Stereoselectivity at sulfur not determined

Figure 3.100 *Diastereoselective oxidation of sulfides to sulfoxides using a VO(acac)₂/ TBHP system.*

Figure 3.101 *Manganese salen catalyst for diastereoselective oxidation of sulfides with hydrogen peroxide.*

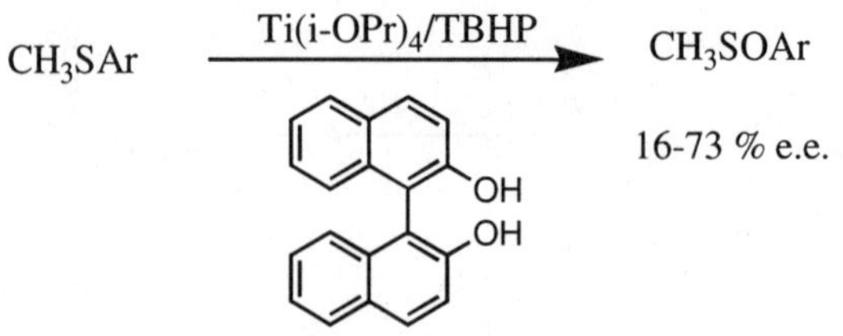

$$CH_3SAr \xrightarrow{\text{Ti(i-OPr)}_4/\text{TBHP}} CH_3SOAr$$

16-73 % e.e.

Figure 3.102 *R-(+)-Binaphthol system for diastereoselective oxidation of sulfides with titanium(IV) isopropoxide and TBHP.*

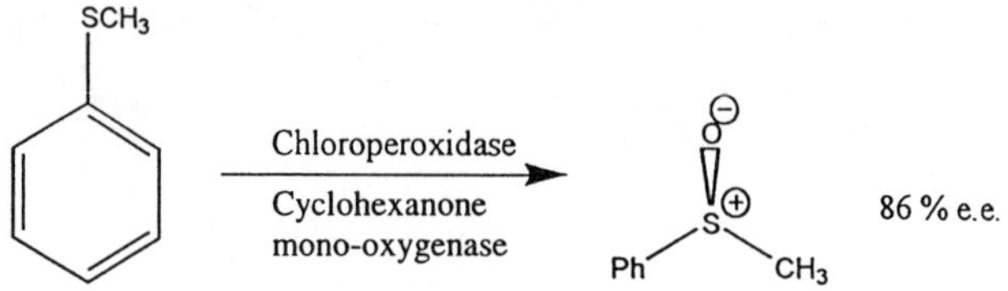

Figure 3.103 *Oxidation of thioanisole with cyclohexanone monooxygenase in the presence of chloroperoxidase.*

Sodium percarbonate (PCS) can also be employed for the oxidation of sulfides to sulfones in acetonitrile assisted by ultrasound.[415]

Oxidation at sulfur by micro-organisms or isolated enzymes has been studied by several workers. However, at present it cannot be recommended as a general synthetic route.[416] Oxidation of a sulfide to a sulfoxide is only stereospecific under certain conditions, and it is often difficult to predict the stereochemistry of the product. *Caldariomyces fumago* has been used in conjunction with chiral peroxides which undergo kinetic resolution during sulfur oxidation. A series of substituted arylmethyl sulfoxides can be prepared with 97–100% d.e.[417] The dramatic effect of substrate structure on enantiomeric excess using cyclohexanone monooxygenase from *Acinetobacter* has been reported for a series of arylalkyl and dialkyl sulfides (3–99% d.e.).[418] It was shown that the substrate structure not only influenced enantiomeric excess of the product, but also its absolute configuration (Figure 3.103).

A number of other inorganic oxidizing agents has also been reported for sulfide oxidation which include [Ru(bpy)$_2$(O)PR$_3$][ClO$_4$],[419] Zn(BiO$_3$)$_2$/AcOH[420] and potassium monopersulfate on wet alumina.[421]

10 Halogenation

Hydrogen peroxide will readily oxidize hydrogen halides and their salts to liberate the corresponding halogen, with the exception of fluorine (Figure 3.104).

$$H_2O_2 + 2HX \longrightarrow X_2 + 2H_2O$$

Figure 3.104 *Liberation of halogens via the oxidation of hydrogen halides.*

The reaction is exothermic and proceeds very quickly provided the hydrogen halide is present in sufficiently high concentration. The minimum concentration of hydrogen halide required increases in the order hydrogen iodide < hydrogen bromide < hydrogen chloride with the critical concentration for HCl being 20%. Mechanistic investigations support the view that the reaction proceeds via an ionic mechanism with hypohalous acid (HOX) as an intermediate.[422,423] However, fairly strong acid conditions are required in order to prevent loss of available oxygen by further reaction of hydrogen peroxide with HOX.

Bromine can be produced in almost quantitative yield, whereas chlorine[424] can be recovered in about 90% yield in the presence of a tungstic acid catalyst.[425] A vanadate catalyst acts in a similar fashion to allow bromine to be generated more efficiently from the hydrogen bromide under mild conditions.[426] The use of acidic catalysts such as sulfuric or phosphoric acid will promote the reaction outlined above, enabling lower strengths of HX to be utilized. The reaction has been known for many years[427] and is widely reported in the literature. It provides a method for the *in situ* generation of halogens and for the recovery of bromine.

In most conventional halogenation processes with elemental halogen only, half the halogen is utilized, with the rest being converted to the hydrogen halide. Using hydrogen peroxide, the halogen can be regenerated from the hydrogen halide, thereby halving halogen consumption, increasing the yield of desired product and producing water rather than acid as the main waste product. The oxidative *in situ* generation of halogen is a technique which fully utilizes the halogen in the same way (Figure 3.105).

The technique has a number of additional advantages compared with preformed halogen:

- The halogen is not used directly and therefore the transport, handling and storage problems associated with elemental halogens are avoided.
- Improved selectivity may be obtained with acid-sensitive substrates due to avoidance of accumulation of hydrogen halide, and in the case of iodination the reverse reaction can be suppressed.

The hydrogen peroxide/hydrogen halide system is suitable for the halogenation of a wide range of aromatic compounds including phenols,[428] phenyl ethers[428] and aromatic amines[429] and amides.[429] The halogenation of aromatic hydrocarbons using this system was first reported for the conversion of codeine halohydrates to the corresponding monochloro and monobromo

$$H_2O_2 + 2HX \longrightarrow X_2 + 2H_2O$$

$$RH + X_2 \longrightarrow RX + HX$$

Figure 3.105 In situ *generation of halogen for the preparation of organo-halogen compounds.*

derivatives.[430] The bromination of perylene using acidified sodium bromide and hydrogen peroxide was reported soon after.[431] Benzene and alkylbenzenes have been shown to undergo halogenation with this system at room temperature.[432] *m*-Xylene is monochlorinated in 90% yield at 20 °C with hydrogen peroxide/hydrogen chloride[433] at a molar ratio of 1:1:6 *m*-xylene/hydrogen peroxide/hydrogen chloride. Dichloro-*m*-xylene can be obtained in similar yield at 40 °C with a 1:3.3:20 molar ratio.[433] The equivalent bromination affords similar results. However, the use of excess hydrogen peroxide/hydrogen bromide at 70 °C gives the tetrabrominated derivative in 80% yield.[434]

Aromatics with powerful activating groups such as hydroxyl are readily halogenated using hydrogen peroxide/hydrogen halide. 2,3,6-Trimethylphenol is chlorinated in the 4-position with hydrogen peroxide/hydrogen chloride in 95% yield. Further oxidation using hydrogen peroxide in acetic acid gives the vitamin E intermediate trimethyl-1,4-benzoquinone.[435] 2,4,6-Trichlorophenol is easily obtained in high yield at 50–60 °C[436,437] and can be further oxidized with hydrogen peroxide in the presence of magnesium chloride catalyst to give tetrachloro-1,4-benzoquinone, chloranil, in 96% yield.[436] Chloranil is used mainly as an intermediate in the production of oxazine dyes and pesticides. It can also be prepared in the absence of a catalyst by the chlorination of hydroquinone using hydrogen peroxide/hydrogen chloride.[438,439] Phenolic methyl ethers have been brominated in the ring using a sodium tungstate/hydrogen bromide/hydrogen peroxide system (Figure 3.106).[440]

Halogenated phenols can be further substituted to give multi-halogenated products in high yield.[441,442] Alkyl-substituted phenols can also be halogenated to give good yields. The antiseptic 4-chloro-3-xylenol is produced using copper(II) chloride catalyst[443] with hydrogen peroxide/hydrogen chloride. The flame retardant tetrabromobisphenol-A is prepared in 95% yield using hydrogen peroxide/hydrogen bromide.[444,445] This compound is particularly useful for the flameproofing of polyesters and epoxy resins. Naphthols can be halogenated using the hydrogen peroxide/hydrogen bromide route; the halogenation is in the activated ring and the products can be further oxidized to the corresponding naphthaquinone. 2-Naphthol can be mono-iodinated to give 1-iodo-2-naphthol in quantitative yield.[446] The selective halogenation of flavanols has also been reported (Figure 3.107).[447]

Like phenols, aromatic amines are readily halogenated, aniline itself being tribrominated in 98% yield. Alkyl-substituted anilines generally give the quinone derivative under the conditions of the halogenation. Quinone formation can, however, be prevented by first forming the acetanilide. This technique has allowed a wide variety of alkyl- and alkoxy-substituted anilines to be halogenated with hydrogen peroxide/hydrogen halide.[448] Such acetylation is not required where electron-withdrawing groups are present, as quinone formation does not occur under such circumstances. The mono- and di-halogenation of nitroanilines has been extensively studied.[449] Aromatics with deactivating substituents such as 4-nitrophenol,[450] nitroaniline,[449] hydroxybenzoic acid,[451] sulfonanilamide[452] and anthranilic acid[453] require more vigorous conditions, and usually require a Lewis acid catalyst.

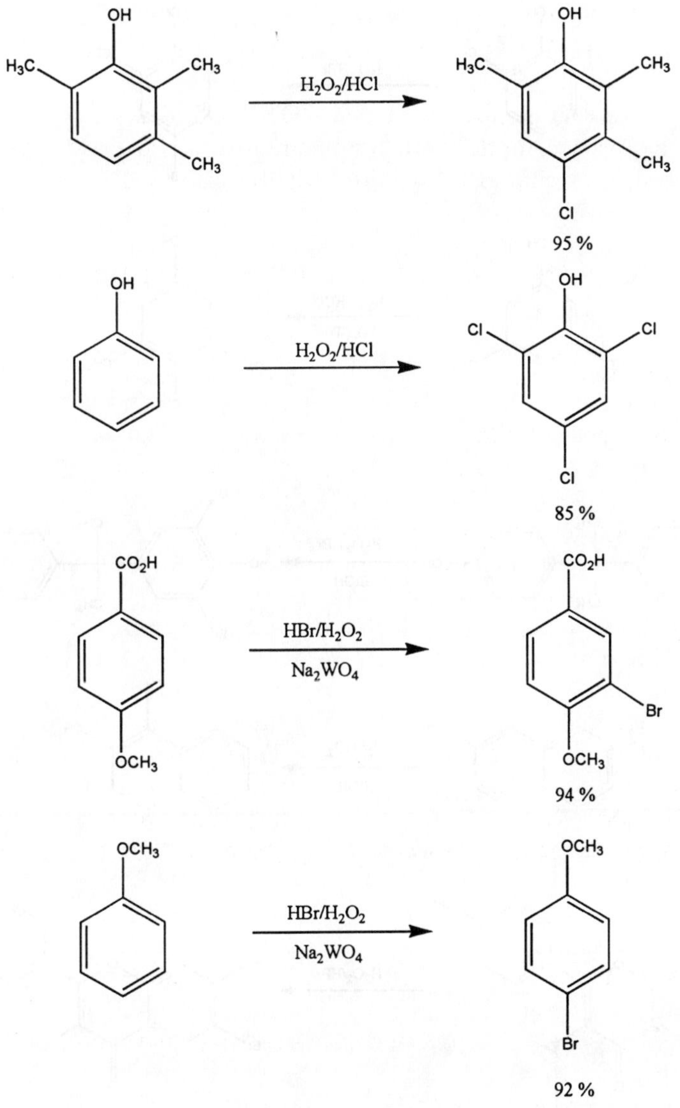

Figure 3.106 *Halogenation of aromatics employing HX/hydrogen peroxide.*

Phthalic anhydride has exclusively electron-withdrawing substituents and can be converted to the tetrabromo derivative in dilute sulfuric acid using an iron(II) iodide catalyst.[454]

The hydrogen peroxide/hydrogen halide system has also been successfully applied in the halogenation of a range of heterocyclic compounds. Pyrrole is tetraiodinated with KI and hydrogen peroxide at 45 °C in 80% yield.[455] The halogenation of substituted pyridines[456] and quinolines[457] have also been reported (Figure 3.108).

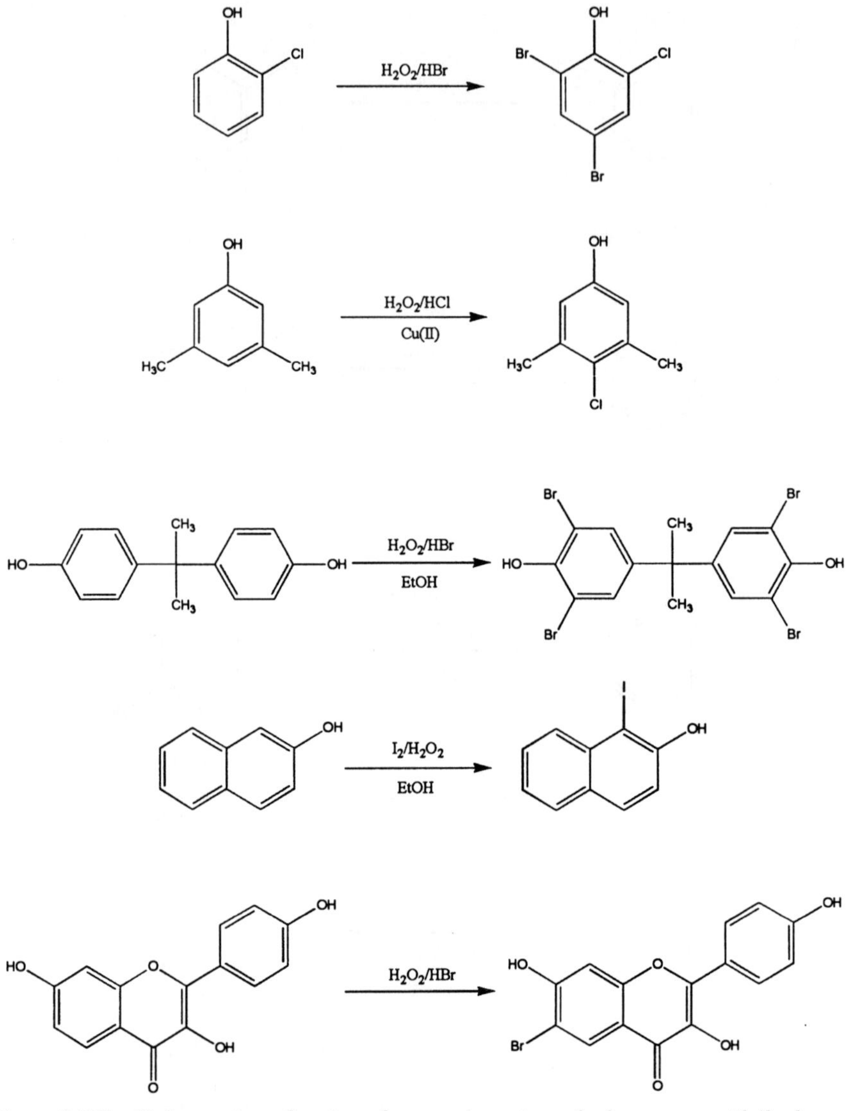

Figure 3.107 *Halogenation of activated aromatics using a hydrogen peroxide/hydrogen halide system.*

The halogenation of alkenes[458] and alkynes[459] using *in situ* generated halogen occurs readily (Figure 3.109). These products have been shown to be useful flame retardent intermediates.

Cyclic alkenes,[460] allyl chloride,[461] and synthetic rubber[462] have been halogenated using this system. In the case of cyclohexene, halogenation occurs readily in a two-phase reaction using a quaternary ammonium salt phase transfer agent. The use of less concentrated hydrogen halide at higher temperature allows the

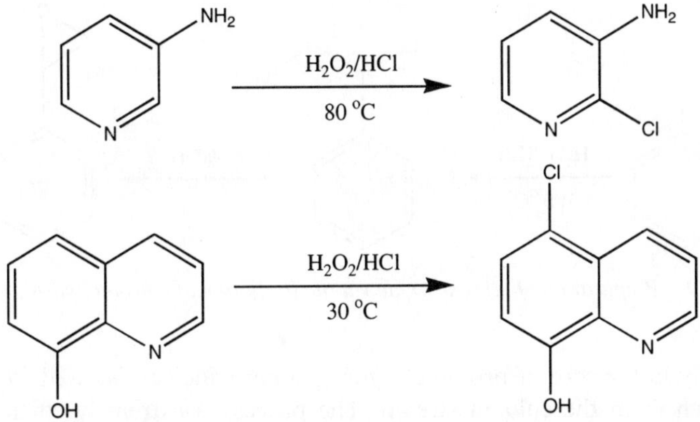

Figure 3.108 *Halogenation of substituted pyridines using hydrogen peroxide/hydrogen halide systems.*

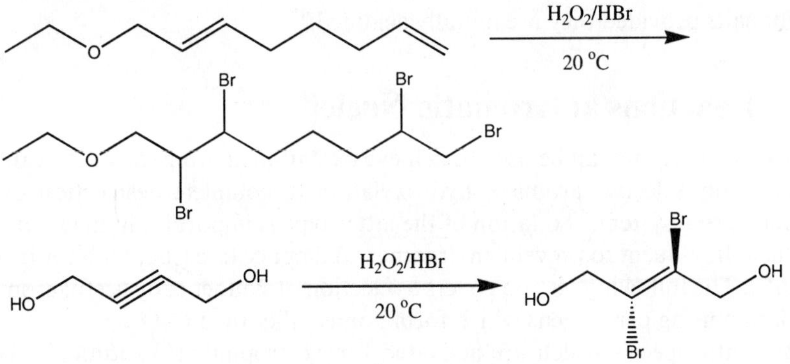

Figure 3.109 In situ *bromination of alkenes and alkynes using hydrogen bromide in the presence of hydrogen peroxide.*

halohydrin to be formed, which can be subsequently treated with alkali to give the epoxide (Figure 3.110).[463,464]

The formation of halohydrins can be promoted by peroxidase catalysts.[465] Recently,[466] it has been shown that photocatalysis reactions of hydrogen peroxide decomposition in the presence of titanium tetrachloride can produce halohydrins. The workers believe that titanium(IV) peroxide complexes are formed *in situ*, which act as the photocatalysts for hydrogen peroxide degradation and for the synthesis of the chlorohydrins from the olefins. The kinetics of chlorohydrin formation were studied, along with oxygen formation. The quantum yield was found to be dependent upon the olefin concentration. The mechanism is believed to involve short-lived di- or poly-meric titanium(IV) complexes.

The hydrogen peroxide/hydrogen halide system is not always applicable due to the presence of water. In cases where the halogen must be used directly,

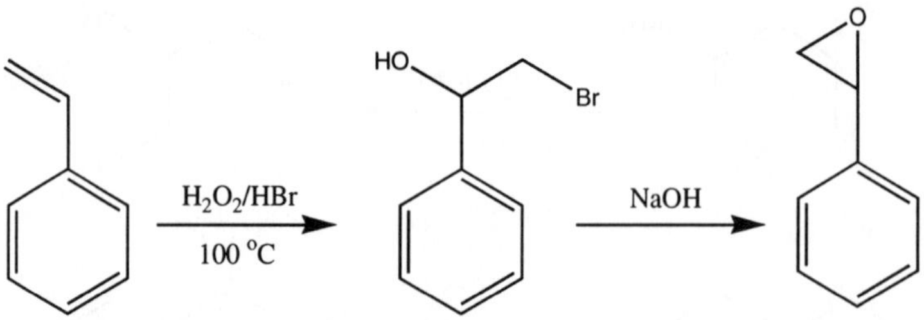

Figure 3.110   *Preparation of styrene oxide via the formation of a bromohydrin.*

particularly in the case of bromine, hydrogen peroxide can be used to recover the halogen from the effluent stream. The process is extremely efficient with bromine being recovered in 95% yield, and does not have the reagent transport and storage problems associated with other bromine recovery technologies. At low concentration of HBr, the reaction may require the input of heat and an acid catalyst to improve recovery.[467] Halogens can also be recovered from their halide salts provided they are initially acidified.[467]

11   Reactions at Aromatic Nuclei

Peroxygen reagents can be used to achieve oxidation at aromatic nuclei ranging from highly selective aromatic hydroxylation to complete destruction of the aromatic ring system. Oxidation of the latter type is important in the treatment of phenolic effluent to prevent environmental damage (see Chapter 5 for further details). The following section covers a selection of synthetically useful aromatic oxidation using peroxygens which form bonds other than C–O.

Aromatic species which are activated for electrophilic substitution may be hydroxylated in the presence of strong acids. The largest single application of this technology is for the hydroxylation of phenol to hydroquinone and catechol using a mixture of perchloric and phosphoric acids as catalysts.[468] As the products are more readily oxidized compared with the substrate, it is important to limit the conversion of the phenol to prevent over-oxidation to tars.

Lewis acids have also been used to catalyse the conversion of alkylbenzenes to the corresponding phenols, however yields and selectivities are generally poor.[469]

Alkaline hydrogen peroxide has not found wide use in the oxidation of aromatic nuclei, although hydroxyquinones can be formed from hydroquinones under strongly alkaline conditions (Figure 3.111).[470]

Hydroxyl radical generated from hydrogen peroxide in the presence of iron(II) salts hydroxylates most aromatic centres, and indeed a phenol hydroxylation process based on this chemistry was operated for nearly a decade.[471] The process ran at similar conversions to the acid-catalysed route mentioned above, however selectivity to dihydroxybenzenes was somewhat lower, and some resorcinol was formed along with catechol and hydroquinone.

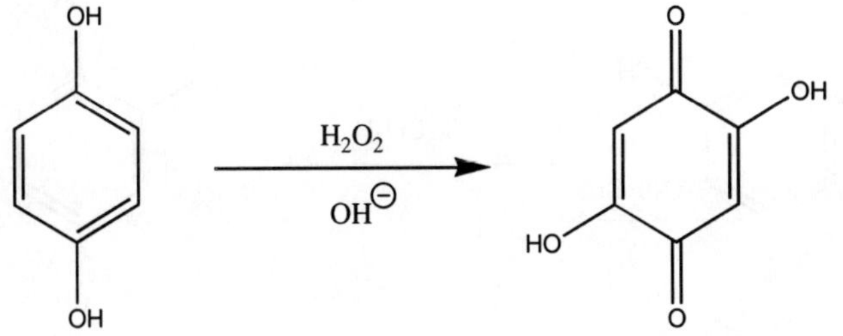

Figure 3.111 *Preparation of hydroxyquinones using alkaline hydrogen peroxide.*

Radical nucleophilc oxidation based on one-electron oxidation, known as the Minisci reaction, is employed for the functionalization of *N*-heterocycles with acidic hydrogen peroxide in the presence of iron(II) salts (Figure 3.112).[472] A range of *N*-heterocycles (pyridines, pyrazines, quinolines, *etc.*) which are activated towards attack by nucleophilic radicals when protonated are suited to this chemistry. The Minisci reaction is suitable for the preparation of carboxylic amides (from formamide), carboxylic esters (from pyruvic esters via a hydroxyhydroperoxide), aldehydes (from 1,3,5-trioxane) and alkylated pyridines (either from carboxylic acids or from alkyl iodides in dimethyl sulfoxide).[473] The latter reaction uses dimethyl sulfoxide as the source of methyl radical (Figure 3.112).

Treatment of quinoline with alkaline hydrogen peroxide results in hydroxylation, and oxidative cleavage of the 'b-ring'. This gives a mixture of products which can be further oxidized to quinolinic acid.[474]

Peracids have found wide application in the oxidation of aromatic species. *In situ* peracids have been used in the presence of strong acids to affect the hydroxylation of phenol in a similar manner to the method using hydrogen peroxide. Trifluoroacetic acid has also been shown to be highly effective for the

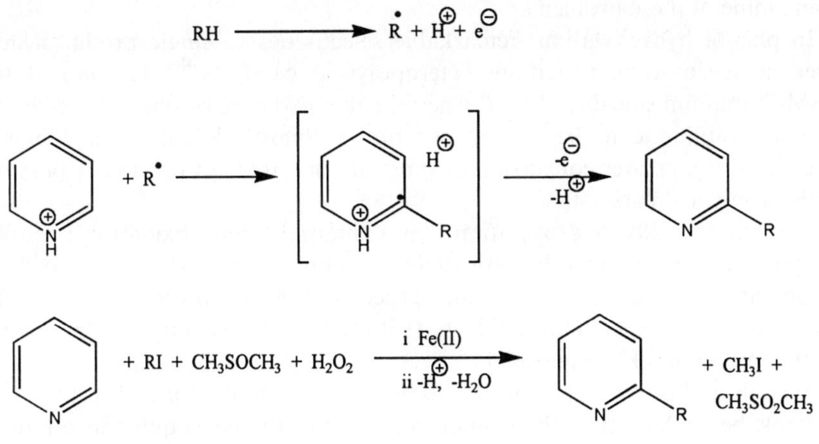

Figure 3.112 *Minisci reaction.*

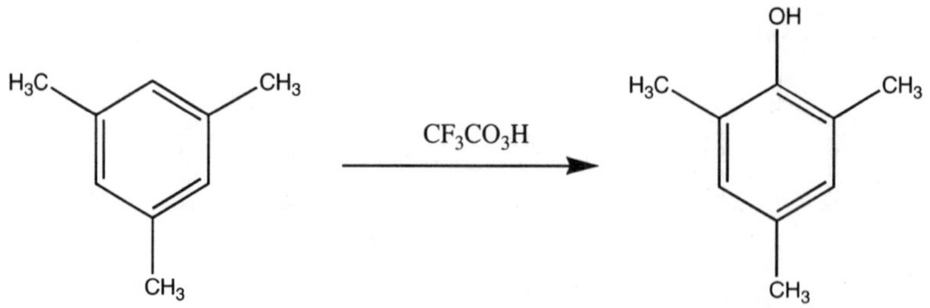

CF_3CO_3H

Figure 3.113 *Hydroxylation of mesitylene with trifluoroperacetic acid.*

hydroxylation of aromatic rings which are more activated than that of toluene, *e.g.* in the conversion of mesitylene to mesitol (Figure 3.113).[475]

If direct products of hydroxylation are required, it is necessary to use an excess of the substrate in order to prevent over-oxidation occurring. However, the further oxidation that can occur may be desirable on occasions. For example, polynuclear aromatic hydrocarbons may be converted to 1,4-quinones using *in situ* peracetic acid.[476] Catechol is readily converted to *cis,cis*-muconic acid via the orthoquinone by peracetic acid in the presence of an iron(III) catalyst.[477] Polycyclic aromatic species can be similarly oxidized to useful products. For example, β-napthol is oxidized to *O*-carboxycinnamic acid.[478] Heteroaromatic species may also react. Substituted furans undergo a variety of oxidations, usually involving ring-opening reactions in the presence of peracids.[479] Quinoline is converted to quinolinic acid using *in situ* peracid in the presence of copper(II) acetate (Figure 3.114).[480]

A wide range of aromatic species including benzene itself has been hydroxylated using hydrogen peroxide with a variety of metal salts. Many of these reactions involve hydroxyl radicals,[481] and not surprisingly, tend to give poor selectivities to specific products. In recent years, a number of complex metal catalysts[482] and heterogenous catalysts[483,484] have been discovered which overcome some of these problems.

In phenol hydroxylation, remarkable selectivities to single products have been achieved using vanadium heteropolyacid catalysts.[485] The use of the ZSM-5 titanium silicalite (TS-1)[483] permits the oxidation of phenol to catechol and hydroquinone to be carried out on an industrial scale with a higher selectivity at a greater conversion of substrate that was not previously possible with strong acid catalysts.

As with peracids, hydroxylation is not the only useful oxidation possible. Dihydroxyarenes are readily converted to quinones using hydrogen peroxide in combination with a wide range of metal species.[486] An interesting example is the use of supported chromium oxide–bis(tributyltin) oxide catalyst in the manufacture of vitamin K_1 (Figure 3.115).[478]

Phenols and phenol ethers may also be converted to quinones directly,[488] an example being 2,3,5-trimethylphenol to 2,3,6-trimethylbenzoquinone, an intermediate in vitamin E production. Ring-opening and degradation reactions are

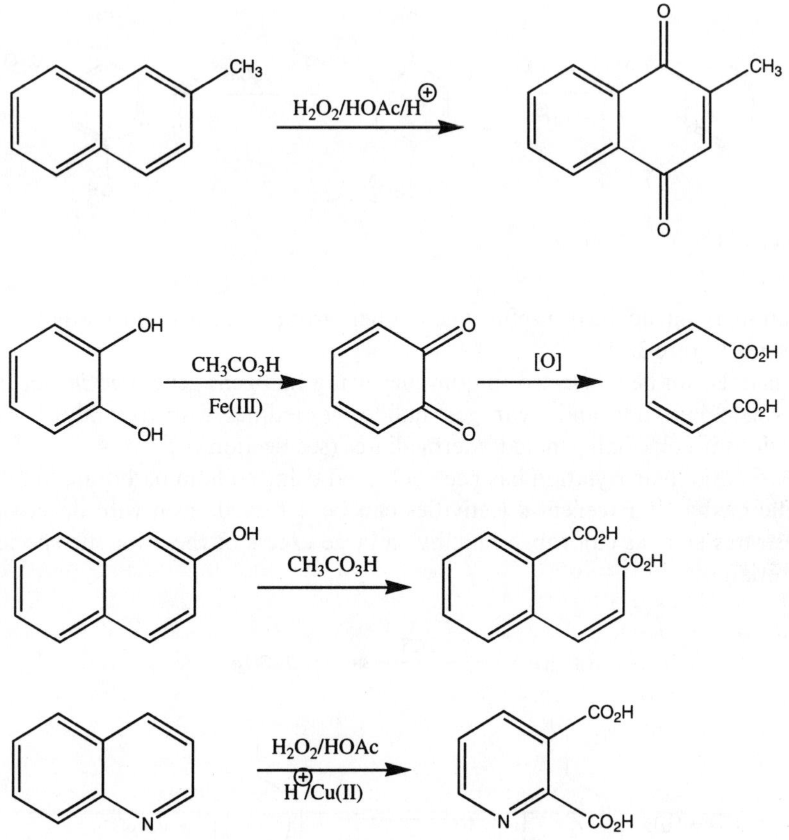

Figure 3.114 *Oxidation of aromatic compounds with peracetic acid.*

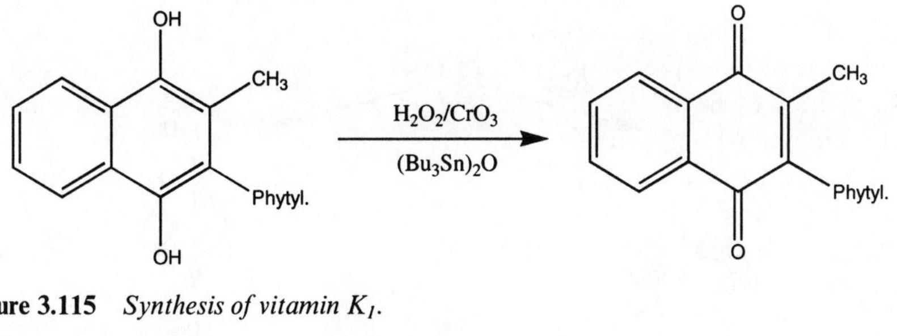

Figure 3.115 *Synthesis of vitamin K_1.*

also possible; for example, the osmium tetraoxide catalysed oxidation of naphthalene to phthalic anhydride.[489]

A number of other aromatic oxidations involving peroxygen species have been documented. Probably the best known of these is the hydroxylation of phenols using alkaline peroxydisulfates, known as the Elbs reaction.[490] This

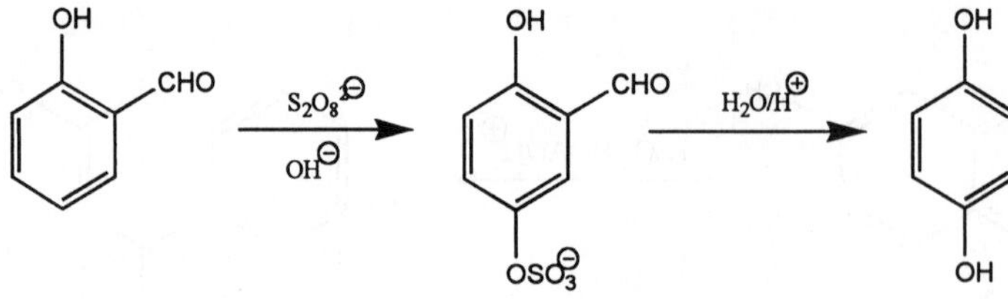

Figure 3.116   *Elbs reaction.*

reaction is particularly useful when other groups sensitive to oxidation are present (Figure 3.116).

Phenols can be converted to quinones using halogens generated *in situ* from hydrogen peroxide and hydrogen halide, sometimes with a catalyst. These reactions involve halogenated intermediates (see Section 10).

Aromatic hydroxylation has been achieved using sodium perborate in highly acidic media.[491] Excellent selectivities can be achieved, even with deactivated substrates such as chlorobenzene, but a large excess of the aromatic species is essential.

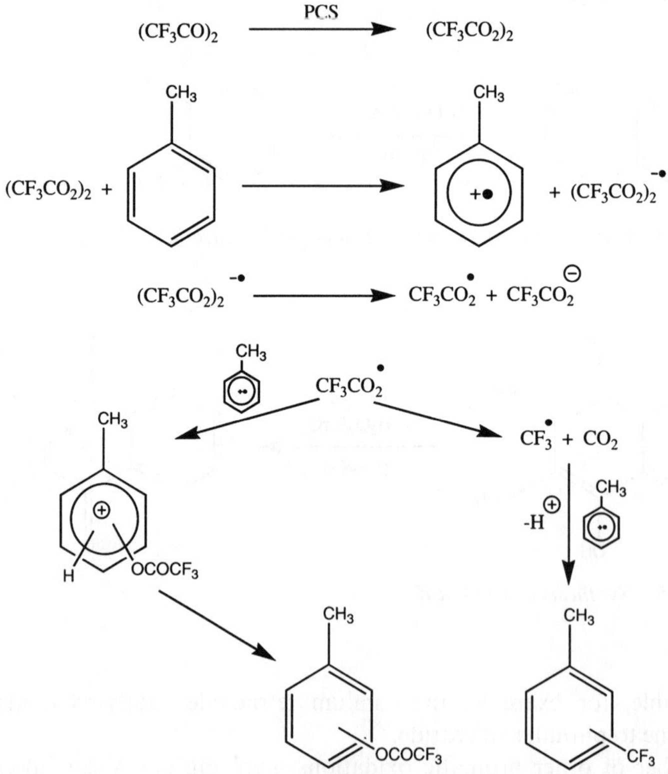

Figure 3.117   *Trifluoromethylation and acetoxylation of toluene in the presence of trifluoroacetic anhydride/hydrogen peroxide.*

Aromatic acyloxylation can be achieved via the decomposition of diacyl peroxides in the presence of aromatic substrates. By careful choice of reaction conditions, either acyloxylation or trifluoromethylation of the ring system can be achieved, using trifluoroacetyl peroxide generated *in situ* from trifluoroacetic anhydride and sodium percarbonate (Figure 3.117).[492]

The Minisci reactions referred to above can also employ *t*-butyl hydroperoxide (TBHP), perdisulfate or dibenzoyl peroxide in the presence of hydrogen peroxide.[493]

12 Conclusion

Based on the above discussion, it is hoped that the reader has gained an insight as to the importance of hydrogen peroxide in the area of organic synthesis, in particular, the use of hydrogen peroxide with metal species for catalytic oxidation. The use of catalytic processes is mainly driven by the need for cleaner processes, and many new and interesting technologies are emerging, especially in fine chemicals where high selectivity to the desired product is crucial. In particular, we expect the design of novel heterogeneous liquid-phase oxidations to play an important rôle in such developments, and such technologies are further expanded upon in the next chapter. Finally, it is hoped that better methods for enantioselective oxidation will be developed, possibly via the utilization of enzymes in organic media.

References

1 N. Mizuno, T. Watanabe, H. Mori and M. Misono, *J. Catal.*, 1990, **123**, 157.
2 L. Casella, E. Monzani, M. Guilotti, F. Glibuch and L. De Giola, *J. Chem. Soc., Dalton Trans.*, 1994, 3203.
3 D.A. Rockcliffe and A.E. Martell, *Inorg. Chem.*, 1993, **32**, 3143.
4 T. Mallet and A. Baiker, *Catal. Today*, 1994, **19**, 247.
5 C. Bronnimann, T. Mallet and A. Baiker, *J. Chem. Soc., Chem. Commun.*, 1995, 1377.
6 M. Reddy, T. Punniyamurthy and J. Iqbal, *Tetrahedron Lett.*, 1995, **36**, 159.
7 J.P. Hague and D.T. Sawyer, 1995, **117**, 5617.
8 T. Funalski, T. Sugimato and S. Yoshida, *Chem. Lett.*, 1982, 1097.
9 R.A. Sheldon, *Bull. Chim. Soc. Belg.*, 1985, **9**, 1450.
10 Ullmann's Encyclopedia of Industrial Chemistry, Vol. 49, p. 531.
11 G. Kolaczinski, R. Mehen and W. Stein, *Fette Sleizen Austrichim*, 1971, **73**, 553.
12 S. Murata, M. Suzuki and R. Noyari, *J. Am. Chem. Soc.*, 1979, **101**, 2738.
13 M. Balzer, F. Lohse and R. Schmid, *Angew. Makromol. Chem.*, 1973, **30**, 349.
14 N. Prilezhaev, *Chem. Ber.*, 1909, **42**, 4811.
15 P.D. Bartlett, *Rec. Chem. Progr.*, 1950, **11**, 47.
16 B.M. Lynch and K.H. Pausaker, *J. Chem. Soc.*, 1955, 1525.
17 R.D. Bach, C.L. Willis and J.M. Domaglas, in 'Applications of MO Theory in Organic Chemistry', ed. I.C. Csizmadia, Elsevier Scientific, Amsterdam, 1977, Vol. 2, p. 221.
18 D.R. Campbell, J.O. Edwards, J. MacLachlan and K.J. Polgar, *J. Am. Chem. Soc.*, 1958, **80**, 5308.
19 G. Berti and F. Botturi, *J. Org. Chem.*, 1960, **25**, 1286.

20 N.N. Schwartz and J.H. Blumbergs, *J. Org. Chem.*, 1964, **29**, 1976.
21 R.D. Bach, C. Canepa, J.E. Winter and P.E. Blanchette, *J. Org. Chem.*, 1997, **62**, 5191.
22 H.C. Brown, J.H. Kawakami and T. Liu, *J. Am. Chem. Soc.*, 1973, **92**, 2209.
23 D. Swern, in 'Organic Peroxides', Wiley Interscience, Vol. 2, 1970.
24 Kirk Othmer's Encyclopedia of Chemical Technology, Wiley Interscience, Vol. 9, p. 259.
25 Solvay Interox, private communication.
26 Kuraray KK, JP 61042504.
27 M.S. Cooper, H. Heaney, A.J. Newbold and W.R. Sanderson, *Synlett.*, 1990, 533.
28 H. Hibbert and C.P. Burt, *J. Am. Chem. Soc.*, 1925, **47**, 2240.
29 S. Danilow and E. Venus-Danilova, *Chem. Ber.*, 1927, **60**, 1050.
30 W.D. Emmons and A.S. Pagano, *J. Am. Chem. Soc.*, 1955, **77**, 89.
31 E.E.G. Werner, *Recl. Trav. Chim. Pays-Bas*, 1948, **67**, 442.
32 E. Klein, W. Rojahn and D. Henneberg, *Tetrahedron*, 1964, **20**, 2025.
33 D. Felix, A. Melera, J. Stibl and E. Kovats, *Helv. Chim. Acta.*, 1963, **46**, 1513.
34 A.R. MacRae and R.C. Hammond, *Biotech. Genetic Engineering Rev.*, 1985, **3**, 193.
35 F. Bjorkling, S.E. Godtfredson and O. Kirk, *J. Chem. Soc., Chem. Commun.*, 1990, 1301.
36 C. Laane, S. Boeren, K. Vos and C. Veeger, *Biotech. Bioeng.*, 1987, **30**, 81.
37 Universitat Jena, DE 232284.
38 D.J. Reif and H.O. House, *Org. Synth. Coll. Vol.*, No. 4, 1963, 860.
39 Peroxid Chemie, EP 55387.
40 Solvay Interox, internal reports.
41 D. Swern, in 'Organic Peroxides', Wiley, New York, 1971, Vol. 2, p. 403.
42 Ciba Geigy, DE 3235664.
43 S.N. Lewis, in 'Oxidation', ed. R.L. Augustine, 1963, Vol. 1, p. 213.
44 W.D. Emmons, *J. Am. Chem. Soc.*, 1954, **76**, 3468.
45 B. Phillips, F.C. Frostide and P.S. Starcher, *J. Org. Chem.*, 1969, **79**, 5982.
46 B. Phillips, P.S. Starcher and B.D. Ash, *J. Org. Chem.*, 1959, **23**, 1823.
47 Interox Chemicals Ltd., GB 1535313.
48 D. Swern, in 'Organic Peroxides', Wiley, New York, 1971, Vol. 2, p. 404.
49 W.D. Emmons and S. Pagano, *J. Am. Chem. Soc.*, 1955, **77**, 2287.
50 Interox Chemicals Ltd., EP 27693.
51 Rhone-Poulenc, FR 2589862.
52 S. Kumar, *J. Org. Chem.*, 1988, **53**, 1329.
53 L.I. Kasvan, *Vopr. Khim. Technol.*, 1986, **80**, 67.
54 S. Medvedev and O. Blokh, *J. Phys. Chem. USSR.*, 1993, **4**, 721.
55 M.S. Malinovski, *Zh. Org. Khim.*, 1966, **2**, 2129.
56 H.O. House and R.S. Ro, *J. Am. Chem. Soc.*, 1958, **80**, 2428.
57 G.B. Payne, *J. Am. Chem. Soc.*, 1958, **80**, 6460.
58 H.O. House and R.L. Wasson, *J. Am. Chem. Soc.*, 1957, **79**, 1488.
59 B. Zwanenburg and J. Ter Weil, *Tetrahedron Lett.*, 1970, 935.
60 H. Newman and R.B. Angler, *Tetrahedron*, 1970, **26**, 825.
61 C.E. Griffith and S.K. Kundi, *J. Org. Chem.*, 1969, **34**, 1532.
62 Dupont de Nemours Co., GB 904877.
63 T.L. Ho and S.O. Lu, *Synth. Commun.*, 1983, **13**, 685.
64 Firmenich SA, EP 181475.
65 M. Miyashita, T. Suzuki and A. Yoshikoshi, *Chem. Lett.*, 1987, 285.
66 G.B. Payne, P.H. Williams and P.H. Deming, *J. Org. Chem.*, 1961, **26**, 659.
67 R.L. Rowland and A. Rodgman, *J. Org. Chem.*, 1964, **29**, 16.
68 Y. Chen and J.L. Reymond, *Tetrahedron Lett.*, 1995, **36**, 4015.
69 W. Adam, R. Curci and J.O. Edwards, *Acc. Chem. Res.*, 1989, **22**, 205.
70 C.W. Jones, J.P. Sankey, W.R. Sanderson, M.C. Rocca and S.L. Wilson, *J. Chem. Res. (S)*, 1994, 114.

71 R.W. Murray and R. Jeyaraman, *J. Org. Chem.*, 1985, **50**, 2847.
72 H. Mimoun, *Tetrahedron*, 1970, **26**, 37.
73 M. Pralus, J.C. LeCoq and J.P. Schrimann, in 'Fundamental Research in Homogeneous Catalysis', ed. M. Tsutsui, Plenum Press, New York, 1989, Vol. 3, p. 327.
74 J.P. Schrimann, BE 863237.
75 S.E. Jacobsen, F. Mares and P.M. Zambri, *J. Am. Chem. Soc.*, 1979, **101**, 6946.
76 J. Itakura, H. Tanaka and H. Ito, *Bull. Chem. Soc. Jpn.*, 1969, **42**, 1604.
77 Interox Chemicals Ltd., EP 68564.
78 E.I. Dupont de Nemours Co., US 3646130.
79 A.W. Hermann, R.W. Fischer, W. Scherer and M. Rauch, *Angew. Chem.*, 1993, **105**, 2775.
80 W.A. Hermann, R.W. Fischer, M. Rauch and W. Scherer, *J. Mol. Catal.*, 1994, **86**, 243.
81 J. Rudolph, K.L. Reddy, J.P. Chiang and K.B. Sharpless, *J. Am. Chem. Soc.*, 1997, **119**, 6189.
82 M. Mugden and D.P. Young, *J. Chem. Soc.*, 1949, 2988.
83 C.G. Allen and A.N. Neogi, *J. Catal.*, 1970, **16**, 197.
84 D. Pratt and R. Lett, *Tetrahedron Lett.*, 1986, **27**, 707.
85 H. Mimoun, *Tetrahedron*, 1970, **26**, 37.
86 G. Amato, *J. Mol. Catal.*, 1986, **37**, 165.
87 B.M. Trost and Y. Masuyama, *Isr. J. Chem.*, 1984, **24**, 134.
88 Montedison SpA, EP 109273.
89 L.I. Simandi, *Inorg. Chim. Acta.*, 1982, **65**, L181.
90 R. Neumann and A.M. Khenkin, *J. Org. Chem.*, 1994, **59**, 7577.
91 C. Venturello, *J. Org. Chem.*, 1983, **48**, 3831.
92 C. Venturello, *J. Mol. Catal.*, 1983, **32**, 107.
93 C.L. Hill and A. Khenkin, *J. Chem. Soc., Mendeleev Commun.*, 1993, 140.
94 K. Sato, A. Masao, M. Ogawa, T. Hashimoto, D. Panyella and R. Noyori, *Bull. Chem. Soc. Jpn.*, 1997, **70**, 905.
95 P. Muller and M. Idamoumaz, *J. Organomet. Chem.*, 1988, **345**, 187.
96 E.P. Tallsi, *J. Chem. Soc., Chem. Commun.*, 1985, 1768.
97 G. Strukel and R.A. Michelin, *J. Am. Chem. Soc.*, 1985, **101**, 7563.
98 B. Meunier, *Bull. Soc. Chim. Fr.*, 1986, 578.
99 D. Mansuy, *Pure Appl. Chem.*, 1987, **59**, 759.
100 T.G. Traylor, *J. Am. Chem. Soc.*, 1993, **115**, 2775.
101 Mitsui Toatsui, JP 6344573.
102 Eniricerche SpA, EP 230949.
103 Solvay Interox Ltd., WO 93/00338.
104 W.F. Brill, *J. Org. Chem.*, 1986, **51**, 1149.
105 T. Katsuki and K.B. Sharpless, EP 46033.
106 B. Hassine, M. Gorsane, F.G. Evard, J. Pecher and R.H. Martin, *Bull. Soc. Chim. Belg.*, 1986, **95**, 547.
107 H. Pluim and H. Wynberg, *J. Org. Chem.*, 1980, **45**, 2498.
108 S. Julia, J.G. Guxier, J. Masana, J. Rocas, S. Colonna, R. Annuziata and H. Molinari, *J. Chem. Soc., Perkin Trans. 1*, 1982, 1317.
109 V. Schrig, K. Hintzer, U. Leyrer, C. Mark, P. Pitchen and H.B. Hagen, *J. Organomet. Chem.*, 1989, **370**, 81.
110 R. Curci, M. Fiorentino and M.R. Serio, *J. Chem. Soc., Chem. Commun.*, 1984, 155.
111 Nippon Shokubai Co., JP 92/236205.
112 T. Ravinden, *J. Appl. Polym. Sci.*, 1988, **35**, 1227.
113 D. Swern, N. Billen and T.J. Scanlan, *J. Am. Chem. Soc.*, 1946, **68**, 1504.
114 Degussa, EP 25940.
115 Degussa, EP 2937840.
116 Ciba Geigy, EP 141775.
117 Chevron, US 3991126.

118 H. Riviere, *Bull. Soc. Chim. Fr.*, 1964, 97.
119 H. Wassermann and N.E. Aubery, *J. Am. Chem. Soc.*, 1956, **78**, 1726.
120 C. Curtin, A. Bradley and Y.G. Hendrickson, *J. Am. Chem. Soc.*, 1956, **78**, 4064.
121 A. Roebuck and H. Adkins, *Org. Synth. Coll. Vol.*, No. 3, 1955, 217.
122 D. Swern, *J. Am. Chem. Soc.*, 1945, **67**, 1769.
123 D. Swern, US 2443280.
124 L.F. Fieser and S. Ragapalan, *J. Am. Chem. Soc.*, 1949, **71**, 3938.
125 Janssen Pharmaceutica, US 4079062.
126 Degussa, DE 3442937.
127 T.M. Luong, H. Schriftman and D. Swern, *J. Am. Oil Chem. Soc.*, 1967, **44**, 316.
128 J.M. Church and R. Bluinberg, *Ind. Eng. Chem.*, 1951, **43**, 1780.
129 G.B. Payne and C.W. Smith, *J. Org. Chem.*, 1957, **22**, 1680.
130 M.A. Beg and I. Ahmed, *J. Catal.*, 1975, **39**, 260.
131 L.I. Kuznetsova, *Izv. Akad. Nauk. SSR., Ser. Khim.*, 1985, 537.
132 L. Cerveny, *Chem. Ztg.*, 1986, **110**, 303.
133 A.A. Akhundova, F.M. Velieva, F.T. Aliev and M. Alimardanov, *Neftekhimya*, 1996, **36**, 237.
134 K.Y. Lee, *Shokubai*, 1996, **38**, 350.
135 E.D. Finashina, V.R. Flid and A.P. Belov, poster presented at the 3rd World Congress on Oxidation Catalysis, San Diego, California, September 21–26, 1997.
136 Eniricherche SpA, EP 226257.
137 N.A. Milas and S. Suismann, *J. Am. Chem. Soc.*, 1936, **58**, 2345.
138 G. Cainella, *Synthesis*, 1989, 45.
139 Exxon Res. Eng., EP 53023.
140 S. Hanessian, in 'Total Synthesis of Natural Products: The Chiron Approach', 1983, Pergamon Press, Oxford.
141 J. Rocek and F.H. Westheimer, *J. Am. Chem. Soc.*, 1962, **84**, 2241.
142 R. Criegee, *Chem. Ber.*, 1931, **64**, 260.
143 E.L. Jackson, *Org. React.*, 1944, **2**, 341.
144 R.D. Temple, *J. Org. Chem.*, 1970, **35**, 1275.
145 P. Wright and J. Abbot, *Int. J. Chem. Kinet.*, 1993, **25**, 901.
146 Continential Oil, DE 2046034.
147 Degussa, DE 2106307.
148 S. Warwal, M. Rusch gen Klass and M. Sojka, Tegungsberich 9204, Proc. DGMK Conference, 'Selective Oxidation in Petro-chemistry', ed. M. Baerns and I. Weitkamp, Goslar, 1992, 161.
149 C. Djerassi and R. Engle, *J. Am. Chem. Soc.*, 1953, **75**, 3838.
150 Atlantic Richfield Co., US 3839375.
151 Degussa, DE 2035558.
152 C. Venturello and M. Ricci, *J. Org. Chem.*, 1986, **51**, 1599.
153 Y. Ishii, K. Yamawaki, T. Ura, H. Yamada, T. Yoshida and M. Ogawa, *J. Org. Chem.*, 1988, **53**, 3587.
154 T. Oguchi, T. Ura, Y. Ishii and M. Ogawa, *Chem. Lett.*, 1989, 857.
155 Ortec Inc., US 5157152.
156 Bayer AG, DE 2201456.
157 Bayer AG, DE2201455.
158 Bayer AG, GB 1426260.
159 Bayer AG, GB 1432217.
160 J. Hoffmann, *J. Am. Chem. Soc.*, 1957, **79**, 503.
161 M. Bochio, V. Conte, E. DiFuria and G. Modena, *J. Mol. Catal.*, 1991, **70**, 159.
162 Nippon Syn. Chem., JP 05/004938.
163 Novamont SpA., WO 93/121064.
164 A.E. Gekhman, N.I. Moiseev and I.I. Moiseev, *Soviet J. Coord. Chem.*, 1992, **18**, 1.
165 B.M. Choudary and P.N. Reddy, *J. Mol. Catal. (A)*, 1995, **103**, L1.
166 G. Barak and Y. Sasson, *J. Chem. Soc., Chem. Commun.*, 1987, 1266.

167 J.L. Courtenay, in 'Organic Synthesis by Oxidation with Metal Compounds', ed. W.J. Mijs, C.R.M.I. de Jonge.
168 S.J.H.F. Arts, F. van Rantwijk and R.A. Sheldon, *J. Carbohydr. Chem.*, 1994, **13**, 851.
169 A. Johnstone, P.J. Middleton, W.R. Sanderson, M. Service and P.R. Harrison, in 'New Developments in Selective Oxidation II', Elsevier Science BV, 1994, 609.
170 Research Corp., US 2414385.
171 Research Corp., US 2437648.
172 M. Inoue, T. Uragakai, M. Kashiwagi and S. Enomoto, *Chem. Lett.*, 1989, 99.
173 M. Shimizu, H. Orita, K. Suzuki, T. Hayakawa, S. Hamakawa and T. Takehira, *J. Mol. Catal.*, 1996, **114**, 217.
174 R.A. Sheldon and J. Dakka, *Catal. Today*, 1994, **19**, 215.
175 R.A. Sheldon, J.D. Chen, J. Dakka and E. Neelman, in 'New Developements in Selective Oxidation II', Elsevier, Amsterdam, 1994.
176 S.B. Kumar, S.P. Mirajaker, G.C.C. Pasis, P. Kumar and R. Kumar, *J. Catal.*, 1995, **156**, 163.
177 W. Adam, A. Corma, A. Martinez and M. Renz, *Chem. Ber.*, 1996, **129**, 1453.
178 Emery Ind., US 2813113.
179 Degussa, DE 2035558.
180 A.L. Henne and P. Hill, *J. Am. Chem. Soc.*, 1943, **65**, 752.
181 Hoechst AG, EP 513600.
182 N. Poklukar, AU 393503.
183 Henkel, WO 93/05007.
184 P.S. Bailey and R.W. Murray, in 'Ozone in Organic Chemistry', Franklin Inst. Press, 1975, 75–129; F.D. Gunstone, *Educ. Chem.*, 1968, **5**, 166; *Long, Chem. Rev.*, 1940, **27**, 437.
185 K.C. Rajanna, K.N. Reddy, U.U. Kumar and P.K.S. Prakash, *Int. Chem. Kinet.*, 1996, **28**, 153.
186 E.G. Rozantzar and M.B. Neiman, *Tetrahedron*, 1964, **20**, 131.
187 J.A. Cella and J.P. McGrath, *J. Org. Chem.*, 1975, **40**, 1860.
188 B. Gareo, *J. Org. Chem.*, 1975, **40**, 1998.
189 H.R. Kim, J.H. Jung, J.N. Kim and E.K. Ryan, *Synth. Commun.*, 1990, **20**, 637.
190 T. Morimoto, H. Hirano, H. Ashiyan and H.E. Gaskira, *Bull. Chem. Soc. Jpn.*, 1987, **60**, 4143.
191 E.J. Corey, E.P. Barette and P.A. Magriotis, *Tetrahedron Lett.*, 1985, **26**, 5855.
192 E.S. Huyzer and G.W. Hawkins, *J. Org. Chem.*, 1983, **48**, 1705.
193 T. Morimoto, M. Hirano, M. Wachi and T. Murakami, *J. Chem. Soc., Perkin Trans. 2*, 1984, 1949.
194 Daicel, JP O1/207255.
195 Atlantic Richfield, US 3997578.
196 B.M. Trost and Y. Masuyama, *Tetrahedron Lett.*, 1984, **25**, 173.
197 S.E. Jacobsen, D.A. Muccigrosso and F. Marco, *J. Org. Chem.*, 1979, **44**, 921.
198 S. Maignien, S. Alt-Mohand and J. Muzart, *Synlett.*, 1996, 439.
199 V. Conte, F. DiFuria and G. Modena, *J. Org. Chem.*, 1988, **53**, 1665.
200 O. Bortalini, V. Conte, F. DiFuria and G. Modena, *J. Org. Chem.*, 1986, **51**, 2661.
201 D.C. Duncan, R.C. Chambers, E. Heat and C.L. Hill, *J. Am. Chem. Soc.*, 1995, **117**, 681.
202 W.P. Griffith, R.G. Morea and H.I.S. Nogueira, *Polyhedron*, 1996, **15**, 3493.
203 G.H. Marcu and A. Bodar, *Inorg. Synth.*, 1985, **23**, 189.
204 G. Barak, J. Dakka and Y. Sasson, *J. Org. Chem.*, 1988, **53**, 3553.
205 H. Hayashi, K. Kikawa, Y. Muri, N. Shigemato, S. Suigyama and K. Kawashiro, *Chem. Lett.*, 1996, **36**, 99.
206 Solvay Interox Ltd., WO 94/21583.
207 Solvay Interox, EP 260054.
208 Solvay Interox, unpublished results.

209 J. Dakka and Y. Sasson, *Bull. Soc. Chim. Fr.*, 1988, 756.

210 Hoechst, EP 326470.

211 C.H. Hassall, *Org. Reac.*, 1957, **9**, 73.

212 D. Swern, in 'Organic Peroxides', Wiley Interscience, Vol. 1, pp. 24 and 240.

213 Solvay Interox Publication, AO.15 'Peroxygen Compounds in Organic Synthesis—Oxidation at Carbon–Oxygen Bonds'.

214 A. Dobrowsky, *Monatsch. Chem.*, 1955, **86**, 325.

215 H.D. Daiken, *Org. Synth*, 1941, Vol. 1, 149.

216 Solvay Interox, GB 2188927.

217 Dequest is a Monsanto tradename.

218 Mykon CIX is a Warwick International tradename.

219 A.V. Wacek and H.O. Eppinger, *Chem. Ber.*, 1940, **73**, 644.

220 Mitsubishi, GB 1431876.

221 J. D'Ans and A. Kriep, *Chem. Ber.*, 1915, **48**, 1136.

222 Denki Kagaku Kogyo, JP 48/103511.

223 I.M. Godfrey, M.V. Sargent and J.A. Elix, *J. Chem. Soc., Perkin Trans. 1*, 1974, 1353.

224 H. Heaney, *Top. Curr. Chem.*, 1993, **164**, 1.

225 A. Nishikara and I. Kubata, *J. Org. Chem.*, 1968, **33**, 2525.

226 C.W. Smith and R.T. Holm, *J. Org. Chem.*, 1957, **22**, 746.

227 L. Syper, *Synthesis*, 1989, 167.

228 B.M. Trost and Y. Masuyama, *Tetrahedron Lett.*, 1989, **45**, 3299.

229 C. Venturello and M Gambaro, *J. Org. Chem.*, 1991, **56**, 5924.

230 M.S. Cooper, H. Heaney, A.J. Newbold and W.R. Sanderson, *Synlett.*, 1990, 533.

231 A. McKillop and D. Kemp, *Tetrahedron*, 1989, **45**, 3299.

232 C.H. Hassall, *Org. Chem. Reac.*, 1957, **9**, 73.

233 P.A.S. Smith, in 'Molecular Rearrangements', ed. P. de Mayo, Wiley Interscience, Vol. 1.

234 S.N. Lewis, in 'Oxidation', ed. R.L. Augestine and M. Dekker, p. 213.

235 H.O. Howie, in 'Modern Synthetis Reactions', 2nd Edition, Benjamin, p. 321.

236 R. Criegee, *Liebigs Annalen*, 1948, **560**, 127.

237 R.R. Sauers, *J. Am. Chem. Soc.*, 1959, **81**, 5824.

238 G.R. Krow, C.A. Johnson, J.P. Guare, D. Kulrak, K.J. Henz, D.A. Shaw, S.W. Szczepanshi and J.T. Carey, *J. Org. Chem.*, 1982, 47, 5239.

239 G. Methan and P.N. Pandy, *Synthesis*, 1975, 404.

240 M. Hindlicky, *Chem. Listy*, 1952, **45**, 380.

241 Solvay Interox Ltd., US 4353832.

242 P.R. Story, D.D. Denson, C.E. Bishop, B.C. Clark and J.C. Farine, *J. Am Chem. Soc.*, 1968, **90**, 817.

243 B.M. Trost and M.J. Bogdanowiczi, *J. Am. Chem. Soc.*, 1971, **93**, 3773.

244 Sagami, JP 61/087677.

245 Y. Ogato, Y. Sawaki and M. Shiroyama, *J. Org. Chem.*, 1977, **42**, 4061.

246 Laporte, GB 1203752.

247 J. Ondracek, *Collect. Czech. Chem. Commun.*, 1989, **54**, 3260.

248 B. Pandey and P.V. Dalvi, *J. Org. Chem.*, 1989, **54**, 2968.

249 Kuhlmann, DE 1643146.

250 Solvay Interox Ltd., unpublished results.

251 J. Mainwald, J.J. Tuffnalls and J.J. Hirst, *J. Org. Chem.*, 1964, **29**, 2914.

252 R. White and W.D. Emmons, *Tetrahedron*, 1962, **17**, 31.

253 J. Neienwald and J.L. Ripoli, *J. Am. Chem. Soc.*, 1967, **89**, 7075.

254 P. Brougham, M.S. Cooper, D.A. Cummerson, H. Heaney and N. Thompson, *Synthesis*, 1987, 1015.

255 W.D. Emmons and G.B. Lucas, *J. Am. Chem. Soc.*, 1955, **77**, 2287.

256 B.D. Hookherjee, R.W. Trenkle and R.R. Patel, *J. Org. Chem.*, 1972, **37**, 3846.

257 W.F. Balley and J.J. Buschaff, *J. Org. Chem.*, 1985, **50**, 3009.

258 A.V. Baeyer and V. Villiger, *Chem. Ber.*, 1899, **32**, 3625.
259 R.W. Murray, *Chem. Rev.*, 1989, **89**, 1187.
260 P.A. Grieco, V. Yokoyama, S. Gilman and Y. Ohfine, *J. Chem. Soc., Chem. Commun.*, 1977, 870.
261 R.T. Taylor and L.A. Flood, *J. Org. Chem.*, 1983, **48**, 5160.
262 Union Carbide Corp, US 4160769.1.
263 L. Syper, *Synthesis*, 1989, 167.
264 IFF, GB 1335930.
265 G.B. Payne and C.W. Smith, *J. Org. Chem.*, 1957, **22**, 1680.
266 ICI, GB 1050846.
267 R. Ugo, G.M. Zanderighi, A. Fusi and D. Carrei, *J. Am. Chem. Soc.*, 1980, **102**, 3745.
268 Allied Chemicals, EP 008496.
269 Mobil Oil, US 4870192.
270 Solvay Interox, WO 4/21624.
271 Nissan Chemicals, JP 51/6851.
272 Y. Zhang, H. Hu, Y. Fang and F. Tao, *Youy Huaxue*, 1996, **16**, 64.
273 Y. Zhang, J. Yu, H. Ai, Y. Fang, G. Xie, F Tao and M. Chen, *Huaxue Tangoao*, 1995, 35.
274 A. McKillop and J.A. Tarbon, *Tetrahedron*, 1987, **43**, 1753.
275 A. Banerjee *et al.*, *Synthesis*, 1989, 765.
276 Y. Ogatoa, K. Tomizawa and T. Ikeda, *J. Org. Chem.*, 1978, **43**, 2417.
277 K. Maruyama, *Bull. Chem. Soc. Jpn.*, 1960, **33**, 1516.
278 Burham, GB 1413475.
279 M. Suzuki, H. Takada and R. Noyari, *J. Org. Chem.*, 1982, **47**, 902.
280 T.A. Cooper and W.A. Waters, *J. Chem. Soc. (B)*, 1967, 687.
281 W. Partenheimer, in 'Catalysis of Organic Reactions', 1990, Chapter 10, p. 321.
282 R.O.C. Normal and J.R. Lindsey-Smith, *J. Chem. Soc.*, 1963, 2897.
283 N. Suzuki and H. Hotta, *Bull. Chem. Soc. Jpn.*, 1967, **40**, 1361.
284 L.S. Mers and W.A. Waters, *J. Chem. Soc.*, 1949, 2427.
285 Laporte Industries, GB 1198261.
286 Interox Chemicals, GB 2246565.
287 Solvay Interox Ltd., WO 93/00319.
288 Solvay Interox Ltd., PCT/GB 94/02080.
289 C.W. Jones, W.R. Sanderson, K. Auty, B.C. Gilbert, B.C. Thomas and S.W. Brown, *J. Mol. Catal. A*, 1997, **117**, 279.
290 A. Sumini and G. Czapski, *J. Chem. Soc., Dalton Trans.*, 1973, 487.
291 V. Najaroyan and R.W. Fressienden, *J. Phys. Chem.*, 1985, **89**, 2330.
292 C.W. Jones, A. Hackett, I. Pattinson, A. Johnstone and L. Wilson, *J. Chem. Res.* 1996, (*S*) 438; (*M*) 2501.
293 T.A. Cooper, A.A. Clifford, D.J. Mills and W.A Walters, *J. Chem. Soc. (B)*, 1966, 793.
294 R. Van Helden, A.F. Bickel and E.C. Kooyman, *Rec. Trav. Chim.*, 1961, **80**, 1237 and 1257.
295 Solvay Interox Ltd., EP 336568; US 4943358; EP 336567.
296 C.W. Jones, N.G. Carter, S.C. Oakes, S.L. Wilson and A. Johnstone, *J. Chem. Technol. Biotechnol.*, 1998, **71**, 111.
297 Solvay Interox Ltd., EP 334511.
298 Solvay Interox Ltd., PCT/GB 93/01112.
299 R.G.R. Bacon, *Chem. Ind.*, 1953, 897.
300 Andrejawa, GB 1365610.
301 C. Walling, *J. Am. Chem. Soc.*, 1978, **100**, 4814.
302 Phillips Petroleum, US 5280001.
303 J.D. Sauer, *Surfactant SCI*, 1990, **34**, 275.
304 Texaco Chem. Co., 5216154.

305 Huls AG, DE 3618352.
306 Ethyl Corp., US 4942260.
307 Ethyl Corp., EP 307184.
308 Atochem, EP 2632638.
309 S. Wawzonak and J.V. Kempf, *Org. Prep. Proced. Int.*, 1972, **4**, 135.
310 Ciba-Geigy, US 289596.
311 Interox SA, EP 233622.
312 H. Mayoshi, *Catal. Rev.*, 1990, **32**, 229.
313 Atochem, EP 399866.
314 Houben–Weyl, 1967, XIV/1,220, X/2, 764.
315 Eiwa, GB 1410446.
316 R. Tan and A. Taurins, *Tetrahedron Lett.*, 1965, 2737.
317 C.C. Cheng and W. Pfleiderer, *J. Med. Pharm. Chem.*, 1966, **9**, 573.
318 P. Brougham, M.S. Cooper, D.A. Cummerson, H. Heaney and N. Thompson, *Synthesis*, 1987, 1015.
319 H. Yamanaka, S. Ogawa and T. Sakamoto, *Heterocycles*, 1981, **16**, 573.
320 G.J. Robke and E.J. Behrman, *J. Chem. Res. (S)*, 1993, 412.
321 G.B. Payne, *J. Org. Chem.*, 1961, **26**, 668.
322 W.D. Emmons, *J. Am. Chem. Soc.*, 1957, **79**, 5528.
323 K.M. Ibne-Rasa and W.D. Emmons, *J. Am. Chem. Soc.*, 1962, **84**, 763.
324 A.M. Islam, E.A. Hassa, I.B. Hannciut and M.Y. Maurad, *J. Chem. UAR*, 1970, **13**, 297.
325 J.N. Brough and B. Lythgoe, *J. Chem. Soc.*, 1954, 4069.
326 K. Bloc, *Synthesis*, 1987, 1084.
327 F.A. Davis, N.F. Abdul-Mallk, S.B. Awas and M.E. Harakal, *Tetrahedron Lett.*, 1981, **22**, 917,
328 F.A. Davis, R.T. Reddy and P.J. Carroll, *J. Am. Chem. Soc.*, 1992, **114**, 1428.
329 F.A. Davis, J. Lamendola and U. Nadir, *J. Am. Chem. Soc.*, 1980, **102**, 2000.
330 P. Duhamil, D. Benard and J.P. Plaquevant, *Tetrahedron Lett.*, 1985, **26**, 6065.
331 D. Enders and A. Plant, *Synlett.*, 1990, 725.
332 Narchim, French Patent 2632954.
333 W.W. Zajac, *Synthesis*, 1988, 808.
334 Pennsalt Corp., US 3274252.
335 A.R. Katritzky and J.M. Lagowski, in 'Chemistry of the Heterocyclic N-Oxides', 1971, Ch. 2, p. 22, Academic Press, London.
336 Sagami Chemical, JP 62/263164.
337 Montedison SPA, EP 201934.
338 Nippon Kayaku, JP 02/124872.
339 S. Murahashi, *J. Chem. Soc., Chem. Commun.*, 1984, 874.
340 S. Murahashi and T. Shiota, *Tetrahedron Lett.*, 1987, **28**, 2382.
341 E. Breur, in 'The Chemistry of Amino Nitroso and Nitro Compounds', ed. S. Patai, 1982, John Wiley, New York, p. 459.
342 E.G. Rozanstev and V.D. Sholle, *Synthesis*, 1971, 190.
343 S. Murahashi, *J. Org. Chem.*, 1990, **55**, 1744.
344 J. Ritz, H. Fuchs and H.G. Perryman, in 'Ullmann's Encyclopaedia of Industrial Chemistry', 5th Edition, Vol. A13, VCH, 1989, p. 527.
345 Bayer AG, GB 1177495.
346 Enichem Anic SPA, EP 564040.
347 Monsanto Co., US 5023369.
348 Monsanto Co., US 5077430.
349 R.W. Murray, R. Jeyaraman and L. Moahn, *Tetrahedron Lett.*, 1986, **27**, 2335.
350 A. McKillop and D. Kemp, *Tetrahedron*, 1989, **45**, 3299.
351 A. McKillop and J.A. Tarbin, *Tetrahedron*, 1987, **43**, 1753.
352 S.M. Mehta and M.V. Vakikawa, *J. Am. Chem. Soc.*, 1952, **74**, 563.
353 W.W. Zajac, *Tetrahedron Lett.*, 1989, **30**, 6495.

354 D. Enders and V. Bhusman, *Z. Naturforsch., Teil B*, 1987, **42**, 1595; A. McKillop and J.A. Tarbin, *Tetrahedron*, 1987, **43**, 1753.
355 D.T. McAllen, *J. Am. Chem. Soc.*, 1951, **73**, 3627.
356 ICI, GB 519617.
357 Bayer, DE 2815964.
358 Wingfoot Corp., US 2024575.
359 A.N. Nathan and M.T. Bogart, *J. Am. Chem. Soc.*, 1941, **63**, 2361.
360 Gilman, in 'Organic Chemistry', John Wiley, New York, 1943, p. 839.
361 UCB, GB 1429925.
362 Sharpless Chemicals, US 2423310.
363 Degussa, DE 2105473.
364 Winglecroft Corp., US 2024567.
365 Vanderbilt Co., US 2509453.
366 Roche, Swiss Patent 200248.
367 M. Gazder and S. Smiles, *J. Chem. Soc.*, 1968, **93**, 1833.
368 T.B. Douglas, *J. Am. Chem. Soc.*, 1946, **68**, 1072.
369 O. Hinsberg, *Chem. Ber.*, 1908, **41**, 2836.
370 D.O. Poak and T.I Walkins, *J. Chem Soc.*, 1950, 445.
371 Laboratoire Lalafone, GB 1519352.
372 D.E. O'Conner and W.I. Lyness, *J. Am. Chem. Soc.*, 1941, **63**, 2939.
373 I. Reddy and S.R. Varma, *J. Chem. Soc., Chem. Commun.*, 1997, 471; P. Lupattelli, R. Ruzziconi, P. Scofato, A. Degl'Innocenti and A.B. Paolobelli, *Synth. Commun.*, 1997, **27**, 441; W. Adam, C.M. Mitchell and C.R. Saha-Moeller, *Tetrahedron*, 1994, **50**, 13121.
374 W.E. Parchem and M.D. Bhavzar, *J. Org. Chem.*, 1963, **28**, 2686.
375 Aquitane, GB 1488818.
376 W.M. Luster and L.E Sutton, *Trans. Faraday Soc.*, 1939, **35**, 495.
377 C.G. Overberger and R.W. Cummins, *J. Am. Chem. Soc.*, 1953, **75**, 4250.
378 D. Swern, in 'Organic Peroxides', Vol. 1, p. 243, Wiley Interscience, 1971.
379 C.R. Harrison and P. Hodge, *J. Chem. Soc., Perkin Trans. 1*, 1976, 2252.
380 Shell Dev., US 2694072.
381 Eli Lilly, GB 1388979.
382 A. Mangia, *Synthesis*, 1978, 361.
383 D.N. Jones and M.J. Green, *J. Chem. Soc. (C)*, 1967, 532.
384 K. Konda, A. Negishi and M. Fukijama, *Tetrahedron Lett.*, 1969, 2461.
385 H. Nawa, *J. Am. Chem. Soc.*, 1960, **82**, 896.
386 J.S. Showell, J.R. Russell and D. Swern, *J. Org. Chem.*, 1962, **27**, 2853.
387 Y. Arai and T. Koizumi, *Reve. Heterochem.*, 1992, **6**, 202.
388 L.A.G.M. Van der Broek, J.H. Colstee and H.C.J. Ottenheym, *Synth. Commun.*, 1989, **19**, 3397.
389 I. Cho and S.Y. Choi, *Makromol. Chem. Rapid. Commun.*, 1991, **12**, 399.
390 D. Batty, D. Crich and S.M. Fartt, *J. Chem. Soc., Perkin Trans. 1*, 1990, 2875.
391 D. Crich and T.J. Richie, *J. Chem. Soc., Perkin Trans. 1*, 1990, 945.
392 L.M. Siemens, F.W. Roltneck and L.S. Trzupek, *J. Org. Chem.*, 1990, **55**, 3507.
393 E.P. Kohler and H. Potter, *J. Am. Chem. Soc.*, 1935, **57**, 1316.
394 R.D. Little and L. Brown, *Tetrahedron Lett.*, 1980, 2203.
395 Schoberi, DE 604016.
396 F. DiFuria and G. Modena, *Pure Appl. Chem.*, 1982, **54**, 1853.
397 O. Bortelini, F. DiFuria and R. Seraglia, *J. Org. Chem.*, 1985, **50**, 2688.
398 Degussa, DE 2504235.
399 Interox SA, French Patent 2616786.
400 Solvay Interox Ltd., UK Patent Application 94/22391.4.
401 M.S. Schultz, H.B. Freyerman and S.R. Buc, *J. Org. Chem.*, 1903, **28**, 1140.
402 L.D. Small, J.H. Bailey and C.J. Cavalito, *J. Am. Chem. Soc.*, 1947, **69**, 1710.
403 GAF Co., US 3005852.

404 N.W. Connan, *Eastman Org. Chem. Bull.*, 1972, **44**, 1.

405 GAF Co., US 3006952.

406 H.J. Reich, F. Chow and S.L. Peake, *Synthesis*, 1978, 299.

407 R. Breitshuh and D. Seebach, *Synthesis*, 1992, 1170.

408 Tokkyo Koho, JP 04/169567.

409 M. Paluki, P. Hanson and E.N. Jacobsen, *Tetrahedron Lett.*, 1992, **33**, 7111.

410 N. Komoatsu, Y. Nishibayashi and T. Sugita, *Tetrahedron Lett.*, 1992, **33**, 5391.

411 W. Adam, Y.Y. Chan, D. Cramer, J. Gauss, D. Schetit and M. Schindler, *J. Org. Chem.*, 1987, **52**, 2800.

412 J.E. Brumwell, N.S. Simpkins and N.K. Terrett, *Tetrahedron Lett.*, 1993, **34**, 1219.

413 M. Shimazaki, M. Takahashi and H. Komatzu, *Synthesis*, 1992, 555.

414 R. Balichi, I. Kaczmavek and P. Nantk-Naimirski, *Liebigs Ann. Chim.*, 1992, 883.

415 T. Ando, D.G. Cork and T. Kimura, *Chem. Lett.*, 1986, 665.

416 R.S. Phillips and S.W. May, *Enzyme Microbail Technal.*, 1981, **3**, 9.

417 H. Fu Kondo, Y. Ichikawa, G.C. Look and C.H. Wong, *J. Org. Chem.*, 1992, **57**, 7265.

418 G. Carrea, B. Redigolo, S. Colonna, N. Gaggero, E. Baltistel and D. Blanchi, *Tetrahedron Asymm.*, 1993, **3**, 1063.

419 J.H. Acquaye, J.G. Muller and J.F. Takeuchi, *Inorg. Chem.*, 1993, **32**, 160.

420 M. Firowzabadi and I. Mohammadour, *Bull. Chem. Soc. Jpn.*, 1992, **65**, 1131.

421 R.P. Greenhaugh, *Synlett.*, 1992, 235.

422 O. Maass and P.G. Hiebert, *J. Am. Chem. Soc.*, 1924, **46**, 290.

423 Ugine Kuhlmann, GB 1463050.

424 Degussa, GB 1332170.

425 K. Kadiz, CZ 170076.

426 V. Conte, F. Di Furia and S. Moro, *Tetrahedron Lett.*, 1994, **35**, 7429.

427 A. Leulier, *Bull. Soc. Chim. Fr.*, 1924, **35**, 1325.

428 P. Lubbecke and P. Boldt, *Angew. Chem.*, 1976, **88**, 641.

429 A. Leulier, *Bull. Soc. Chim. Fr.*, 1930, **47**, 730.

430 E. Speyer and E. Rosenfeld, *Chem. Ber.*, 1925, **58**, 1110.

431 Uniroyal, DE 2110210.

432 Toya Soda Mfg., JP 51/068534.

433 D.K. Abdullaev and O.M. Mamedova, *Issled Obl. Sint. Polim. Monomermykh Prod.*, 1977, 23.

434 Degussa, DE 2227439.

435 Merck, DE 2460361.

436 H. Lubbecke and P. Boldt, *Tetrahedron*, 1978, **34**, 1577.

437 Chem. Fabrik. Kalk, DE 2852645.

438 Hoechst AG., EP 278377.

439 Rhone Poulenc, EP 326456.

440 P. Bezodis, J. Hanson and P. Petit, *J. Chem. Res. (S)*, 1996, 334.

441 Chem. Fabrik. Kalk, DE 2828645.

442 Nippon Kayaku, JP 62/223140.

443 Nippon Kayaku, JP 56/131535.

444 Ugine Kuhlmann, FR 2274586.

445 Ethyl Co., US 5237112.

446 I.E. Marsh, *J. Chem. Soc.*, 1927, 3164.

447 M. Ichikawa, *Org. Prep. Proc. Int.*, 1985, **17**, 56.

448 GAF Co., US 2649482.

449 A. Borovanski and S. Cizmarik, CZ 136315.

450 A.M. Islam, *J. Chem. UAR*, 1970, **13**, 297.

451 A. Leulier and L. Pinet, *Bull. Soc. Chim. Fr.*, 1927, **41**, 1362.

452 M.K. Seikel, *J. Am. Chem. Soc.*, 1940, **62**, 1214.

453 BASF, DE 2750292.

454 Degussa, DE 1399065.

455 Olin Co., US 4008278.
456 H.J. Den Hertog, *Recl. Trav. Chim. Pays-Bas*, 1951, **70**, 353.
457 A. Cerulis, USSR 433146.
458 Chem. Fabrik. Kalk, DE 2855184.
459 Chem. Fabrik. Kalk, DE 3133577.
460 T.C. Ho, B.G.B. Gupta and G.A. Olah, *Synthesis*, 1977, 676.
461 Chem. Fabrik. Kalk, DE 2913277.
462 Chem. Fabrik. Kalk, DE 2855184.
463 Esso, GB 867737.
464 Degussa, DE 2160613.
465 Cetus Co., US 4284723.
466 S.Y. Kuchmii, A.V. Korzhak and A.I. Kryukov, *Ukr. Khim. Zh. (Russ. Ed.)*, 1995, **61**, 21.
467 R. Johnson and K.M. Reeve, *Speciality Chemicals*, 1992, 292.
468 J. Varagnat, *Ind. Chem. Eng.*, 1976, **15**, 213.
469 A.H. Haines, in 'Methods for the Oxidation of Organic Compounds; Alkanes, Alkenes, Alkynes and Arenes', Academic Press, 1985, Chapter 5.
470 R.G. Jones and H.A. Shonle, *J. Am. Chem. Soc.*, 1945, **67**, 1034.
471 F. Minisci and P. Maggioni, *Chim. Ind.*, 1979, **61**, 834.
472 F. Minisci, *J. Org. Chem.*, 1986, **51**, 536.
473 F. Minisci, *Tetrahedron Lett.*, 1989, **30**, 4569.
474 American Cyanamid, EP 331899.
475 C.A. Buehler and H. Hart, *J. Am. Chem. Soc.*, 1963, **85**, 2177.
476 R.T. Arnold and R. Larson, *J. Org. Chem.*, 1940, **5**, 250.
477 A.J. Pandell, *J. Org. Chem.*, 1976, **41**, 3992.
478 G.A. Page and D.S. Tarbell, *Org. Synth.*, 1954, **34**, 8.
479 S.B. Gingerich and P.W. Jenkins, *Adv. Oxygenated Processes*, 1990, **2**, 117.
480 Dupont, US 2371691.
481 C. Walling, *J. Am. Chem. Soc.*, 1978, **100**, 4814.
482 Solvay Interox Ltd., GB 2269588.
483 Enichem., EP 200260.
484 Solvay Interox Ltd., WO 94/14740.
485 S.W. Brown, A. Hackett, A. Johnstone, A.M. King, K.M. Reeve, W.R. Sanderson and M. Service, *Tagungsbericht*, 9204 Proc. DGMK. Conference, 1992, 339.
486 B.C. Kar, *J. Ind. Chem. Soc.*, 1937, **14**, 291.
487 M. Inoue, T. Uragaki and S. Enomoto, *Chem. Lett.*, 1986, 2075.
488 W. Adam, W.A. Hermann and J. Lin, *J. Org. Chem.*, 1994, **59**, 8281.
489 Research Corp., US 2395638.
490 L.F. Fieser, *Org. Reac.*, 1942, **1**, 129.
491 G.K.S. Prakash, N. Krass, Q. Wang and G. Olah, *Synlett.*, 1991, 39.
492 Solvay Interox Ltd., EP 0700885.
493 T. Carrona, A. Cittero, L. Grossi, F. Minisci and K. Ogawa, *Tetrahedron*, 1976, **32**, 2741.

CHAPTER 4

Heterogeneous Activation and Application of Hydrogen Peroxide

1 Introduction

The 1990 Environmental Protection Act necessitates a greater need for clean technology, and the priorities of the chemical industry today includes clean technology, re-cycling, waste treatment, disposal and environmental monitoring. As part of this picture there is great interest in the replacement of conventional reagents and catalysts for supported species.[1] Indeed, supported catalysts are widely used throughout industry due to their enhanced activity, stability, economic benefits, ease of recovery and resistance to poisoning.[2] In order to appreciate the reasons for the present interest in supported and heterogeneous catalysts, it is important to examine the advantages and disadvantages of heterogeneous and homogeneous catalysts.

The major disadvantage of homogeneous catalysts is the difficulty of separating the often relatively expensive catalyst from the products at the end of the reaction. Often distillation is inevitable, and unless it is extremely efficient, it may result in loss of catalyst and hence render the process uneconomic. In contrast, a heterogeneous catalyst can be removed by coarse filtration techniques. In heterogeneous systems, reaction has to occur at catalytically accessible sites, *i.e.* some sites will remain unused during the reaction. By contrast, all the sites on a homogeneous catalyst are theoretically available as catalytic centres, hence, these catalysts are potentially more efficient in terms of the concentration required to catalyse a given amount of reaction. The problem is somewhat negated via the theory of restricted dimensionality, *i.e.* most surfaces can be represented as two dimensions, whereas the homogeneous system is three dimensional. Consequently, provided that adsorption processes are not rate limiting, the kinetics of the reaction may be faster in a heterogeneous system.

Homogeneous catalysts also have the advantage over heterogeneous systems in that the former are highly reproducible, because they have a definite stoichiometry and structure; by contrast, the structure of a heterogeneous catalyst is heavily dependent on both its method of preparation and its antecedents prior to preparation. Furthermore, homogeneous catalysts have only one type of active site, and therefore will often be more specific than a heterogeneous catalyst, in which several types of active site may be present in the form of surface defects. Such defects are extremely difficult to control, hence any defects present can make systematic design and improvement very difficult.

As mentioned above, the initial driving force for preparing heterogeneous systems is currently to turn an environmentally aggressive homogeneous assemblage into a relatively benign one. However, experience has also shown that the presence of both support and catalyst can have synergistically beneficial effects. Firstly, the support may not merely be an inert backbone, but may play a positive rôle leading to preferred orientations of the substrate at the catalytic site.[3] Secondly, organic functional groups covalently bound to the surface of inorganic species or polymers are subject to special constraints which may alter their chemical reactivity relative to the analogous small molecules.[4] Thus the chemical properties of supported catalytic species can be different to their homogeneous counterparts. Thirdly, supporting a metal complex catalyst on what is effectively a multidentate ligand may alter the stereochemistry around the metal ion in a beneficial way.[5] Fourthly, supporting the catalyst may alter the equilibrium position between the metal ion and the surrounding ligands.[6] Finally, it is sometimes possible to stabilize catalytically active centres, often through site isolation, which prevents dimerization of the catalyst to an inactive form.[7] In addition, attachment of the catalyst to a support can sometimes provide protection against poisons for the catalytic species.[8]

The aim of this chapter is to illustrate the benefits of employing heterogeneous catalysts with hydrogen peroxide for the liquid-phase oxidation of organic substrates. Two broad classes of heterogeneous system will be discussed:

- redox zeolites;
- non-crystalline heterogeneous catalysts.

2   Redox Zeolites

There are two major problems associated with selective oxidation catalysis with soluble oxo-metal complexes. Firstly, the propensity of certain oxo-metal species such as titanyl ($Ti^{IV}=O$) towards oligomerization to inactive μ-oxo complexes. Secondly, many complexes contain organic ligands which can be oxidatively destroyed during the oxidation transformation of interest. These problems can, in principle, be circumvented by site isolation of a discrete oxo-metal species in an inorganic matrix, which functions as a thermodynamically stable ligand.[9] One approach to isolating metal ions in stable inorganic matrices is to incorporate them, via isomorphous substitution, into the framework of molecular sieves such as silicates, zeolites and aluminophosphates (AlPOs).

Sheldon and co-workers have described such materials as redox molecular sieves.[10,11] The materials have several advantages over conventional metal-supported catalysts. For example, unlike amorphous materials such as alumina, silica or titania, molecular sieves have a regular micro-environment with homogeneous internal structures, which consist of well defined cavities and channels. They provide the possibility of performing shape-selective catalysis, and by fine tuning the size, and hydrophobic/hydrophilic character of the redox cavity can be turned into 'tailor made' oxidation catalysts. The molecular sieve can also be viewed as a second solvent that extracts the substrate out of the bulk solvent. High silica molecular sieves (*e.g.* silicalite) possess hydrophobic cavities, and hence are suitable for oxidation of organic substrates with aqueous hydrogen peroxide. In contrast, aluminophosphates have hydrophilic cavities that render them unsuitable for reactions with aqueous hydrogen peroxide. They can, however, be successfully employed with alkyl hydroperoxides or molecular oxygen.

Table 4.1 illustrates a number of redox molecular sieves which have been successfully prepared and employed as oxidation catalysts with hydrogen peroxide, alkyl hydroperoxides and molecular oxygen.

Whilst a whole host of redox zeolites have been prepared and employed with hydrogen peroxide for the oxidation of organic substrates, by far the most studied is a titanium-containing silicalite known as TS-1. For this reason, the bulk of this section will discuss this material in terms of its preparation, oxidation ability with aqueous hydrogen peroxide and the mechanistic pathways believed to be functioning during the oxidation of organic substrates.

The discovery of TS-1 in 1983 at Enichem in Italy,[12] opened up active research into the area of titano-silicates. TS-1 is a ZSM-5 type molecular sieve with MFI topology. The discovery of TS-1 is interesting as it can be used in

Table 4.1 *Pore size and structural topographic species of various redox molecular sieves*

Structure type	Framework structure	Pore size (Å)	Metals incorporated
MFI	ZSM-5	5.6 × 5.5	Ti, Zr, V, Fe
MEL	ZSM-11	5.1 × 5.5	Ti, V, Fe
EUO	ZSM-50	5.7 × 4.1	—
ZSM-48	ZSM-48	5.4 × 4.1	Ti
TON	ZSM-22	5.5 × 4.4	Fe
MTT	ZSM-23	5.2 × 5.5	Fe
FER	ZSM-35	5.5 × 4.2	—
FAU	X, Y	7.4 × 7.4	Ti, Fe
MOR	Mordenite	6.7 × 7.0	Ti, Fe
MTW	ZSM-12	5.7 × 6.1	Fe
BEA	Beta	7.6 × 6.4	Ti
NCL-1	NCL-1	7.0 × 7.3	V
MCM-41	MCM-41	40–100	Fe

conjunction with aqueous hydrogen peroxide. Since water is detrimental to conventional catalysts, organic peroxides have to be used as oxygen sources, and these can give rise to toxic side products. The hydrophobic nature of the TS-1 micropores favours diffusion of non-polar substrates to the active site. The hydrophobic micropores of the catalyst are assumed to exclude water from its internal voids, and thus protect the active site from deactivation. The titanium atoms in TS-1 are tetrahedrally co-ordinated to four atoms in the framework, and it is generally considered that the hydroperoxides are accommodated in the titanium co-ordination sphere in the first step of the catalytic conversion.[13]

TS-1 has been synthesized by three methods. The first involves the hydrolysis of tetraethylorthosilicate and tetraethylorthotitanate with tetrapropylammonium hydroxide and water.[14] The preparation does not require mineralizing alkali metals; in fact these can prove detrimental to the catalytic functioning of the material. The hydrolysis is followed by hydrothermal crystallization carried out under autogeneous pressure and agitation at about 175 °C. The second method whilst similar to the first involves the reaction of aqueous hydrogen peroxide with an aqueous solution containing hydrolysed titanium alkoxide which results in the formation of peroxytitanates. Thirdly, the synthesis of TS-1 can be carried out with co-precipitated amorphous TiO_2–SiO_2 in an aqueous solution of tetrapropylammonium hydroxide. Precautions are necessary to avoid precipitation of TiO_2 as a separate phase, as it can act as a catalyst poison in the subsequent oxidation reactions.

TS-1 functions as a very efficacious catalyst in the oxidation of a large number of organic substrates with aqueous hydrogen peroxide (Figure 4.1).[15–18]

TS-1 exhibits some quite remarkable activities and selectivities in the epoxidation of many substrates. For example, relatively unreactive olefins such as ethene and allyl chloride are epoxidized in high yield under mild conditions using methanol as the solvent.[19] As a result of the shape selectivity effect, larger, more reactive olefins such as cyclohexene are epoxidized very slowly (Table 4.2).

Whilst the discovery of TS-1 is a seminal milestone in heterogeneous catalysis, the smaller pore size of the zeolite framework restricts its use, even for some

Table 4.2 *Epoxidation of olefins with aqueous hydrogen peroxide catalysed with TS-1*

Olefin	Temperature (°C)	Time (min)	Hydrogen peroxide (% conversion)	Epoxide yield (%)
Propylene	40	72	90	85
Pent-1-ene	25	60	94	86
Hex-1-ene	25	70	88	79
Oct-1-ene	45	90	81	74
Cyclohexene	25	90	10	< 1
Allyl chloride	45	30	98	90
Allyl alcohol	45	35	81	63

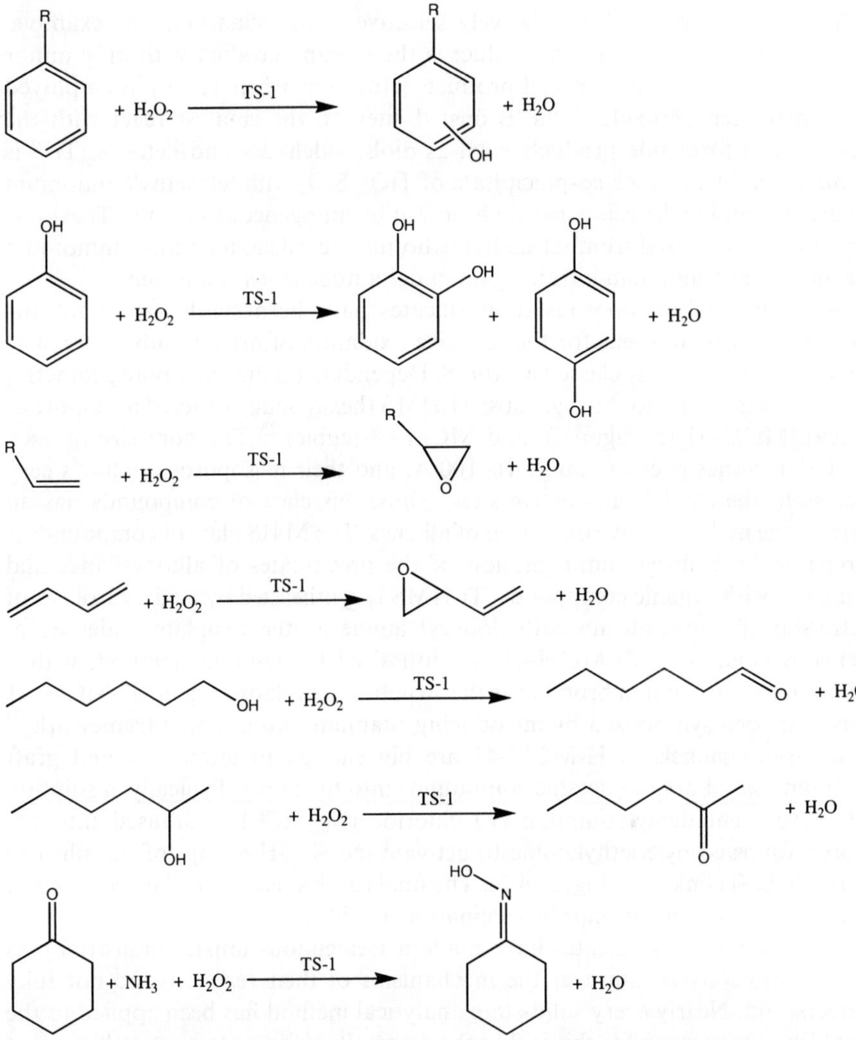

Figure 4.1 *Organic functional group oxidation in the presence of hydrogen peroxide and TS-1.*

relatively small alkenes. Conformationally rigid, as well as larger substrates, cannot approach the catalytically active sites. Hence, new materials with larger pores have been sought. An interesting material worth discussing is titanium-doped zeolite beta (Ti-β).[20] Ti-β is a high-silica zeolite with a three-dimensional pore system containing twelve-membered ring apertures. The pore size is of the order of 7 Å $\times$ 7 Å. The material shows high selectivity and catalytic activity with branched and cyclic alkenes to epoxides with aqueous hydrogen peroxide. Ti-β is hydrophilic in nature and good catalytic activity has been found with organic hydroperoxides. The Ti-β has a Si/Al ratio of greater than 200, which affords a catalyst which is relatively selective in its oxidation. For example, during epoxidation the main product is the oxirane product with only minor contributions from ring-opened products. However, when Ti-Al-β is employed with hydrogen peroxide,[21] the Brönsted sites on the catalyst react with the epoxides to form side products such as diols, aldehydes and ketones. Ti-β is synthesized by mixing a co-precipitate of TiO_2–SiO_2 with tetraethylammonium hydroxide and hydrogen peroxide followed by autogeneous heating. Ti-Al-β is generally synthesized from tetraethylorthotitanate, silica, tetraethylammonium bromide and aluminium nitrate again under autogeneous conditions.

A number of mesoporous titanosilicates have been synthesized[22] for the activation of peroxygens for the selective oxidation of organic substrates. The materials are generally classed as M41S. Depending on the basic pore geometry, they are classified into three groups: Ti-HMS (hexagonally ordered mesoporous silica), MCM-41 (hexagonal), and MCM-48 (cubic).[23] The pore size in such materials varies over the range 16–100 Å, and their mesoporosity allows easy access to the catalytically active sites. Thus, this class of compounds has an advantage in the selective oxidation of alkenes. The M41S class of compounds is prepared by hydrogel impregnation of the precipitates of alkoxysilanes and titanates with organic compounds. Ti-HMS is synthesized by acid hydrolysis of tetraisopropylorthosilicate with dodecyl amine as the template under auto-geneous conditions. Ti-MCM-41 is synthesized by a similar method, with a quaternary ammonium bromide as the structure-templating agent. Ti-MCM-41 has also been synthesized by introducing titanium into a zeolite framework.[24] The large channels of H-MCM-41 are big enough to intercalate and graft titanium-based organometallic compounds into the pores. Typically, a solution of dicyclopentadienyl titanium(IV) chloride (Cp_2TiCl_2) is diffused into the pores, followed by triethylamine to activate the Si–OH groups of the silica to form Si–O–Ti linkages (Figure 4.2). The final catalyst is obtained on removal of the cyclopentadienyl groups by calcination at 550 °C.

The porous titanosilicates have made a tremendous impact industrially as oxidation catalysts, however the mechanisms of their reactions are not fully understood. Nearly every solid-state analytical method has been applied to the problem. One reason for this is that the active titanium centres are only present in very low concentrations. Consequently, whilst there is a continuing search for new heterogeneous titanosilicates for selective oxidation, there has also been an interest in the preparation of homogeneous models for such materials, in an

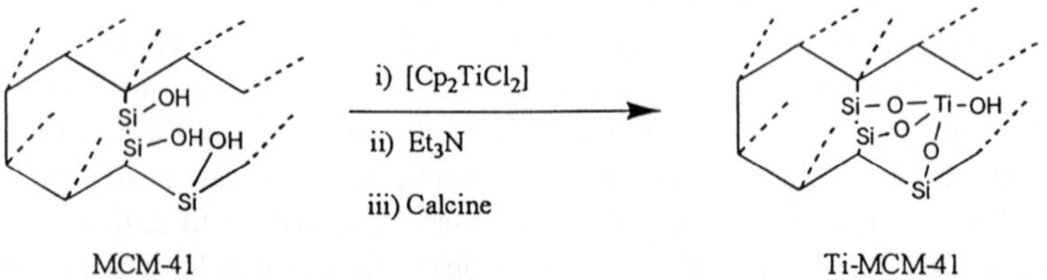

MCM-41   Ti-MCM-41

Figure 4.2 *Preparation of Ti-MCM-41 from MCM-41 in the presence of a titanium-containing organometallic precursor.*

attempt to understand their complex chemistry, and the nature of the active site. Stable organosilanetriols and various titanium-containing organometallic precursors as building blocks are used to investigate such chemistry. With this strategy, it has been possible to prepare a series of molecular titanosilicates with cage structures containing titanium centres in the formal oxidation state of (IV) with tetrahedral (TiO_4), trigonal bipyrimidal (TiO_5), and octahedral (TiO_6), co-ordination environments.[25] The molecules can be viewed as single-phase titanosilicates having a high titanium content. The crystal structure of a titanium-rich titanosilicate [$Cs_3HTi_4O_4(SiO_4)_34H_2O$] synthesized under autogeneous conditions has been reported.[26] The structure was found to resemble the mineral pharmacosiderite.

Numerous studies investigating the nature of the titanium environment in TS-1 have been documented. Based on the unit cell expansion, it is understood that the titanium atoms in TS-1 are incorporated into the zeolite framework.[27] The Ti sites are believed to be tetrahedrally co-ordinated.[28] Further, EXAFS studies on Ti-Al-β suggest that the co-ordination number of the titanium varies depending on whether the samples are as made, calcined, hydrated or dehydrated.[29] XPS has shown that TS-1 has an identical binding energy to that of Ti(IV) in co-precipitates of TiO_2–SiO_2, suggesting that the oxidation state of the titanium in TS-1 is (IV).[30] Davis and co-workers have investigated the selective oxidation of n-octane with amorphous TiO_2–SiO_2 and TS-1 in the presence of aqueous hydrogen peroxide and organic hydroperoxides.[31] They also investigated the epoxidation of hex-1-ene. The aim of their work was to investigate the mechanism of alkane activation on TS-1 by determining the stereoselectivity of this catalyst for *cis-* and *trans-*1,3-dimethylcyclopentane oxidation. The paper proposed a mechanism for the oxidation of alkanes and compared it with the suggested mechanism for alkene epoxidation. Figure 4.3 illustrates their proposed interaction between TS-1, hydrogen peroxide and an alkyl hydroperoxide.

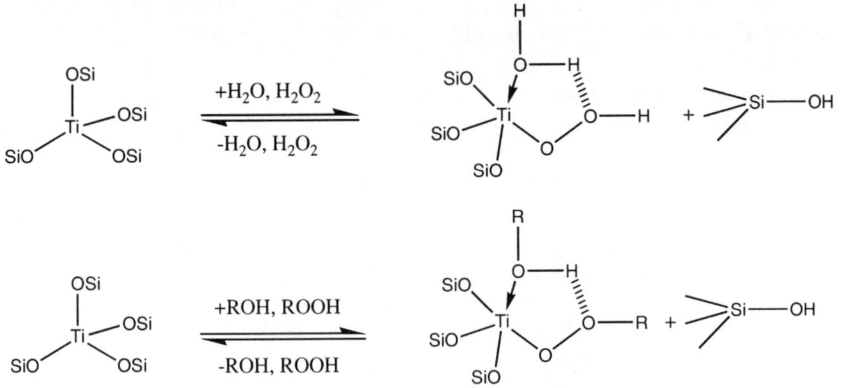

Figure 4.3 *Possible interaction between TS-1 and aqueous hydrogen peroxide or an alkyl hydroperoxide.*

The workers proposed that alkyl hydroperoxides and aqueous hydrogen peroxide interact with TS-1 in a similar manner, forming titanium alkyl peroxo complexes and titanium peroxo complexes, respectively. However, the titanium alkyl peroxo complexes were not active because the substrate could not enter the void due to steric effects. Consequently, no activity was possible for either alkane hydroxylation or alkene epoxidation. Comparison with TiO_2–SiO_2/alkyl hydroperoxide for alkane and alkene oxidation indicated that this material was active because the oxidation took place on the surface and not in the pores. Figures 4.4 and 4.5 show the possible mechanisms in operation for the oxidation of alkenes and alkanes with a TS-1/hydrogen peroxide system.

Stereoselectivity differences were found between alkane and alkene oxidation in the presence of TS-1, which suggested that the oxidations proceeded via different mechanisms. Stereo-scrambling was present during alkane oxidation on TS-1, without any radical clock rearrangement, suggesting that the radicals formed may have had a very short lifetime or that their movements were restricted such that no rearrangement could occur.

The work of Davis was, however, unable to distinguish which oxygen was attacked on the titanium peroxo complex when the alkene co-ordinates. Therefore, Neurock and Manzer conducted a theoretical study of the mechanism of alkene epoxidation using TS-1 with aqueous hydrogen peroxide.[32] The workers concluded that their calculations to predict both the structure and relative bands in the IR spectra for TS-1 were in good agreement with experimental data. The calculations indicated that the oxygen closest to the titanium centre was the active site for alkene attack. The result was the direct formation of

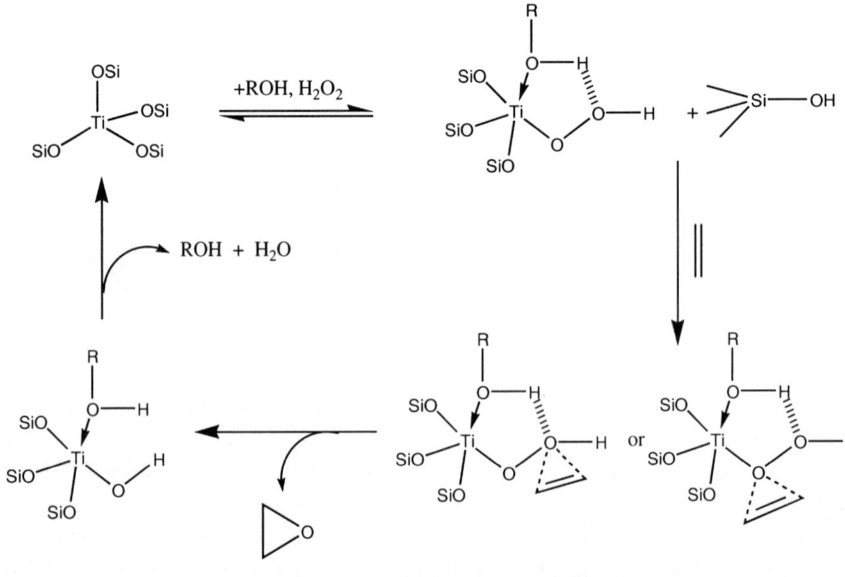

Figure 4.4 *Possible mechanism of alkene epoxidation on TS-1.*

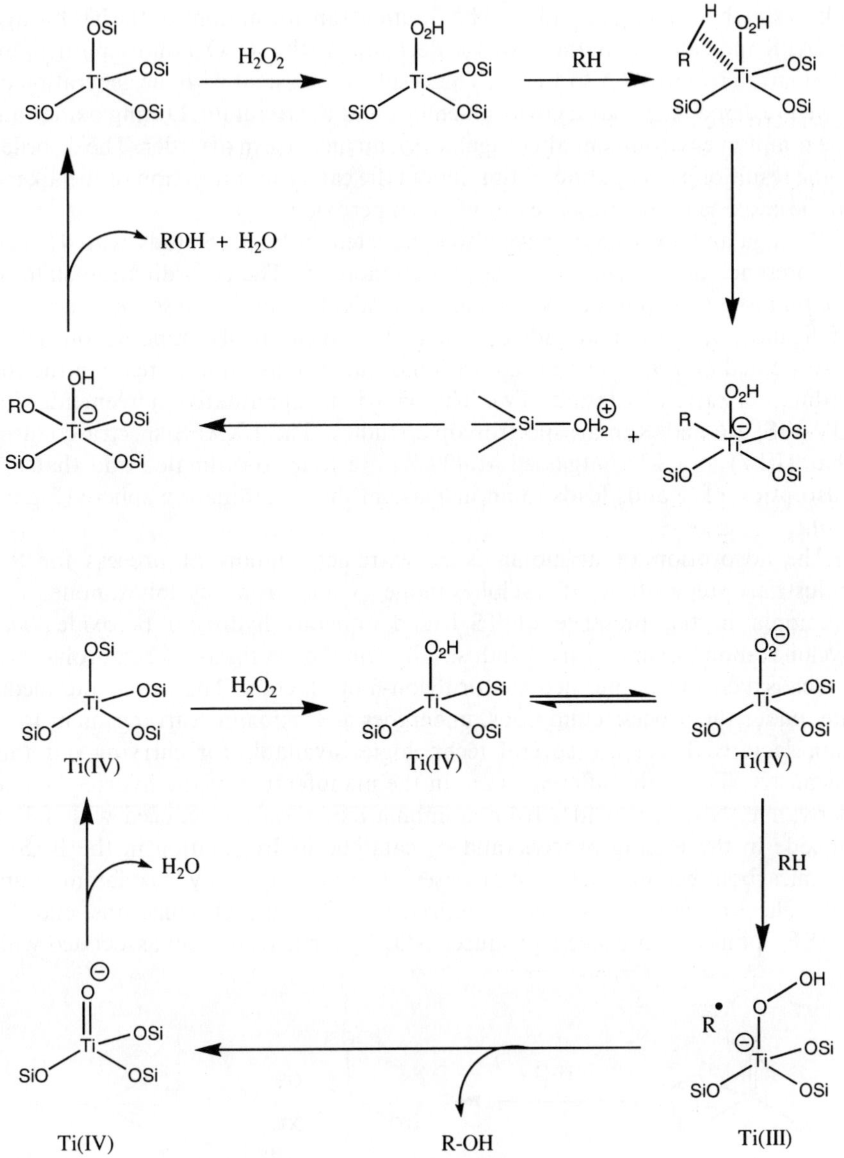

Figure 4.5 *Possible mechanisms operating for alkane oxidation on TS-1.*

water, the regeneration of the tetrahedral titanium centre and a weakly bound epoxide product. Substituent effects suggested that the reaction became more exothermic and released the epoxide more readily upon increasing the electron-withdrawing nature of the substituent.

Sankar and co-workers have used X-ray absorption fine structure (EXAFS) measurements with a synchrotron source to investigate the local environment around the titanium-centred active site in TS-1, during the epoxidation of alkenes by hydrogen peroxide.[33] The Fourier transformations of the Ti–K-edge EXAFS indicated that there was a decrease in the Ti–O bond length upon calcination from 1.87 Å to 1.80 Å. This result is consistent with the generation of a well-ordered tetrahedral environment around the titanium. During oxidation, the titanium environment alters again with an increase in disorder. The disorder is the result of participation of titanium in the catalytic conversion of the alkene to the epoxide in the presence of hydrogen peroxide.

Bordiga and co-workers have also conducted an EXAFS study with TS-1 in the presence of adsorbates, water and ammonia.[34] The co-ordination state of the titanium was found to be strongly dependent upon the presence or absence of ligands. In particular, adsorption and desorption of ammonia on TS-1, investigated *in situ*, gave clear indication of the increase and decrease of the co-ordination state of titanium(IV). The data is in quantitative agreement with UV-VIS, IR and Raman spectroscopic studies. The EXAFS spectra confirm that Ti(IV) in TS-1, outgassed at 400 K, are four co-ordinated and that the adsorption of ligands leads to an increase in the co-ordination sphere (Figure 4.6).

The adsorption of ammonia is an extremely important process for the industrial preparation of cyclohexanone oxime from cyclohexanone and ammonia in the presence of TS-1 and aqueous hydrogen peroxide. The cyclohexanone oxime is used industrially for the synthesis of ε-caprolactam, which is a chemical intermediate for nylon-6 production. The oxime intermediate, under the process conditions, undergoes a Beckmann rearrangement to ε-caprolactam. There are several technologies available for carrying out this chemistry. The main differences are in the manufacture of the hydroxylamine derivative. Nitrogen oxides from ammonia oxidation are reduced with sulfur dioxide in the Rashig process, and by catalytic hydrogenation in the BASF/Stamicarbon technologies.[35] In all cases, the product is a hydroxylammonium salt. The Rashig process is characterized by SO_2 and NO_x emissions, and the BASF/Stamicarbon process produces NO_x. A further problem associated with

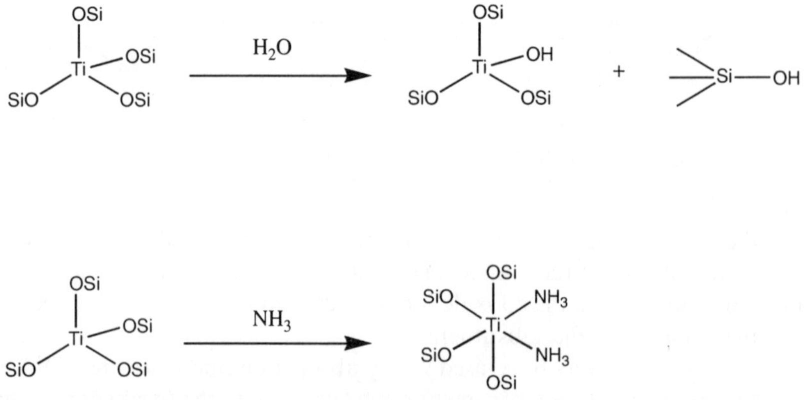

Figure 4.6   *Co-ordination of water and ammonia on TS-1.*

such processes for the production of ε-caprolactam is the large by-product formation of ammonium sulfate, which is a critical environmental and economic issue.

The problems associated with the processes outlined above for the production of ε-caprolactam encouraged workers at Enichem to develop a 'once-through' synthesis of cyclohexanone oxime that would satisfy the following. Firstly, introduction of nitrogen into the cyclohexanone molecule via the direct oxidation of ammonia to the correct oxidation state without over-oxidation to nitrogen oxides. Secondly, the use of aqueous hydrogen peroxide as the oxidant source. Thirdly, low or no by-production of sodium sulfate. Finally, high yield and selectivity together with a simplification of the whole process. The Enichem workers investigated the use of TS-1 with hydrogen peroxide for the cyclohexanone ammoximation and compared this with a range of catalysts (Table 4.3).[36]

The results obtained with the different catalysts indicate that both TiO_2 supported on silica and TS-1 perform well. However, TS-1 was the superior catalyst. The Enichem process for this step in the ε-caprolactam process greatly simplifies the current technology (Figure 4.7).

The ammoximation of cyclohexanone using a TS-1 and hydrogen peroxide system provides a dramatic example of how commercial processes can be improved not only on an environmental basis but also in terms of process simplification leading to cost savings. The main advantages of the Enichem technology are that the current industrial processes can be easily adapted to accommodate the changes and that there is a significant reduction of ammonium sulfate by-product, together with the complete elimination of NO_x and SO_2 emissions.

The discovery of TS-1 and its catalytic properties in the oxidation of organic

Table 4.3 *Cyclohexanone ammoximation with different catalysts*[a]

$$NH_3 + H_2O_2 \xrightarrow{[CAT]} NH_2OH + H_2O$$

Catalyst	Conversion of cyclohexanone (%)	Selectivity to oxime (%)
None	53.7	0.6
SiO_2 amorphous	55.7	1.3
SiO_2 crystalline	59.4	0.5
TiO_2–SiO_2	66.8	85.9
TS-1	99.9	98.2

[a] Work carried out by Enichem SpA.

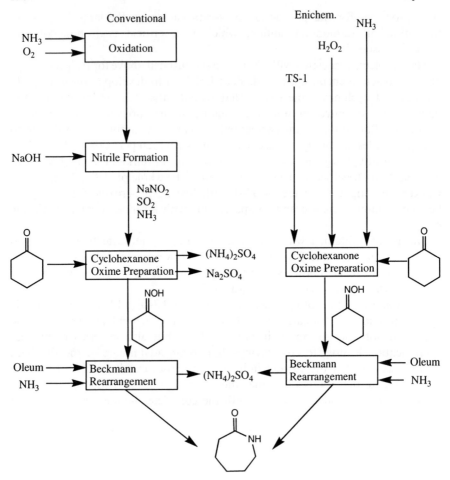

Figure 4.7 *Comparison of conventional and Enichem processes for the production of ε-caprolactam.*

compounds with hydrogen peroxide has opened the way to new technological possibilities and has demonstrated that hydrogen peroxide can be an extremely selective oxidizing agent when coupled with titanosilicates. Before leaving the section, there are other silicalite materials worthy of note, namely tin silicalite, chromium silicalite and ZSM-5.

Tin silicalite materials with MFI (ZSM-5) topology have been prepared by several workers.[37,38] Larger pore size tin silicalites of ZSM-12 (MTW topology, consisting of unidimensional channels with 12-membered ring pore openings of 5.7 Å × 6.1 Å in size) were prepared by Ramaswamy and co-workers.[39] The material was synthesized hydrothermally using a hexamethylene bis(benzyl-dimethylammonium hydroxide) templating agent. The tin silicalite has an IR absorption band at 970 cm^{-1} and a charge-transfer band at 250 nm in the UV-VIS spectrum indicating the presence of Si–O–Sn units with Sn(IV) centres in a

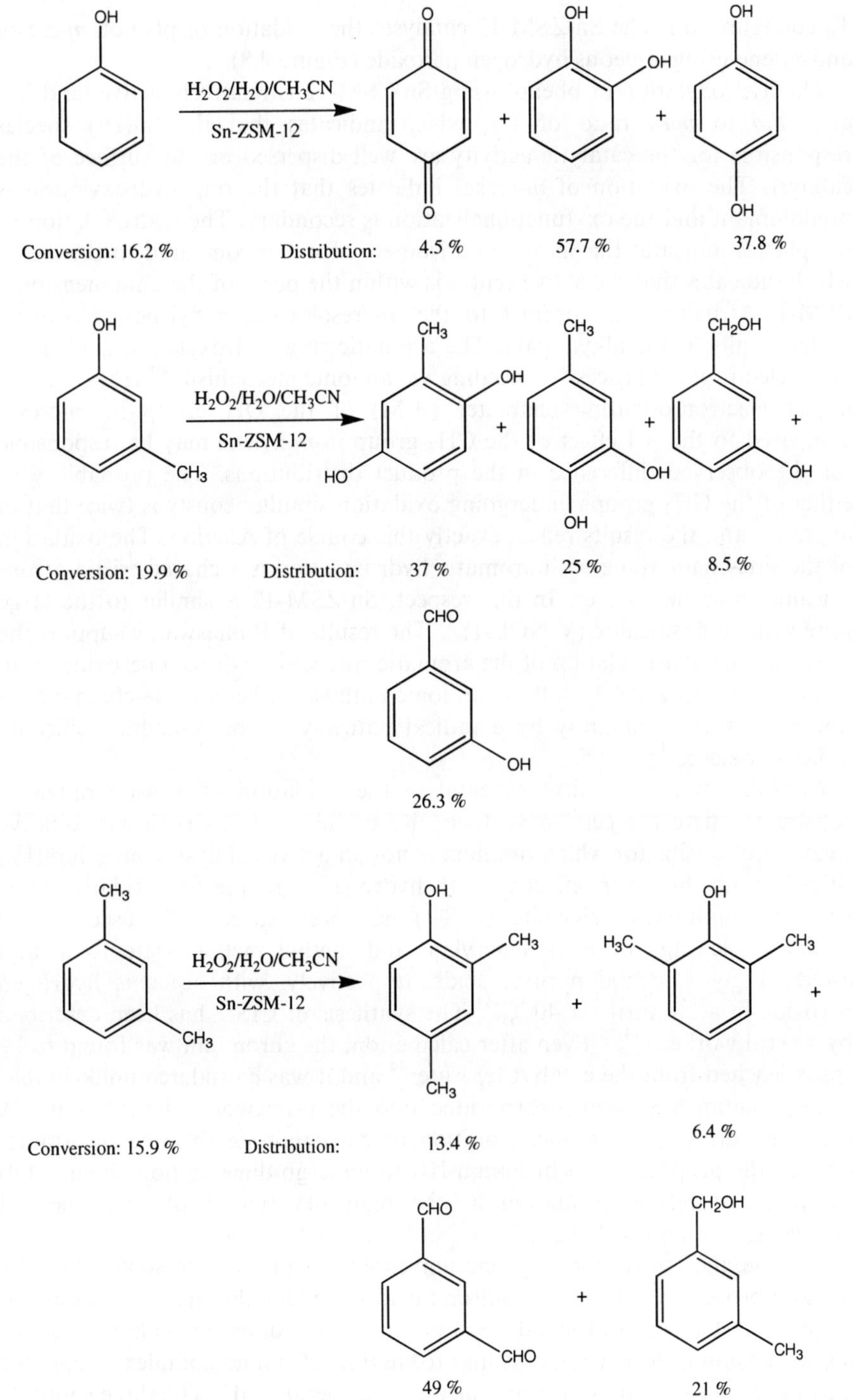

Figure 4.8 *Oxidation of arenes in the presence of aqueous hydrogen peroxide and Sn-ZSM-12.*

T_d configuration. The Sn-ZSM-12 catalyses the oxidation of phenol, *m*-cresol and xylene using aqueous hydrogen peroxide (Figure 4.8).

The hydroxylation of phenol using Sn-ZSM-12 is relatively active, and has an *ortho* to *para* ratio of 1.4, which indicates that the Sn(IV) species responsible for the catalytic activity are well dispersed on the surface of the catalyst. The oxidation of *m*-cresol indicates that the ring hydroxylation is predominant and the oxyfunctionalization is secondary. The hydroxylation of the phenol indicates that more *para* isomer is formed compared to the *ortho*, which indicates that the active centre is within the pores of the unidimensional ZSM-12 structure. In contrast to the *m*-cresol case, *m*-xylene is oxidized preferentially at the alkyl chain. The aromatic ring hydroxylation is likely to be an electrophilic process proceeding via an ionic mechanism.[40] However, the higher electron-donating character ($+M$) of the OH group in *m*-cresol compared to the $+I$ effect of the CH_3 group in *m*-xylene may be responsible for the observed difference in the product distributions. The probability of either of the CH_3 groups undergoing oxidation simultaneously is twice that in *m*-cresol, and the results reflect exactly this course of reaction. The oxidation of the side-chain along with aromatic hydroxylation is a characteristic of tin-silicalite molecular sieves. In this respect, Sn-ZSM-12 is similar to the large pore vanadium silicalite (V-NCL-1).[41] The results of Ramaswamy support the view that the hydroxylation of the aromatic ring with hydrogen peroxide as an oxidant over Sn-ZSM-12 follows an ionic pathway, whereas side-chain oxidation occurs predominantly by a radical pathway, as on vanadium silicalite molecular sieves.[42]

Although titanium silicalites catalyse the oxidation of a wide range of substrates, there are certain synthetically useful reactions such as oxidative cleavage of olefins for which titanium is not an active catalyst. Chromium(IV) silicalites are, however, effective with hydrogen peroxide for oxidative cleavage. Chromium(IV) silicalite (CrS-1) has been successfully used in the oxidative cleavage of methyl acrylate and methyl methacrylate to methyl esters of glyoxilic and pyruvic acids, respectively, with aqueous hydrogen peroxide in acetonitrile at 40 °C.[43] The synthesis of CrS-1 has been described by several workers.[44,45] Even after calcination, the chromium was found to be easily leached from the catalyst by water[44] and it was considered unlikely that the chromium had been incorporated into the framework of the zeolite. A major problem with the incorporation of chromium in the silicalite framework is the propensity of chromium(III) to undergo dimerization (Figure 4.9) via hydroxo bridge formation at the high pH typical of hydrothermal synthesis.

Sheldon and co-workers have circumvented this problem to some extent by three approaches:[46] the use of sulfuric acid to reduce the pH, by addition of ammonium fluoride and by addition of ammonia. Ammonia stabilizes monomeric chromium(III) species via the formation of amine complexes, and the fluoride effects dissolution of the silica at near-neutral pH. The three catalysts that were synthesized were evaluated in the oxidative cleavage of styrene with 35% *m/m* hydrogen peroxide in 1,2-dichloroethane at 70 °C (Table 4.4). The

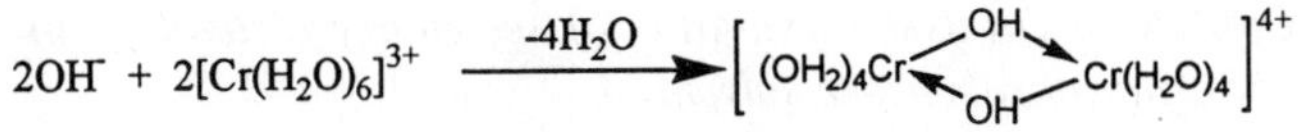

$$2OH^- + 2[Cr(H_2O)_6]^{3+} \xrightarrow{-4H_2O} \left[(OH_2)_4Cr \underset{OH}{\overset{OH}{<}} Cr(H_2O)_4 \right]^{4+}$$

Figure 4.9 *Dimerization of chromium(III) species under basic conditions.*

major products were benzaldehyde and 1,2-diphenylethane-1,2-diol, together with minor amounts of styrene oxide and phenylacetaldehyde. The latter results from the rearrangement of styrene oxide and is the major product of TS-1 catalysed oxidation of styrene with hydrogen peroxide.[47]

The Sheldon group found that the highest conversions were observed with CrS-1 prepared by the ammonia method. CrS-1 made by the fluoride method gave a slightly lower styrene conversion and roughly the same selectivities. The catalyst prepared by the sulfuric acid method gave substantially lower styrene conversions, but the selectivity to benzaldehyde was high.

Aluminophosphates (AlPO$_4$) were discovered in 1982[48] and a large amount of research has been directed towards the incorporation of various elements into the framework of these molecular sieves.[49] A particular area of study is the oxidation of primary and secondary alcohols to the corresponding carbonyl compounds, which are useful synthetic intermediates. Traditionally, alcohol transformations are performed with stoichiometric chromium(VI) reagents.[50] However, due to environmental problems associated with chromium-containing effluent, attention has focused on the use of chromium in conjunction with oxidizing agents such as *tert*-butyl hydroperoxide.[51] Sheldon and co-workers

Table 4.4 *Chromium silicalite cleavage of styrene with aqueous hydrogen peroxide[a]*

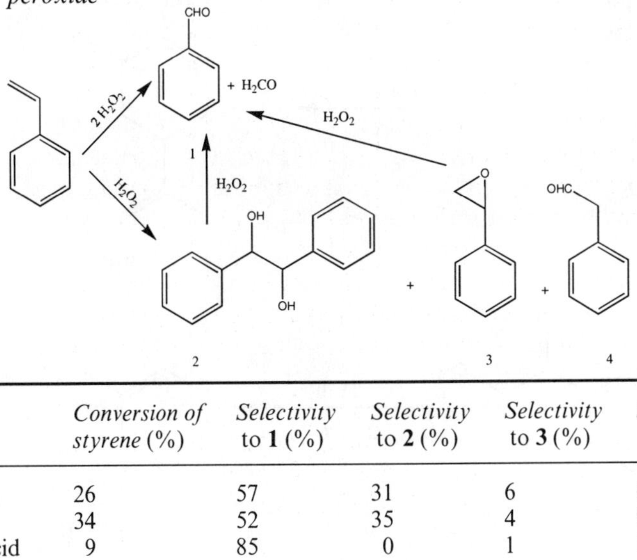

Synthesis method	Conversion of styrene (%)	Selectivity to 1 (%)	Selectivity to 2 (%)	Selectivity to 3 (%)	Selectivity to 4 (%)
Fluoride	26	57	31	6	6
Ammonia	34	52	35	4	9
Sulfuric acid	9	85	0	1	1

[a] 70 °C, 1,2-dichloroethane, 4 h.

Table 4.5 *Cr-APO-5 catalysed oxidation of secondary alcohols with TBHP at 85 °C in chlorobenzene solvent*

Substrate	Product	Conversion (%)	Selectivity on substrate (%)	Selectivity on TBHP (%)
α-Ethylbenzyl alcohol	Propiophenone	77	100	91
α-Methylbenzyl alcohol	Acetophenone	77	96	89
Cyclohexanol	Cyclohexanone	72	85	73
Carveol	Carvone	62	94	66
1,2-Diphenylethane-1,2-diol	α-Hydroxy acetophenone	54	73	40

have prepared chromium-substituted AlPO$_4$s in conjunction with TBHP for alcohol oxidation.

The material synthesized by the group containing chromium in the +3 oxidation state, Cr-APO-5, was green; however, upon calcination in air at 550 °C, the catalyst turned yellow, and the majority of the chromium was converted to the +6 oxidation state. Table 4.5 summarizes some of Sheldon's results. The alcohol oxidation selectivities were found to be good to excellent with respect to both the substrate and TBHP. Carveol undergoes chemoselective oxidation of the alcohol group to give carvone, without any attack on the double bonds. Diols containing secondary and primary alcohol groups can also be selectively oxidized (Figure 4.10).

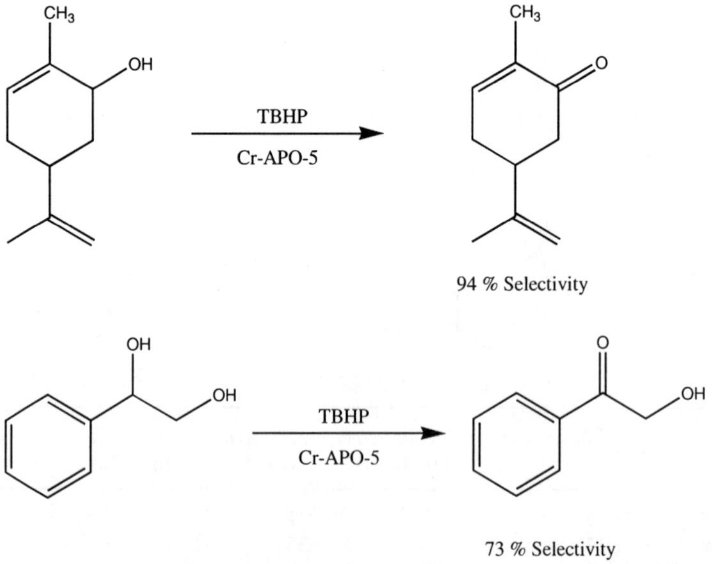

Figure 4.10 *Chemoselective oxidation of alcohols with Cr-APO-5 in the presence of TBHP.*

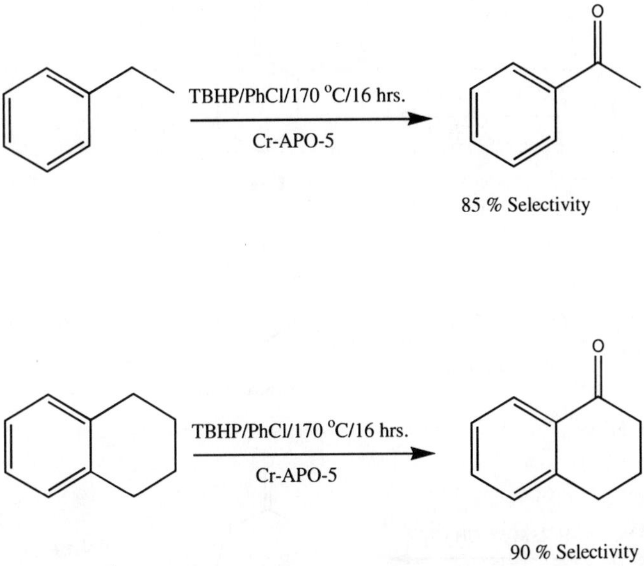

Figure 4.11 *Oxidation of hydrocarbons with Cr-APO-5 and TBHP.*

Cr-APO-5 and TBHP systems have also been employed for the oxidation of hydrocarbons. For example, ethylbenzene and tetralin have been selectively oxidized under mild conditions employing this system (Figure 4.11).

Whilst the majority of the discussion thus far has been concerned with metallo-substituted redox molecular sieves, it is important to note that protonated zeolite forms can also be employed for selective oxidation with aqueous hydrogen peroxide. An excellent example of this is the study conducted by the Mobil Oil Corporation.[52] Their work has shown that a number of protonated zeolites such as H-ZSM-5 or zeolite-β can be used with hydrogen peroxide to catalyse the oxidation of cyclic ketones to lactones or the ω-hydroxycarboxylic acids (Figure 4.12).

3   Non-crystalline Heterogeneous Catalysts

The final section in this chapter will discuss several areas of current research, which have attempted to use non-crystalline materials such as metallosilicates or supported oxidation catalysts on high surface area solids. The definition of non-crystalline employed here refers to the support system and not necessarily the active catalytic centre.

The previous section discussed the use of crystalline metal silicalites as catalysts for the liquid-phase oxidation of organic substrates with peroxygens. The unfortunate feature of many of the crystalline catalysts is their relatively difficult synthesis, requiring specialized expertise, equipment and the use of proprietary knowledge. Also the small pores of the zeolite limit access to the catalytic sites by large organic substrates, which are often more prevalent in the

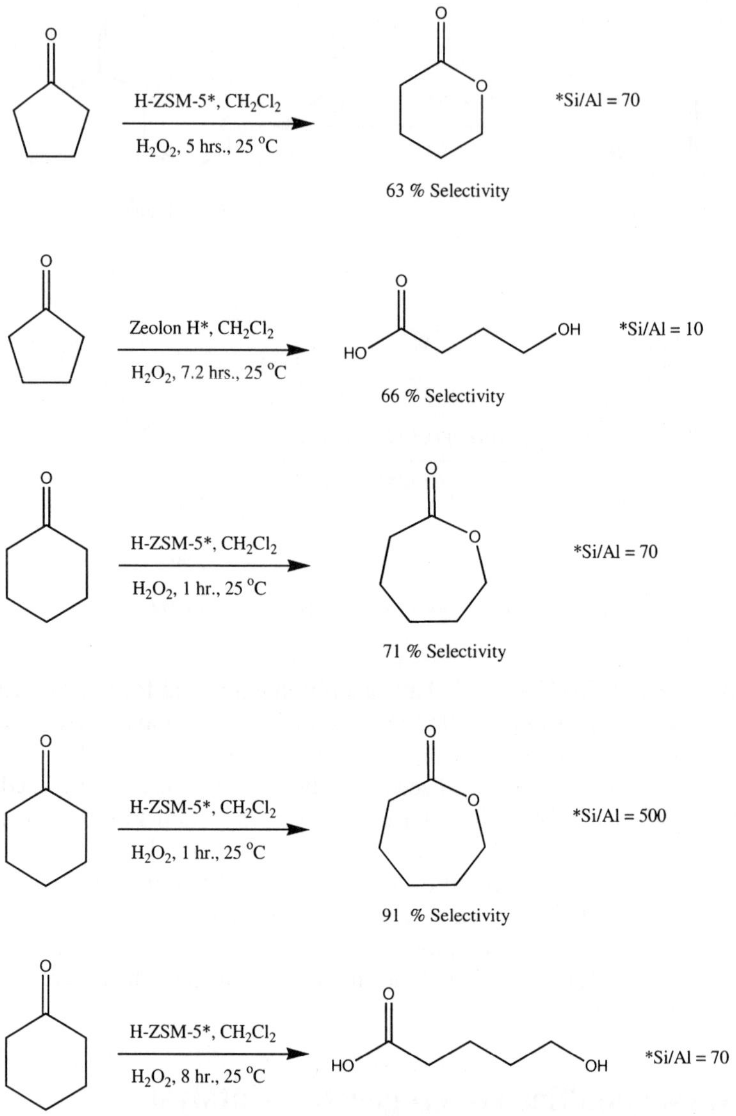

Figure 4.12 *Oxidation of cyclic ketones with aqueous hydrogen peroxide in the presence of protonated zeolites.*

synthesis of fine chemicals. In response to such limitations, Neumann has synthesized a range of amorphous metallosilicalite xerogels, MO_x–SiO_2,[53] by the application of the sol-gel method[54] and used them as catalysts in oxidation of various organic substrates (Figure 4.13). The required metallo-silicalites are prepared by dissolving tetraethylorthosilicate in acidified aqueous ethanol. The solution is heated at 60 °C for about 2 h. The appropriate metal isopropoxide is then added, and the resultant gel is left in an open beaker allowing slow

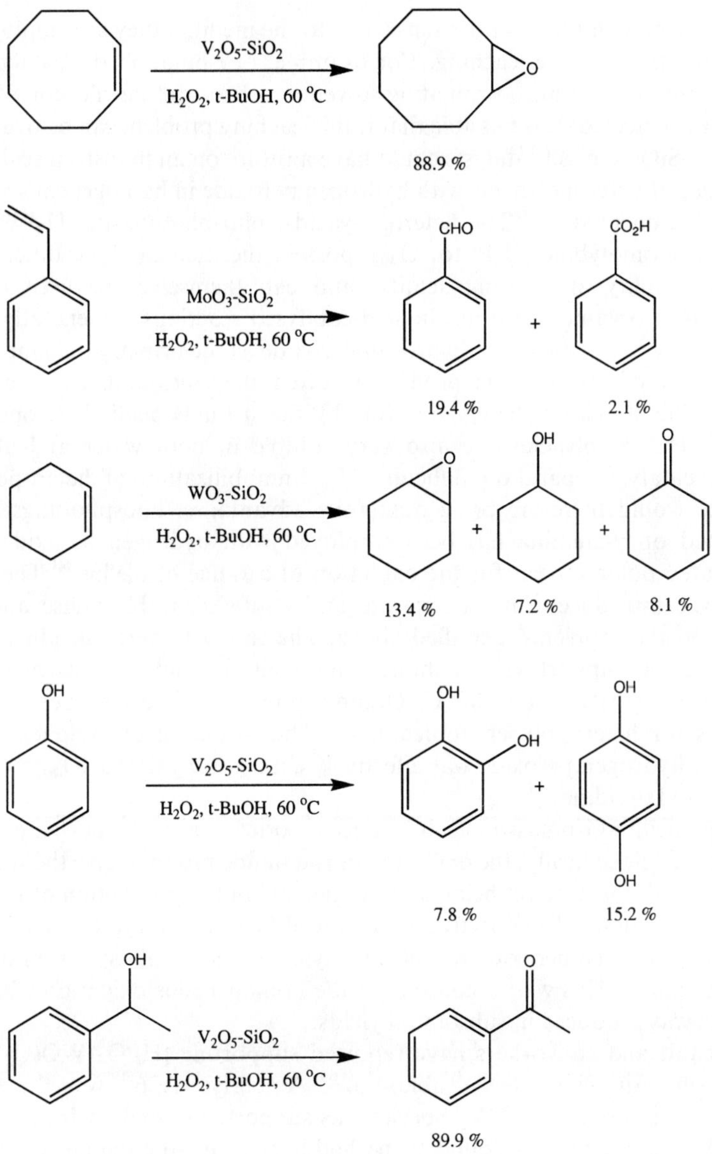

Figure 4.13 *Use of MO_x–SiO_2 amorphous catalysts for the oxidation of organic substrates with aqueous hydrogen peroxide.*

evaporation of the solvent. The brittle MO_x–SiO_2 xerogel generated is ground and dried at 100 °C. The catalysts are easily characterized by IR spectroscopy which indicates a band at 940–960 cm^{-1}, attributable to the M–O–Si vibration.[55]

Whilst the metal containing xerogels show reasonable catalytic activity with hydrogen peroxide, they are still inferior to the redox zeolites mentioned earlier.

One problem with them is their sensitivity to the medium they are employed in, as regards metal oxide leaching. For example, in *t*-butanol, the catalysts are relatively stable, but in co-ordinating solvents such as acetonitrile, considerable leaching is detected. It is possible that if the leaching problem can be overcome, the MO_x–SiO_2 xerogel catalysts could have a future on an industrial scale.

The use of heteropolyacids with hydrogen peroxide in homogeneous systems is well documented.[56,57] The heteropolyacids, phosphotungstic ($H_3PW_{12}O_{40}$) and phosphomolybdic ($H_3PMo_{12}O_{40}$), possess the dual catalytic functions of oxidizing ability and strong acidity, and can themselves be used for the oxidation of organic substrates in acid-catalysed reactions.[58] Generally, when a heteropolyacid is used with hydrogen peroxide as the primary oxidant for the epoxidation of alkenes, the product is often the *trans*-diol, as a result of electrophilic attack on the epoxide ring by the strongly acidic heteropolyacid anion.[59] Heteropolyacids are also very soluble in both water and alcohol, making catalyst separation difficult. The immobilization of heteropolyacid catalysts would therefore bring numerous advantages. Phosphotungstic acid supported on γ-alumina has been employed with hydrogen peroxide in an oxygenated polar solvent for the oxidation of a range of olefins.[60] The heavy metal was introduced into a separate and easily separable phase and thus eliminated the problems identified above. The catalysts were introduced onto the inorganic support via alcoholic impregnation, and then calcined at a temperature greater than 300 °C. Organic polymers have also been used as supports for heteropolyperoxometallates. The oxidation of cyclohexane-1,2-diol with hydrogen peroxide was effectively catalysed by $[PMo_{12}O_{40}]^{3-}$ bound to polyvinyl pyridines.[61]

The penicillin-G-potassium salt has been oxidized using supported polyoxometallates.[62] Specifically, the oxidation of the sulfide group to give the sulfoxide was achieved, the product being an intermediate in the production of *Cephalosporin* antibiotics.[63] The selective oxidation of benzenethiol to benzenesulfonic acid with hydrogen peroxide was effectively catalysed by phosphotungstic acid on γ-alumina.[62] This was in contrast to the homogeneous oxidation where the disulfide was produced in substantial yields.

Bregeault and co-workers have reported supporting $[HPO_4\{WO(O_2)_2\}_2]^{2-}$ species on resins and silica (Table 4.6).[64] Amberlyst A26® was the macroreticular resin used. The 'PW$_2$' species was supported onto dehydrated porous silica. The catalysts were found to be highly selective for the epoxidation of limonene by hydrogen peroxide.

A different strategy has been applied by Neumann and co-workers with $[PO_4\{WO(O_2)_2\}_4]^{3-}$ for use in the epoxidation of olefins with hydrogen peroxide.[65] Here they created insoluble, silicate-based particles capable of selective adsoption of the olefin substrate and possessing catalytic centres at the particle and adsorbed-substrate–water interface. Silicate xerogels with covalently attached phenyl rings were prepared, and after connection of quaternary ammonium cations to the xerogel the anionic heteropolyperoxometallate was introduced as the catalytic centre. The catalytic particles were reasonably effective in the oxidation of cyclooctene with hydrogen peroxide.

Table 4.6 *Epoxidation of limonene with supported heteropolyacid species*

Catalyst	Substrate conversion (%)	Yield of **1** (%)	Yield of **2** (%)
PW$_2$/A-26®ᵃ	95	94	Trace
PW$_2$-silicaᵇ	100	95	0
Homogeneous 'PW$_2$'ᶜ	71	< 1	67

ᵃ $[HPO_4\{WO(O_2)_2\}_2]^{2-}$; ᵇ $[HPO_4\{WO(O_2)_2\}_2]^{2-}$; ᶜ $[Bu_4N]_2[HPO_4\{WO(O_2)_2\}_2]$.

Since the catalytic system does not require any other organic solvents, its future development and its ecological benefits are of interest.

Gelbard has used peroxotungstates supported on polypyridine polymers in the epoxidation of cyclohexene with hydrogen peroxide.[66] Polypyridine polymers were also used to support heteropolyperoxometallates for use in the oxidation of alcohols with hydrogen peroxide.[67] The tetranuclear complex [cetylpyridinium chloride][PO$_4$\{WO(O$_2$)$_2$\}$_4$] supported on polypyridine was found to be an effective catalyst for the oxidation of cyclohexanol to cyclohexanone, also with hydrogen peroxide.

Several heterogeneous catalysts have been developed for the hydroxylation of alkanes under mild conditions.[68,69] One of them is the bi-catalytic system, which combines the ability of palladium to convert hydrogen and oxygen to hydrogen peroxide, with the capability of the iron ions to activate the hydrogen peroxide to hydroxylate hydrocarbons.[70] Iron oxide and palladium supported on silica have been used as efficient catalysts for the oxidation of cyclohexane to the alcohol and ketone, via the *in situ* generation of hydrogen peroxide in an acetone solvent.[71]

A series of iron(III) complexes, *cis*-[Fe(tetraamine)(OH)$_2$]$^{2+}$, where tetraamine is a tetradentate amine such as *N,N'*-bis(2-picolyl)ethylenediamine, bound electrostatically to poly-L-glutamate or dextran sulfate, has been studied as catalysts for the decomposition of hydrogen peroxide into water and oxygen.[72] The catalyst performance varies with the mode of binding and nature of the support, thus demonstrating the importance of the local environment around the catalytically active site.

Polystyrene-supported vanadyl acetate catalyses the oxidation of a range of substrates, including dimethyl sulfide to the sulfone, di-n-butylthioether to the sulfoxide, and cyclohexene to the epoxide.[73] A series of trinuclear carboxylate complexes [M$_3$O(O$_2$CR)$_6$(H$_2$O)$_3$]$^{n+}$ (M = V, Cr, Mo, Mn, Ru, Co or Rh) bound electrostatically to a cation-exchange resin catalyses the oxidation of cyclohexene by cumene hydroperoxide.[74] The activity depends on the metal employed in the order Co < V < Rh < Ru < Cr < Mn < Mo.

Metalloporphyrin-type catalysts for alkane hydroxylation and alkene epoxidation have been supported on a diverse range of materials, including organic polymers and inorganic solids such as silica, alumina, magnesia, clays and zeolites. Organic polymers have the disadvantage that they are susceptible to oxidation, and can therefore act as competitive substrates leading to low yields of the desired product.[75] In contrast, the inorganic supports are inert to oxidation and have rigid structures and frameworks which are of potential benefit for metalloporphyrin-based oxidation catalysts. The most common approach to preparing supported metalloporphyrins is to combine a pre-formed porphyrin with a solid support. A second approach is to form a highly cross-linked co-polymer of a metalloporphyrin and a suitable monomer, thus rendering the catalyst insoluble.[76] Taylor and Mansuy have polymerized a fluorinated porphyrin by nucleophilic substitution of the *para*-fluorines by sodium sulfide (Figure 4.14).[77]

In cytochrome P-450 mediated oxidations, substrate specificity is controlled by the 'hand in glove' substrate binding pocket in the protein above the active site of the enzyme. In theory, these features can be realized by inserting metalloporphyrins between the layers of smectite clays, such as montmorillonite, or by trapping them in zeolite cages (Figure 4.15). The clay or zeolite then serves the dual function of supporting the metalloporphyrin and restricting substrate access to the catalyst. Iron and manganese porphyrins have been synthesized within a zeolite supercage.[78] The catalysts were, however, found to be poor for cyclohexane hydroxylation by hydrogen peroxide, probably due to poor substrate access.

Ionic meso-tetraarylporphyrins can be bound to a range of charged supports including cross-linked polystyrene based anions,[78] cations,[79] exchange resins, cross-linked and quaternized poly(4-vinylpyridine)[80] and the oxides of silica, alumina and magnesia.[81] Metalloporphyrins may be attached to supports by co-ordination of a ligand on the support to the metal centre. Such catalysts mimic

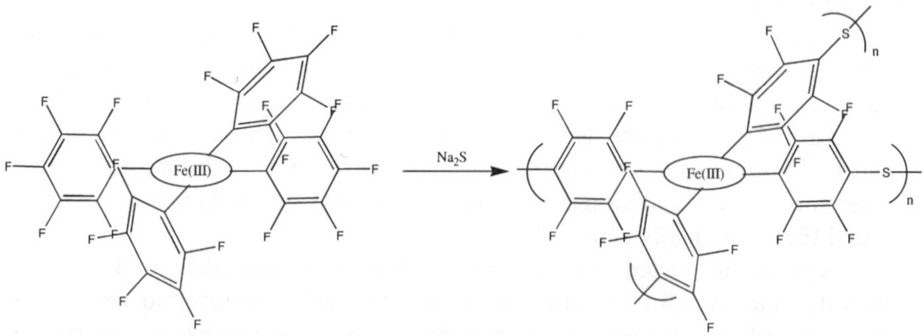

Figure 4.14 *Polymerization of a fluorinated porphyrin.*

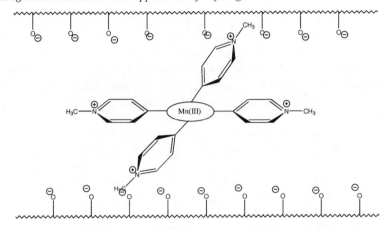

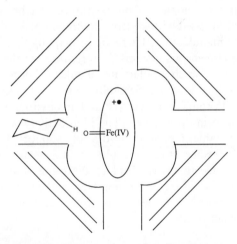

Figure 4.15 *Encapsulated porphyrins in clay and zeolite structures.*

haemoproteins in which the axial ligand serves to anchor the iron porphyrin prosthetic group and to mediate the activity of the metal ion. A few reports have been published of non-polar metalloporphyrins adsorbed directly on silica being used as oxidation catalysts, where it is assumed that bonding is by ligation of the surface silanol groups to the metal centre.[82,83] However, the more common practice is to modify the silica surface with a suitable ligand such as imidazole,[84] pyridine or its *N*-oxide,[85] trialkylamines, thiols and carboxylate groups.[86]

Covalent binding has also been used for attaching a metalloporphyrin to a support. The strong bond formed eliminates the problem of leaching found with co-ordinatively bound catalysts and immobilizes the porphyrin on the surface preventing aggregation. The main advantage of covalent binding, however, is the variable nature of the linker unit, allowing a greater variety of supports to be used and the environment around the metalloporphyrin to be structurally manipulated. The structural groups that have been used as linkages include amine,[87] amide,[88] ester,[89] ether,[90] ketone[90] and quaternized ammonium ions.[91] The supports used include flexible organic polymers such as poly(methyl methacrylate), polystyrene and polymerized isocyanide, as well as inorganic solids such as silica and clays.

Groves has recently supported a polyfluorinated ruthenium porphyrin catalyst for selective hydrocarbon oxidation.[92] The oxidant employed was 2,6-dichloropyridine *N*-oxide prepared via the oxidation of the parent pyridine with peracetic acid. The catalyst was highly active showing turnover numbers of greater than 2500. The use of such supported porphyrin systems with hydrogen peroxide has to date been rather scant. However, because of the advantages that could be gained from coupling supported porphyrins with hydrogen peroxide for the selective oxidation of organic substrates, it is likely that a plethora of active research will soon be reported.

Clearly, the major developments of the future will be the application of heterogeneous catalysts used with aqueous hydrogen peroxide in fine chemical and pharmaceutical manufacture. There are principally three main reasons for employing such heterogeneous systems. Firstly, because the catalyst has high activity and long life which implies no significant deactivation by either poisons, or as a result of leaching. Secondly, because it gives rise to very high selectivity. Finally, it produces both a cost and an environmental benefit to the particular oxidation being performed. The high activity will be promoted by having a thermally stable, mechanically durable support. Inorganic supports that clearly merit further study include zeolites, clays and glasses. The importance of high selectivity will lead to far more studies of the detailed three-dimensional nature of the active site. Hence, the support will not merely be an insoluble support but rather a material that contributes to the total environment of the active site.

4   Conclusion

The future of hydrogen peroxide lies in its ability to be applied under zero effluent conditions (not including water as a by-product). Therefore, the use of hydrogen peroxide coupled with solid surfaces is crucial if the true potential of the reagent is to be realized. The use of zeolite type materials such as TS-1 has made an important contribution in respect of zero effluent processes industrially. However, further research is required, particularly in the area of supported metal catalysts on inorganic surfaces for the activation of hydrogen peroxide. It will also be important for researchers to seriously look at the suppression of metal catalysts leaching from the surface of the support. The future for heterogeneous catalysed hydrogen peroxide looks bright, and an area worthy of further study.

References

1 A.J. Butterworth, J.H. Clark, P.H. Walton and S.J. Barlow, *J. Chem. Soc., Chem. Commun.*, 1996, 1859.
2 G. Bond, *Acc. Chem. Res.*, 1993, **26**, 490.
3 W. Heitz, *Adv. Polym. Sci.*, 1977, **23**, 1.
4 S. Mazur, Pjayalekishmy, J.T. Anderson and T. Matusinovic, *Am. Chem. Soc., Symp. Ser.*, 1982, **192**, 43.
5 D.P. Harrison and M.F. Rose, *Ind. Eng. Chem. Fund*, 1967, **6**, 161.
6 A.T. Whitehurst, *Adv. Chem. Ser.*, 1974, **132**, 240.
7 F.H. Jardine, *Prog. Inorg. Chem.*, 1981, **28**, 63.
8 B.M. Trost and E. Keinan, *J. Am. Chem. Soc.*, 1978, **100**, 7779.
9 M. Mayoshi, K. Kikawa, Y. Murai, N. Shigenmato, S. Sugiyama and K. Kawashioro, *Chem. Lett.*, 1996, **36**, 99.
10 R.A. Sheldon, *Chem. Tech.*, 1991, 566.
11 R.A. Sheldon, *Topics Curr. Chem.*, 1993, **164**, 21.
12 B. Notari, *Catal. Today*, 1993, **18**, 163.
13 R. Landau, G.A. Sullivan and D. Brown, *Chemtech.*, 1997, 602.
14 Enichem SPA, US 4656016.
15 A. Esposito, M. Taramasso, C. Neri and F. Buonomo, UK 2116974.
16 C. Neri, A. Esposito, B. Anfosi and F. Buonomo, EP 100119.
17 C. Neri, B. Anfosi and F. Buonomo, EP 100118.
18 A. Esposito, C. Neri and F. Buonomo, US 4480135.
19 M.G. Clerei, G. Bellussi and U. Romano, *J. Catal.*, 1991, **129**, 159.
20 M.A. Camblar, M. Constantini, A. Carma, L. Gilbert and P. Esteve, *J. Chem. Soc., Chem. Commun.*, 1996, 1339.
21 T. Sato, J. Dakka and R.A. Sheldon, *J. Chem. Soc., Chem. Commun.*, 1994, 1887.
22 C.T. Kresge, M.E. Leonowicz, W.J. Roth, J.C. Vertilli and J.S. Beck, *Nature*, 1992, **359**, 710.
23 W. Zhang, M. Fraba, J. Wang, P.T. Tanev, J. Wong and T. Pinnavia, *J. Am. Chem. Soc.*, 1996, **118**, 9164.
24 J. Haggin, *Chem. Eng. News*, 1995, **73**, 6.
25 A. Voigt, R. Murugavel, M.L. Montero, H. Wessel, F.Q.J. Liu, H.W. Roesky, I. Uson and T. Albers, *Angew. Chem., Int. Ed. Engl.*, 1997, **36**, 1001.
26 W.T.A. Harrison, T.E. Gier and G.D. Stucky, *Zeolites*, 1995, **15**, 408.
27 G. Perego, G. Bellussi, C. Corno, M. Taramasso, F. Buonomo and A. Esposito, *Stud. Surf. Sci. Catal.*, 1986, **28**, 129.
28 M. Boccuti, K. Rao, A. Zecchina, G. Leofanti and G. Petroni, *Stud. Surf. Sci. Catal.*, 1989, **48**, 133.
29 M.A. Camblor, A. Carma and J.P. Pereinte, *J. Chem. Soc., Chem. Commun.*, 1993, 1557.
30 S.M. Mukhopdhyay and S.H. Garofalini, *J. Non-Cryst. Solids*, 1990, **126**, 202.
31 C.B. Khouw, C.B. Dartt, J.A. Labinger and M.E. Davis, *J. Catal.*, 1994, **149**, 195.
32 M. Neurock and L.E. Manzer, *J. Chem. Soc., Chem. Commun.*, 1996, 1133.
33 G. Sankar, F. Rey, J.M. Thomas, G.N. Greaves, A. Corma, B.R. Dobson and A.J. Dent, *J. Chem. Soc., Chem. Commun.*, 1994, 2279.
34 S. Bordiga, F. Boscherini, S. Coluccia, F. Genoni, C. Lamberti, G. Leofanti, L. Merchase, G. Petrini, G. Vlaic and A. Zecchina, *Catal. Lett.*, 1994, **26**, 195.
35 J.N. Armor, in 'Catalysis of Organic Reactions', Ed. J.R. Kosak, Dekker, New York, 1984, p. 409.
36 R. Roffia, G. Leofanti, A. Cesana, M. Mantegazza, M. Padovan, G. Petrini, S. Tonti and P. Gervasutti, in 'New Developments in Selective Oxidation', Ed. G. Centi, Elsevier, Amsterdam, 1990, p. 43.
37 F.G. Dwyer and E.E. Jenkins, US 3941871.
38 M. Constantini, J.L. Guth, A. Lopez and J.M. Popa, EP 466545.

39 N.K. Mal, R. Bhaumik, R. Kumar and A.V. Ramaswamy, *Catal. Lett.*, 1995, **33**, 387.
40 J.S. Reddy, S. Sivasanker and P. Ratnasamy, *J. Mol. Catal.*, 1992, **71**, 373.
41 K.R. Reddy, A.V. Ramaswamy and R. Ratnasamy, *J. Chem. Soc., Chem. Commun.*, 1992, 1613.
42 A.V. Ramaswamy and S. Sivasanker, *Catal. Lett.*, 1993, **22**, 239.
43 Mitsui Toatsu Chemicals, JP 0356439.
44 J.S.T. Mambrin, E.J.S. Vichi, H.S. Pastore, C.U. Davanzo, H. Vargas, E. Silva and O. Nakamura, *J. Chem. Soc., Chem. Commun.*, 1991, 922.
45 U. Cornaro, P. Jiru, Z. Tvaruzkova and K. Habersberger, in 'Zeolite Chemistry and Catalysis', Ed. P. Jacobs, Elsevier, Amsterdam, 1991, p. 165.
46 R.A. Sheldon, J.D. Chen, J. Dakka and E. Neelman, in 'New Developments in Selective Oxidation II', Eds. Corberan and S.V. Bellon, Elsevier Science, 1994, p. 515.
47 U. Romano, A. Esposita, F. Maspero, C. Neri and M. Clerici, *Chim. Ind. (Milan)*, 1990, **72**, 610.
48 S.T. Wilson, B.M. Lok, C.A. Messina, T.R. Cannon and E.M. Flanagan, *J. Am. Chem. Soc.*, 1982, **104**, 1146.
49 B.M. Lok, C.A. Messini, R.L. Patton, R.T. Gajek, T.R. Cannon and E.M. Flanagan, *J. Am. Chem. Soc.*, 1984, **106**, 6092.
50 G. Cainelli and G. Cardilo, in 'Chromium Oxidations in Organic Chemistry', Springer Verlag, Berlin, 1984.
51 J. Muzart, *Chem. Rev.*, 1992, **92**, 113.
52 Mobil Oil Co., US 4870192.
53 R. Neumann, M. Chava and M. Levin, *J. Chem. Soc., Chem. Commun.*, 1993, 1685.
54 J. Livage, M. Henry and C. Sanchez, *Prog. Solid State Chem.*, 1988, **18**, 259.
55 A. Zecchina, G. Spoto, S. Bordiga, A. Ferrero, G. Petrini, G. Leofani and M. Padovan, *Stud. Surf. Sci. Catal.*, 1991, **69**, 251.
56 J.R. Briggs, A.M. Harrison and J.M. Roberts Jr., *Polyhedron*, 1985, **5**, 281.
57 O. Bortolini, F. di Furia, G. Modena and R. Seraglia, *J. Org. Chem.*, 1985, **50**, 2688.
58 E. Matveev and I.V. Kozhevnikov, *Kinet. Katal.*, 1980, **21**, 1189.
59 Y. Matoba, H. Inoue, Y. Ishii, S. Hamanka and H. Ogawa, *Rep. Asahi Glass Found. Ind. Technol.*, 1987, **30**, 439.
60 Solvay Interox Ltd., WO 93/003388.
61 M. Ogawa, H. Tanaka and Y. Ishii, *Sekiyu Gakkaishi*, 1993, **36**, 27.
62 S.W. Brown, A. Johnstone, C.W. Jones, A.M. Lee, S.C. Oakes and S.L. Wilson, *Recl. Trav. Chim. Pays-Bas*, 1996, **156**, 147.
63 A. Mangia, *Synthesis*, 1978, 361.
64 L. Salles, C. Aubrey, R. Thouvenot, F. Robert, C. Doremieux-Morin, G. Chottard, M. Ledon, Y. Jeannin and J.M. Bregeaut, *Inorg. Chem.*, 1994, **33**, 871.
65 R. Neumann and H. Miller, *J. Chem. Soc., Chem. Commun.*, 1995, 2277.
66 G. Gelbard, F. Breton, M.T. Charreyre and D. Dong, *Makromol. Chem., Macromol. Symp.*, 1992, **59**, 353.
67 M. Ogawa, H. Tanaka and Y. Ishii, *Sekiyu Gakkaishi*, 1993, **36**, 27.
68 N. Heron, G.D. Stuky and C.A. Tolman, *J. Chem. Soc., Chem. Commun.*, 1986, 1521.
69 Y. Kurusu and D.C. Neckers, *J. Org. Chem.*, 1991, **56**, 1981.
70 N. Herron and C.A. Tolman, *J. Am. Chem. Soc.*, 1987, **109**, 2837.
71 K.W. Jun, K.W. Lee, E.K. Shim and N.S. Cho, *Appl. Catal. A*, 1993, **96**, 269.
72 M. Barteri, M. Farinella and B. Pipisa, *Biopolymers*, 1977, **16**, 2569.
73 S. Bhaduri, A. Ghosh and H. Khwaya, *J. Chem. Soc., Dalton Trans.*, 1981, 447.
74 T. Szymanska-Buzar and J.J. Ziolkowski, *J. Mol. Catal.*, 1981, **11**, 371.
75 D.C. Sherrington, *Pure Appl. Chem.*, 1988, **60**, 401.
76 J.H. Fuhrop, S. Besecke, W. Vogt, J. Ernst and S. Subramanian, *J. Am. Chem. Soc.*, 1977, **178**, 1621.
77 T.G. Traylor, Y.S. Byun, P.S. Traylor, P. Battioni and D. Mansuy, *J. Am. Chem. Soc.*, 1991, **113**, 7821.
78 L. Barley, J.P. Lallier, P. Battioni and D. Mansuy, *New J. Chem.*, 1992, **16**, 71.

79 D.R. Leonard and J.R. Lindsay-Smith, *J. Chem. Soc., Perkin Trans. 2*, 1990, 1917.
80 S. Campestrini and B. Meunier, *Inorg. Chem.*, 1992, **31**, 1999.
81 P. Battioni, J.P. Lallier, L. Barloy and D. Mansuy, *J. Chem. Soc., Chem. Commun.*, 1989, 1149.
82 S. Fukuzumi, S. Mochizuki and T. Tanaka, *Isr. J. Chem.*, 1987, **28**, 29.
83 F. Pautet and M. Daudon, *Tetrahedron Lett.*, 1991, **32**, 1457.
84 T. Tatsumi, M. Nakamura and M. Tominga, *Chem. Lett.*, 1989, 419.
85 C. Gilmartin and J.R. Lindsay-Smith, *J. Chem. Soc., Perkin Trans. 2*, 1995, 243.
86 J. Nener, M.Sc. Thesis, University of York, 1994.
87 K. Marujama, H. Tamiaki and S. Kawabata, *J. Chem. Soc., Perkin Trans. 2*, 1986, **2**, 543.
88 D. Wohle and J. Gitzel, *Makromol. Chem. Lett., Rapid Commun.*, 1988, **9**, 229.
89 L.D. Rollman, *J. Am. Chem. Soc.*, 1975, **97**, 2133.
90 J.A.S. Razenberg, A.W. van der Made, J.W.H. Smeets and R.J.M. Nolte, *J. Mol. Catal.*, 1985, **31**, 271.
91 H.S. Hilal, C. Kim, M.L. Sito and A.F. Schreiner, *J. Mol. Catal.*, 1991, **64**, 133.
92 J.T. Groves, M. Bonchio, T. Carofiglio and K.V. Shalyaev, *J. Am. Chem. Soc.*, 1996, **118**, 8961.

CHAPTER 5

Environmental Application of Hydrogen Peroxide

1 Introduction

The unique properties of peroxygens make them particularly suited to providing environmental protection in all areas of everyday life, including industry where noxious and environmentally hazardous effluents are routinely produced. From pollution prevention to abatement and cure, peroxygens can provide an effective solution to environmental problems. This chapter provides an introduction to the use of peroxygens in controlling pollution problems arising in solid, aqueous and gaseous wastes. The types of peroxygens which have been used for environmental applications include hydrogen peroxide, peracetic acid, sodium percarbonate, calcium peroxide and Caro's acid. Table 5.1 illustrates the range of industries and pollutants in which peroxygens can be applied for noxious control. Peroxygens offer environmentally acceptable solutions to pollution control through chemical oxidations, oxygen supply and disinfection.

Peroxygen species, with their active oxygen content, are powerful oxidizing agents capable of oxidatively destroying many of the most commonly occurring pollutants in solid, aqueous and gaseous wastes to give innocuous or easily biodegradable products. The oxidative power of hydrogen peroxide in particular can be further increased by application in conjunction with specific metal catalysts, UV irradiation or ozonation. Such conditions are known as advanced oxidation processes (AOPs), which are very powerful and can be used to reduce toxicity or chemical oxygen demand (COD) arising from recalcitrant organic species. Thus through oxidation, peroxygens and AOPs can be used to improve the biodegradability of a waste stream.

The active oxygen content of peroxygens can also be utilized as a source of oxygen to enhance microbial degradation of polluting species. Hydrogen peroxide and other peroxygens are used in the treatment of overloaded biological treatment units, in storm water discharges and in remediation of contaminated soil. Hydrogen peroxide is applied to foul sewers to prevent the release of toxic hydrogen sulfide gas by the maintenance of aerobic conditions.

Table 5.1 *Range of industries employing peroxygens for pollutant control*

Pollutant	Agriculture	Animal processing	Chemicals	Food	Horticulture	Metals processing	Mining	Municipal waste	Pharmaceutical	Pulp and paper	Refinery	Soil remediation	Tanneries	Textile	Timber	Water (potable)	Water (waste)
Anaerobicity	θ◆	◆	θ◆					θ	θ◆	θ◆	θ◆	θ◆	θ◆	θ◆			θ◆
Chlorine	◆	θ◆	θ◆			θ			θ			θ◆	θ	θ		θ	
Chromate			θ◆			θ			θ▽	θ▽	θ▽	θ◆	θ◆▽	θ▽			θ▽
COD			θ▽	▽					θ▽	θ▽			θ▽	θ▽			▽
Colour			▽						θ▽	θ▽	θ		▽	▽			▽
Cyanide			θ◆	θ		θ◆	θ		θ◆		θ						θ◆
Formaldehyde			θ						◆θ						θ		
Hydrocarbons			▽			▽			θ▽		▽	▽					
Hydrogen sulfide		◆	θ						θ◆	θ◆	θ◆		θ◆	θ◆			θ◆
Hypochlorite			θ						θ▽					θ			
Microbial	θ◆		θ	◆	◆	θ	◆	◆	θ◆	◆	θ		◆			θ	θ◆
Nitrogen oxides									θ◆								
Nitrite			θ	θ		θ			θ◆		θ		θ				
Odour	θ◆	θ◆	θ◆	θ				θ◆	θ◆	θ◆	θ◆		θ◆			θ	θ◆
Organo-cyanides			θ			▽			θ◆								
Organo-halogens			θ▽						θ▽	▽		▽			▽		
Phenolics			θ◆▽				θ		θ◆▽	θ	θ▽	θ◆			◆		
Sulfite			θ◆	θ		θ	θ		θ◆		θ◆	▽		θ			
Sulfur dioxide			θ	θ		θ			θ		θ	▽		θ			
Thiocyanate			θ◆				θ		θ◆		θ◆			θ			
Thiols		▽							θ▽		θ						
Thiosulfate			θ						θ▽								

θ = Hydrogen peroxide. ◆ = Other peroxygens. ▽ = Advanced oxidation processes.

The mild antimicrobial activity of hydrogen peroxide is applied in curing the problem of sludge bulking and controlling algae in lakes and reservoirs. Percarboxylic acid compounds are extremely powerful biocides used for disinfection of raw or partially treated sewage effluents and sewage sludge. Percarboxylic acids can also be used for pre- and post-treatment of industrial cooling and waste water and for microbial control of industrial circuit and process waters.

The following areas will be covered in this chapter:

- Advanced oxidation processes (AOPs)
- Fenton's treatment
- Control of cyanide and NO_x effluent streams
- Control of reduced sulfur
- Contaminated site remediation
- Waste water treatment

2 Advanced Oxidation Processes

AOPs, which involve the *in situ* generation of highly potent chemical oxidants such as the hydroxyl radical (OH$^\bullet$), have emerged as an important class of technologies for accelerating the oxidation and hence destruction of a wide range of organic contaminants in pollution solids, water and air.[1] The processes include UV irradiation [either direct irradiation of the contaminant or photolytic oxidation mediated by hydrogen peroxide (UV/H_2O_2) and/or ozone (UV/O_3)],[2] heterogeneous photocatalysis using semiconductor catalysts (UV/TiO_2),[3] electron beam irradiation,[4] X-rays, γ-ray radiolysis, non-thermal electrical discharge,[5] supercritical water[6] and ultrasonic irradiation.[7] These technologies involve widely different methods of activation as well as oxidant generation, and can potentially utilize a number of different mechanisms for organic destruction. All of these processes, however, are electrically driven, and share the common denominator for hydroxyl radical chemistry (at least in part). Of the above AOPs we shall concentrate on the ones which employ peroxygens as an integral part of their make-up.

Bolton and co-workers have derived rate expressions for the majority of AOPs.[8] Most AOPs can be modelled by the following simple equations:

$$A \rightarrow Z^\bullet \qquad \qquad Rate_1 = cP/V$$
$$Z^\bullet + C \rightarrow Products \qquad Rate_2 = k_c[Z^\bullet][C]$$
$$Z^\bullet + S_1 \rightarrow Products \qquad Rate_3 = k_{S1}[Z^\bullet][S_1]$$
$$Z^\bullet + S_2 \rightarrow Products \qquad Rate_{3'} = k_{S2}[Z^\bullet][S_2]$$
$$Z^\bullet + S_i \rightarrow Products \qquad Rate_{3i} = k_{Si}[Z^\bullet][S_i]$$

where

c = constant which depends on the system and the technology;
P = electrical power input (kW);
V = treated volume (dm^3);

A = initiating compound, *e.g.* hydrogen peroxide;
Z = reactive intermediate (OH$^{\bullet}$);
C = organic contaminant;
S_1, S_2, S_i = scavengers for the Z$^{\bullet}$ radical;
Rate$_1$ = rate of formation of Z$^{\bullet}$;
Rate$_2$ = rate of reaction of Z$^{\bullet}$ with C;
Rate$_3$, Rate$_{3'}$, Rate$_{3i}$ = Rate of reaction of Z$^{\bullet}$ with scavengers.

The application of steady analysis by Bolton and co-workers yielded an overall rate law of:

$$\text{Rate} = \frac{cPk_c[C]/V}{k_c[C] + \sum k_{Si}[S_i]}$$

The uncomplicated mechanism means that simple overall kinetics are observed, which are either zero or first order in contaminant. For example, if [C] is high then zero-order kinetics are observed, whereas if [C] is low, then first-order kinetics are followed. The demarcation between high and low concentration of pollutant is often very dependent upon the AOP employed.

The work described in Bolton's paper was seminal because it allowed widely different AOP technologies to be compared by comparison of their electrical energy consumption within the two rate regimes. This work allows potential users to have a standardized base for comparison of AOPs. Of course there are economic factors (chemicals operation/maintenance, capital, *etc.*) that go in to a cost analysis, however the Bolton paper allows a more complete picture to be drawn.

Several different peroxygen-based systems exist, all of which are compatible with biological treatment. The peroxygen treatment can be used up- or down-stream of the biotreatment unit, either as a pre-treatment to reduce toxicity and improve biodegradability or as a post/polishing treatment to remove remaining biorecalcitrant species from the effluent. All the peroxygen-based treatments can also be used as stand-alone processes for the total treatment of a waste stream. The most common AOPs are:

- Fenton's treatment (see Section 3)
- UV/H$_2$O$_2$
- O$_3$/H$_2$O$_2$
- UV/O$_3$/H$_2$O$_2$ combinations

Hydroxyl radicals are produced from hydrogen peroxide via different pathways and to different efficiencies depending on the nature of the catalyst (AOP system) involved. Figure 5.1 demonstrates hydroxyl radical production efficiencies for the three most commonly applied technologies: Fenton's treatment, UV/H$_2$O$_2$ and O$_3$/H$_2$O$_2$.

Whilst all of the above systems are capable of breaking down biorecalcitrant

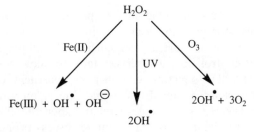

Figure 5.1 *Hydroxyl radical production via Fenton's treatment, UV/hydrogen peroxide and ozone/hydrogen peroxide.*

species to innocuous products, the best system for a particular application will be dependent upon several factors including:

- nature of the polluted medium (solid, aqueous or gaseous);
- total load of oxidizable species;
- existing and target pollutant levels;
- availability of secondary treatment or pre-treatments, *e.g.* biotreatment;
- effluent characteristics, including flow rate, flow regularity, temperature and pH;
- discharge constraints.

The number of treatment options available ensures that almost all toxic, environmentally persistent or otherwise undesirable pollutant species can be destroyed over a wide range of contaminant concentrations and pH ranges in solid, liquid and gaseous phases.

UV/H_2O_2 involves the use of ultraviolet wavelength energy, emitted from specially developed UV lamps, to catalyse production of high powered hydroxyl radicals from cleavage of the hydrogen peroxide molecule. Oxidation of organic species can be via formation of different intermediates to those involved in Fenton's treatment (see Section 3), a result of differences in the reaction pathways, *e.g.* direct photocatalysis of organic species by short wavelength UV light. Unlike Fenton's treatment, which is normally carried out under acidic conditions, UV/H_2O_2 treatment can be carried out over a wide pH range. The optimum pH for the treatment will, however, be dependent upon the nature of organic and inorganic compounds present.

The quantity of hydrogen peroxide needed to achieve the desired pollutant removal and degree of oxidation is affected by the presence of other oxidizable organics, hence the hydrogen peroxide requirement is effluent specific. UV/H_2O_2 can be used to treat water-, air- and solid-borne pollutants. In the latter two cases, the medium is washed with water to abstract the pollutant and the wash waters are passed through the UV/H_2O_2. UV/H_2O_2 systems must be designed to suit the needs of a particular effluent and the specific requirements of each manufacturer. Systems can be designed to cope with both dilute and more concentrated waste waters (CODs of over several thousand) over a range of flow rates up to tens of cubic metres per hour. A typical UV/H_2O_2 treatment requires storage and design facilities for chemicals (hydrogen peroxide and alkalis for any pH adjustment), a UV lamp system, including power source and control box, a reservoir tank to allow constant flows of effluent through the UV reaction chamber at consistent quality and an upstream suspended solids removal unit (depending upon water quality). As the economics of UV/H_2O_2 systems are reliant upon the efficiency and the spectral output of the lamp, it is important to discuss proposed treatments with a UV system supplier.

Ozone reacts with hydrogen peroxide in solution to form highly oxidizing radicals and ionic species including hydroxyl radical. The exact nature of the reactions which occur and the levels and nature of the resultant intermediate oxidizing species formed, *e.g.* HO_2, O^{2-}, O_3, HO_3, HO_2^-, O_2^-, O_2, O_3 and $OH^{\bullet}$, are dependent upon the effluent characteristics (types and concentrations of organic and inorganic species, pH, temperature), the concentration of ozone in solution and the concentration ratio of hydrogen peroxide to ozone applied. The nature of the final oxidation products will also be effluent/condition dependent, hence even for the same pollutant, the optimum treatment conditions in terms of, for example, O_3 concentration, H_2O_2/O_3 mole ratio, *etc.*, will vary for each case.

The production of hydroxyl radicals is generally in the order of two radicals per three molecules of oxidant consumed. In terms of overall oxidant consumption per mole of hydroxyl radicals produced, ozone peroxide combinations are therefore not as efficient as UV/H_2O_2. Since O_3/H_2O_2 does not rely upon transmission of UV light through the effluent for hydrogen peroxide activation, it does have advantages over UV/H_2O_2 in applications where the effluent is particularly turbid and pre-filtration is not an option. O_3/H_2O_2 is normally applied to treatment of water-borne pollutants but can be used, like UV/H_2O_2, for treating soil- or air-borne pollutants, where the pollutants are first transferred to the aqueous phase. O_3/H_2O_2 is particularly well established in the USA as a technology for reducing contamination from organo-chlorines in potable water, and its popularity is growing throughout Europe as demand for high quality water increases. It has also been applied to treatment of turbid and highly coloured leachates from waste tips and other contaminated sites. O_3/H_2O_2 is most economical when applied to water with low pollutant and oxidizable substance loading and is particularly suited to coping with very high flow rates such as those encountered in potable water treatment works.

Ozone generation can be from air or liquid oxygen; the latter can be manufactured off-site or brought in as required. When air is used, it must first be thoroughly dried to allow efficient ozone production by the electrical discharge system. A typical O_3/H_2O_2 treatment requires storage and dosing facilities for chemicals (hydrogen peroxide, acids/alkalis required for pH adjustments, and liquid oxygen if required), the ozone manufacturing system, including power sourcing (possibly air dryers and control units), reaction chambers, an off-gas treatment unit for residual ozone destruction, reservoir tanks or lagoon areas to allow constant feed quality or recirculation facilities as necessary and an upstream suspended solids removal system, depending upon the water quality. The reactor chambers consist of contactor towers for the incorporation of ozone gas into the liquid phase. Ozone gas is introduced at the bottom of the tower via a sparge pipe; water flow may be co- or counter-current in relation to gas flow. Hydrogen peroxide is injected into the water flow via a static mixer as it enters the contactor. As the economics of O_3/H_2O_2 systems are reliant upon the efficiency of the ozone production unit and the gas/liquid mass transfer system, it is important to discuss proposed treatments directly with an ozone system supplier.

Combined treatments involving the use of O_3/H_2O_2 and UV radiation find applications similar to the well established parent technologies. There are, however, advantages to such systems, for example in the treatment of waste water with highly variable quality. The UV lamps which tend to be used in such systems are known as low pressure lamps and emit single wavelength radiation rather than the broad spectrum higher energy radiation emitted from medium pressure sources. In cases where the target pollutant oxidation is enhanced by exposure to short wavelength UV radiation, standard medium-pressure UV/H_2O_2 treatment is generally more efficient.

UV/H_2O_2 has been used in the remediation of ground water from several contaminated sites in North America and Europe. For example, a US food processing plant, whose well water contained trichloroethylene (TCE), installed a UV/H_2O_2 treatment process. TCE levels were reduced from around 4000 $\mu g/l$ to less than 0.83 $\mu g/l$ on a 14 m^3/h flow rate, using a H_2O_2 concentration of 50 ppm and a 30 kW UV lamp. O_3/H_2O_2 has been widely used for the removal of trace contaminants, *e.g.* solvents, pesticides, herbicides, from potable water. In 1991, the Los Angeles Department of Water and Power installed a full scale O_3/H_2O_2 treatment plant to reduce TCE and PCE (perchloroethylene) levels in water supplied to the city of Los Angeles.[9] In bench-scale tests, the combination of H_2O_2 and O_3 reduced by between 56% and 64% the amount of O_3 required to reduce the TCE and PCE levels by 95%. Similar results were found in scale-up studies.[10] Since the successful installation of the plant in Los Angeles, several other existing potable water ozonation plants in the USA and Europe have been converted to O_3/H_2O_2 facilities.

3 Fenton's Treatment

The technology behind Fenton's treatment dates back over a hundred years to 1894 when M.J.H. Fenton reported that ferrous ion promoted the oxidation of tartaric acid with aqueous hydrogen peroxide.[11] Ferrous-catalysed hydrogen peroxide at acidic pH has since come to be known as Fenton's reagent and is widely used in both oxidative treatment of industrial effluents and in the manufacture of several types of polymers and polyelectrolytes. Subsequently, it was shown that the hydroxyl radical is the oxidizing species in this system,[12] and is formed according to:

$$Fe(II) + H_2O_2 \rightarrow Fe(III) + OH^- + OH^\bullet$$

$k = 76.5 \ M^{-1} s^{-1}$.

The use of Fenton's reagent for waste water treatment is relatively new compared to its use in mechanistic investigations in organic chemistry.[13-15] Fenton's reagent is able to destroy phenols (Figure 5.2), chlorinated phenols and herbicides in water media, as well as reducing chemical oxygen demand in municipal waste.

The use of Fenton's reagent as an oxidant has been applied to the removal of soil contamination.[16,17] It has been shown that pentachlorophenol and tri-fluoralin are extensively degraded[18] while hexadecane and dieldrin are only partially transformed in a soil suspension at an acidic pH.[19]

The application of Fenton's reagent as an oxidant for waste water treatment is attractive due to the fact that iron is a highly abundant and non-toxic element,

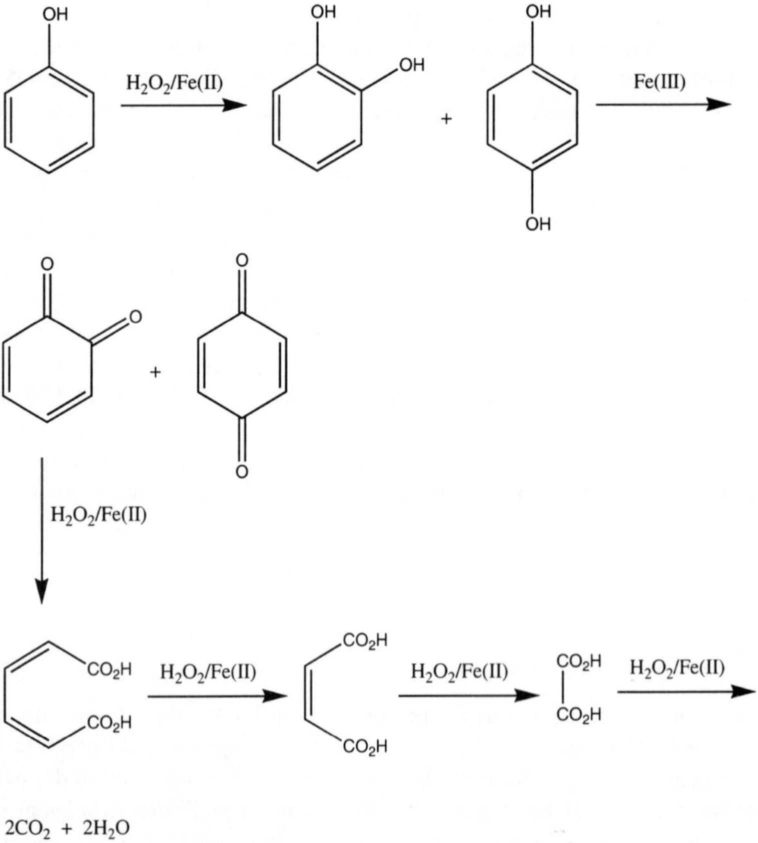

Figure 5.2 *Oxidation of phenol in the presence of Fenton's reagent.*

$$Fe(III) + H_2O_2 \rightleftharpoons \overset{\oplus}{H} + Fe(OOH)^{2\oplus}$$

$$Fe(OOH)^{2\oplus} \longrightarrow OH_2^\bullet + Fe^{2\oplus}$$

$$Fe^{2\oplus} + H_2O_2 \longrightarrow Fe^{3\oplus} + \overset{\bullet}{OH} + OH^\ominus$$

$$Fe^{2\oplus} + HO_2^\bullet \longrightarrow Fe^{3\oplus} + HO_2^\bullet$$

$$\overset{\bullet}{OH} + H_2O_2 \longrightarrow H_2O + HO_2^\bullet$$

Figure 5.3 *Fenton's mechanism with iron(II) salts.*

and hydrogen peroxide is easy to handle and breaks down to environmentally benign products. It does, however, require a stoichiometric amount of Fe(II). Hydrogen peroxide decomposes catalytically in the presence of Fe(II) and generates hydroxyl radicals in the process.[12] Thus Fe(II) in Fenton's reagent can be replaced by Fe(III).[20] The mechanism proposed for free Fe(II) ion involves the hydroxyl ($^\bullet$OH) and hydroperoxyl radicals (HO$_2$$^\bullet$) (Figure 5.3).

The rate constant for the reaction of Fe(II) with hydrogen peroxide is high and Fe(II) is oxidized to Fe(III) in a few seconds in the presence of excess hydrogen peroxide. For this reason, it is believed that the majority of the waste destruction catalysed by Fenton's reagent is simply a Fe(III)–H$_2$O$_2$-catalysed destruction processes.

The Fe(II)/Fe(III)–H$_2$O$_2$ system has its maximum catalytic activity at a pH of 2.8–3.0. Any increase or decrease in the pH sharply reduces the catalytic activity of the metal ion. At high pH, the ferric ion precipitates as ferric hydroxide, whilst at low pH, the complexation of Fe(III) with hydrogen peroxide is inhibited. To overcome this problem, Sun and co-workers have used Fe(III) chelates in place of Fe(II)/Fe(III).[21] Sun has shown that a variety of herbicides and pesticides can be transformed and practically mineralized by Fe(II) chelates at neutral pH.

The rate of degradation of organic pollutants with Fenton reagents is accelerated strongly by irradiation with UV-VIS radiation.[22] If desired, organic pollutants can be mineralized completely by a Fe(II)/Fe(III)/H$_2$O$_2$ system. The increased efficiency of Fenton's systems is believed to be attributed to the following:

- Photoreduction of ferric ion. Lunck and co-workers observed the enhanced rate of photo-oxidation of salicyclic acid by hydrogen peroxide in the presence of Fe(III) as well as the increased rate of photodecomposition of hydrogen peroxide in the presence of transition metal ions.[23] The ferrous ion reacts with hydrogen peroxide, generating a second hydroxyl radical and ferric ion, and the cycle continues.
- Efficient use of light quanta. The absorption spectrum of hydrogen peroxide does not extend beyond 300 nm and has a low molar absorption coefficient beyond 250 nm. The absorption spectrum of ferric ion,

however, extends to the neat UV-VIS region and has a relatively large molar absorption coefficient, thus making more efficient use of the lamp output when polychromatic light is employed. Consequently, photo-oxidation and mineralization can even proceed by irradiation with visible light.

- Photolysis of Fe(III) organo-intermediate chelates. The initial oxidation of organic pollutants generates oxygenated intermediates which can react with Fe(III) to form complexes. These complexes are also photo-reactive and produce CO_2, organic radicals and ferrous ions on irradiation. This results in an increased rate of destruction of organic pollutants as the reaction progresses.

The photo-reactivity of Fe(III) carboxylates has long been recognized and photo-decarboxylation[24] and deamination[25] of amino acid complexes have been known for some time. Ferrioxylate is the best known photo-active example of a Fe(III) polycarboxylate complex. Irradiation of ferrioxalate in acidic solution generates Fe(II) and CO_2 (Figure 5.4).

Safarzadeh-Amiri and co-workers have shown that hν/ferrioxalate/H_2O_2 is a highly efficient and powerful oxidant for organic pollutants in contaminated ground water.[26] The high efficiency of the process, as compared with other UV-driven processes, is attributed to ferrioxalate absorbing in the UV-VIS region over 250–480 nm (resulting in a more efficient use of the lamp output) and the quantum yield of Fe(II) formation being high and independent of irradiation wavelength in the range 250–480 nm. Thus the yield of the hydroxyl radical is expected to be high. The photo-oxidation of organic pollutants in water, mediated by iron and iron chelates, is initiated by the hydroxyl radical which, with dissolved oxygen, eventually converts organic pollutants to CO_2, H_2O and mineral acids. The detailed mechanism of the complex reaction pathways have, however, not yet been elucidated.

A typical Fenton's treatment installation requires facilities for storing and dosing chemicals (hydrogen peroxide, ferrous catalyst, acid and alkali for pH adjustment), a reaction tank, a settlement tank or centrifugal pump for solids removal and a UV-VIS lamp if photo-Fenton's is to be employed. The use of Fenton's and photo-Fenton's systems are likely to grow in use over the next decade as environmental legislation becomes more widespread, and as the systems become more efficient.

$$[Fe(C_2O_4)_3]^{3-} \xrightarrow{\ h\nu\ } [Fe(C_2O_4)_2]^{2-} + C_2O_4^{2-}$$

$$C_2O_4^{2-} + [Fe(C_2O_4)_3]^{3-} \longrightarrow [Fe(C_2O_4)_2]^{2-} + C_2O_4^{2-} + 2CO_2$$

$$C_2O_4^{2-} + O_2 \longrightarrow O_2^{-\cdot} + 2CO_2$$

Figure 5.4 *Breakdown of ferrioxalate species under irradiation.*

$$CN^{\ominus} + H_2O_2 \longrightarrow CNO^{\ominus} + H_2O$$

$$CNO^{\ominus} + 2H_2O \longrightarrow NH_3 + HCO_3^{\ominus}$$

Figure 5.5 *Oxidative destruction of cyanide wastes in the presence of hydrogen peroxide.*

4 Cyanide and NO$_x$ Control

Peroxygens are used widely throughout industry for the safe and effective treatment of cyanide in process and waste waters. Under alkaline conditions, cyanide is efficiently oxidized to the much less toxic cyanate by peroxygens such as hydrogen peroxide, Caro's acid and sodium percarbonate.[27] The cyanate slowly hydrolyses to ammonia and bicarbonate (Figure 5.5).

Peroxygen-treated water can usually be safely discharged following ammonia removal, if necessary via ammonia recovery or biological nitrification. The use of hydrogen peroxide for the treatment of highly concentrated cyanide wastes is very effective (several thousand ppm to below 100 ppm). The rate of cyanide removal with hydrogen peroxide varies from hours down to minutes depending on the nature of the cyanide (whether complexed or free), the treatment conditions (pH and temperature) and the presence of other components in the waste water. In most cases, reaction rate and extent of cyanide removal can be enhanced by the addition of soluble metal catalysts, such as copper (5–10 ppm Cu). It is advantageous to work at pH 9–10, just above the pK_a of HCN. This is in contrast to chloride oxidation which, although more rapid, must be run at pH 11–12 in order to prevent the build up of and possible release of cyanogen chloride.

Caro's acid is an alternative to metal-catalysed hydrogen peroxide for the treatment of low level cyanide waste or metal-complexed cyanides. As peroxymonosulfuric acid is more stable than hydrogen peroxide in the presence of certain transition metals, and under conditions of elevated temperature, it is the preferred peroxygen for the treatment of metal plating waste waters. To prevent emission of toxic hydrogen cyanide, Caro's acid treatment is carried out in conjunction with addition of alkali to maintain a high pH. Caro's acid is usually prepared on-site as required. Sodium percarbonate can also be employed in cyanide detoxification. This is particularly useful for emergency treatment, *e.g.* for spillage in natural waters when solid oxidants can provide storage and handling advantages over liquid alternatives.

Cyanides occur widely in industrial effluents, both from organic processes, such as the cyanohydrin route to methyl methacrylate[28] and the manufacture of acrylonitrile,[29] and from inorganic processes, such as gold mining, electroplating, case hardening, and coking.[30] The coking effluents contain a cocktail of noxious chemicals including phenols, sulfides and thiocyanates. Thiocyanates under alkaline conditions are oxidized to cyanate and sulfate by hydrogen peroxide.

$$2CNS^- + 5H_2O_2 \rightarrow 2CNO^- + 2SO_4^{2-} + 10H^+$$

In many cases, particularly in effluent from hardening shops, electro-plating plants, *etc.*, nitrite and cyanide co-exist in effluents. Nitrite is an undesirable component of waste water since it is an easily oxidizable chemical and adds to the biological oxygen demand (BOD) of the waste water, causing oxygen depletion downstream of the discharge. In most circumstances, nitrite discharge above 20 ppm is prohibited. Peroxygens readily oxidize nitrite to nitrate under acidic conditions, *i.e* pH < 5, for example, with hydrogen peroxide:

$$HNO_2 + H_2O_2 \rightarrow HNO_3 + H_2O$$

Under mildly acidic conditions, *e.g.* pH 3–4, the reaction can take minutes to hours to complete, whereas under highly acidic conditions (pH < 1), the reaction is complete in seconds. Where the effluent is to be treated under conditions of elevated temperature or the effluent contains significant levels of transition metals, Caro's acid is the preferred oxidant.

In the case of a mixed cyanide/nitrite waste stream, it is advisable to treat the effluent first at alkaline pH to remove cyanide before reducing the pH to the acid range for nitrite treatment. This reduces the risk of toxic hydrogen cyanide laden waste streams.

Nitrogen oxides (NO_x) can be treated with hydrogen peroxide, and options exist for both its cure and prevention. Gas streams containing NO_x arise from organic nitration processes and from metal pickling, notably stainless steel, copper, and brass. Such streams can be treated using alkali, but this has three drawbacks. Firstly, the nitric oxide component is poorly absorbed compared to nitrogen dioxide. Secondly, the alkali can absorb the carbon dioxide, thus reducing the amount of alkali available for reaction. Finally, the mixed sodium nitrite/nitrate/carbonate solution produced may itself require treatment for disposal as an effluent if recovery of the solid salts are not appropriate to the manufacturing operation. A cleaner alternative is the use of hydrogen peroxide in nitric acid as a scrubbing solution. Not only does this absorb nitric oxide more efficiently, but the sole product which is recovered, nitric acid, is usually recycled (Figure 5.6).[31] The scrubbing reactions have been successfully modelled using computer software to allow scrubber design to be readily and reliably specified.

The prevention of NO_x can sometimes be achieved by the addition of hydrogen peroxide directly to the process. Clearly, such addition is not always possible, especially in organic reactions, and safety rules must be rigorously observed (see Chapter 1). However, a salient example of its success is in stainless steel pickling. This is carried out using mixtures of nitric and hydrofluoric acids, where the nitric acid is reduced to NO_x as the steel surface is removed. Direct injection of hydrogen peroxide into the pickling bath oxidizes the NO_x in solution, thus preventing gaseous emissions and substantially reducing the net consumption rate of the nitric acid.[32]

$$2NO_2(g) \rightleftharpoons N_2O_4(g) \rightleftharpoons N_2O_4(aq.)$$

$$N_2O_4(aq.) + H_2O \longrightarrow HNO_2 + HNO_3$$

$$NO + NO_2 \rightleftharpoons N_2O_3(g) \rightleftharpoons N_2O_3(aq.)$$

$$N_2O_3(aq.) + H_2O \longrightarrow 2HNO_2$$

$$NO(g) \rightleftharpoons NO(aq.)$$

$$NO(aq.) + H_2O_2 \rightleftharpoons NO_2 + H_2O$$

$$3NO_2 + H_2O \longrightarrow 2HNO_3 + NO$$

$$HNO_2 + H_2O_2 \longrightarrow HNO_3 + H_2O$$

Figure 5.6 *Reactions taking place during NO_x scrubbing.*

5 Control of Reduced Sulfur Species

In solution, sulfides and polysulfides exist in an equilibrium between their basic and protonated forms. Since the equilibrium position is dependent upon the pH of the solution (Figure 5.7), this nature of the reaction with hydrogen peroxide is also influenced by pH.[33] Under acidic to neutral conditions, sulfidic species are oxidized predominantly to sulfates. The key reactions that occur are outlined in Figure 5.8.

The pH at which the oxidation occurs therefore affects the quantity of hydrogen peroxide required. For example, under neutral conditions a mass ratio of between 1.1:1 and 1.3:1 (H_2O_2/S) is required, whereas, at alkaline pH this ratio is increased to 4.25:1. Polysulfides react with hydrogen peroxide in a similar way to sulfides, *i.e.* forming sulfate or elemental sulfur, depending on the pH of the oxidation.

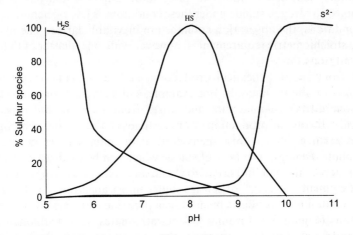

Figure 5.7 *Effect of pH on hydrogen sulfide equilibrium.*

Acid pH $H_2S + H_2O_2 \longrightarrow S° + 2H_2O$

Neutral pH $\overset{\oplus}{H} + \overset{\ominus}{HS} + H_2O_2 \longrightarrow S° + 2H_2O$

Alkaline pH $\overset{\ominus}{S^2} + 4H_2O_2 \longrightarrow \overset{\ominus}{SO_4} + 4H_2O$

Figure 5.8 *Reactions of hydrogen sulfide with hydrogen peroxide at various pH values.*

Absorption and oxidation in alkaline hydrogen peroxide is an efficient and cost-effective way of removing hydrogen sulfide down to very low levels from gas streams. The hydrogen sulfide is removed from the gaseous phase into the liquid phase by dissolution in alkali at pH 10–12, and is then oxidized to the sulfate form by hydrogen peroxide. The pH operating range is narrow due to the necessity to ensure fast absorption of the hydrogen sulfide by acid–base neutralization, whilst keeping the catalysed decomposition of the hydrogen peroxide to a minimum. The presence of hydrogen peroxide ensures that the dynamic equilibrium established between the sulfide in the gas and the liquid phases is moved towards the latter, enabling much lower gaseous sulfide levels to be achieved compared with using the same quantities of alkali alone. The sulfate produced can safely be discharged as it is stable and has no oxygen demand.

In cases where the gas to be treated contains sufficiently high levels of a flammable component, *e.g.* a digester gas (which can contain high levels of methane) which can cause hazardous operating problems, particularly if oxygenated, a two-stage process can be used to eliminate the risk. Here, the hydrogen sulfide is absorbed into the alkali alone and then oxidized in a remote reactor away from the flammable gas. Occasionally the gas may contain elevated levels of carbon dioxide or other acidic species which consume alkali and thus increase the treatment costs. Precipitation of salts may also be a problem in such circumstances. In these cases, absorption and oxidation of sulfide under neutral conditions to give sulfur can be employed. Since the absorption of hydrogen sulfide under these conditions is less efficient than under alkali conditions, the absorption columns are inevitably larger. The hydrogen peroxide stoichiometric requirement is, however, only one quarter of that under alkali conditions.

Apart from specific chemical production, sulfide effluents arise from the processing of natural materials. For example, sulfide is used for the de-hairing of hides in leather manufacture and arises from the xanthate process for cellophane or rayon production from cellulose. Edible oil refining with activated earth gives rise to a spent earth containing sulfides among other contaminants. Biological sources of sulfide which can be oxidized by hydrogen peroxide include tip (landfill) leachate,[34] animal carcass rendering,[35] municipal effluent (*e.g.* rising drains),[36] or sewage sludge under anaerobic conditions.[37]

The oxidation of thiosulfates by hydrogen peroxide proceeds through a series of reactions (Figure 5.9) forming first tetrathionates, then trithionates, then sulfites, and finally sulfates.

$$2S_2O_3^{2\ominus} + H_2O_2 \longrightarrow S_4O_6^{2\ominus} + 2OH^{\ominus}$$

$$S_4O_6^{2\ominus} + 3H_2O_2 \longrightarrow S_3O_6^{2\ominus} + SO_4^{2\ominus} + 2H_2O + 2H^{\oplus}$$

$$S_3O_6^{2\ominus} + H_2O_2 + H_2O \longrightarrow 3SO_3^{2\ominus} + 4H^{\oplus}$$

$$SO_3^{2\ominus} + H_2O_2 \longrightarrow SO_4^{2\ominus} + H_2O$$

Figure 5.9 *Oxidation of thiosulfates by hydrogen peroxide.*

In direct oxidation with hydrogen peroxide, the extent of oxidation is pH dependent. The quantity of hydrogen peroxide consumed, and the extent to which the COD of the waste water is lowered, is therefore also pH dependent. Under acidic to neutral oxidation with hydrogen peroxide, only the tetrathionate stage is reached, whereas under alkaline conditions, oxidation proceeds to the sulfate. If it is desired, metal catalysts can be applied with hydrogen peroxide under acid to neutral conditions to allow complete oxidation to sulfate. Most thiosulfate waste streams from refinery sources are alkaline. In such systems, the reaction would normally proceed through to sulfate requiring a minimum hydrogen peroxide dose of 1.21:1 mass ratio for $H_2O_2/S_2O_3^{2-}$. Copper salts have been employed with hydrogen peroxide for the removal of thiosulfate at an enhanced rate with a reduced hydrogen peroxide requirement.[38] Polythionates are intermediate species in the oxidation of thiosulfate, *e.g.* trithionate. As indicated above, oxidation with hydrogen peroxide proceeds rapidly to sulfate under alkaline conditions. However, under acidic to neutral conditions, oxidation only occurs in the presence of certain metal catalysts. Regardless of pH, sulfites react with hydrogen peroxide to form sulfates. The reaction is very fast, and requires no catalyst.

Thiols (or mercaptans) are the sulfur analogues of alcohols, containing an SH group. They are toxic, volatile, and flammable, and have a disagreeable odour that increases with relative molecular mass. They react with hydrogen peroxide in alkaline conditions to form disulfides. Disulfides, particularly high relative molecular mass species, are considerably less soluble in water than their parent thiols, and may separate from treated liquor as an oily layer. In some cases, the oil layer formed may be separated via skimming and disposed of separately. For safety reasons, it is often preferable to prevent formation of this second phase. In some cases, a soluble metal catalyst can be added,[39] which enhances oxidation through to the water soluble sulfonic acid (Figure 5.10).

$$2RSH + H_2O_2 \longrightarrow RSSR + H_2O$$

$$2OH^{\ominus} + RSSR + 5H_2O_2 \xrightarrow{M^{n\oplus}} 2RSO_3^{\ominus} + 6H_2O$$

$$M^{n\oplus} = Fe(II), Cu(II).$$

Figure 5.10 *Oxidation of thiols to sulfonic acid salts using metal activated hydrogen peroxide.*

Using a molar ratio of 5:1 peroxide to pollutant is generally enough to control toxic odours. The soluble metal catalyst can be copper or iron used in conjunction with a chelating agent to prevent the catalyst precipitating out of solution. A tungstate catalyst with hydrogen peroxide at alkaline pH has been employed for thiol oxidation without the necessity of employing a chelating agent.[40] Dialkyl sulfides can also be oxidized with hydrogen peroxide and an iron catalyst at pH 2. Offensive smelling dialkyl sulfides are deodorized to their corresponding practically odourless sulfoxides (R_2SO) and sulfones (R_2SO_2). As with hydrogen sulfide, volatile organic sulfides can be removed from gas streams by absorption in alkali and concurrent or subsequent oxidation with hydrogen peroxide, possibly using a homogeneous catalyst to increase the oxidation rate where necessary.

6 Contaminated Site Remediation

Remediation of contaminated sites and groundwater requires technologies which can safely destroy persistent and problem pollutants in an environmentally and economically acceptable way. Peroxygen-based technologies can very often answer this need, providing highly effective solutions to pollution problems. In contaminated land remediation, peroxygens are applied as both chemical oxidants, to break down toxic, hazardous or ecologically persistent pollutants to biodegradable components, and as a source of molecular oxygen, to enhance the natural aerobic break down of pollutants by micro-organisms in the soil.

A number of treatment technologies are applied in contaminated site remediation:

- Pump and treat
- Direct oxidation *in situ* or *ex situ*
- Biological treatment *in situ* or *ex situ*

In pump and treat technology, contaminated ground waters or artificially introduced wash waters are pumped to the surface for treatment. The surface treatment system can consist of a variety of technologies, including bio-reactors, filtration and oxidation methodologies. The clean ground/wash water is then re-introduced into the soil. The transfer of the pollutant from the soil matrix to the surface treatment system, and the time required for site remediation is therefore governed by the solubility of the pollutant in the ground/wash water. Speed of remediation can be improved in some cases by the use of surfactants to improve the pollutant mass-transfer rate. Hydrogen peroxide is used in a number of pump and treat systems, as a chemical or a supplier of molecular oxygen, for the surface treatment system.

Direct oxidation of organic contaminants in the soil can be performed either *in situ* or *ex situ* with hydrogen peroxide alone or with the addition of iron salts to increase the oxidation power of the treatment (Fenton's reagent). Direct oxidation is normally applied in order to partially oxidize the contaminants present, thus improving their natural biodegredation in the soil. Once this is accomplished, *in situ* bioremediation occurs at a greatly enhanced rate and reduced cost. Direct oxidation can allow for shorter treatment times than pump and treat or simple bioremediation treatments.

Biological treatment can take a number of different forms including land-forming, pump and treat bioreactors, and *in situ* bioremediation. In order for bioremediation to be effective, the contaminant must be biodegradable. Equally important is the fact that the levels of contaminants should not be so high as to be toxic to the micro-organisms. If toxicity exists, direct oxidation may be used to reduce the toxicity prior to biological treatment. Both hydrogen peroxide as a liquid peroxygen, and calcium peroxide as a solid peroxygen, are applied to enhance biological treatment via the supply of oxygen to aerobic micro-organisms.

Landforming consists of ploughing the appropriate micro-organisms, nutrients, and sometimes additional waters into the soil to form a compost.[41] The mixture provides the appropriate environment for the micro-organisms to degrade the contaminants. A number of factors impact on the success of this type of treatment including type and extent of contamination, moisture level, pH of the soil, and nitrogen and phosphate content. Weekly tilling or disking can be used to introduce oxygen into the system, although calcium peroxide can be used to replace the need for tilling or disking. The calcium peroxide works by slowly releasing oxygen into the composting soil.

Bioreactor systems, used in pump and treat methods, can experience limitations in oxygen transfer which can slow the rate of degradation of pollutants. Hydrogen peroxide or sodium percarbonate can be applied as a supplemental oxygen source for bioreactor systems, improving the bioreactor efficiency when dealing with situations of limited oxygen supply.

In nature, bioremediation occurs slowly because of the low population of micro-organisms with degradation ability. Degradation is also hindered because of environmental conditions such as nutrient levels or available oxygen. For bioremediation to be commercially viable, site managers must stimulate the natural biodegradation of hazardous compounds to yield practical remediation rates. By assisting nature through the supply of nutrients and hydrogen peroxide as an oxygen source, the remediation time frame can be drastically shortened. *In situ* bioremediation offers the advantage of being used where other technologies will not work, such as at sites that cannot be evacuated. *In situ* treatment avoids the cost of excavation as well as freight for off-site treatment. An added advantage is that the soil and ground water can be treated in a one-step process with minimal equipment. Additionally, the treatment can easily follow the contamination plume in the groundwater. For *in situ* bio-remediation to be effective, the components must also degrade readily. Several factors are important, including contaminant characteristics, hydro-geology, soil microbiology, and soil and groundwater chemistry.[42] Where contaminants are not readily biodegradable, direct oxidation may be useful in reducing the toxicity of various contaminants prior to biodegradation.

There are several examples of soil remediation technologies which have been applied to real situations, some of which are discussed below.

Creosote and pentachlorophenol contamination. The Champion Superfund site in Montana, USA, chose *in situ* bioremediation to treat the upper aquifer. Hydrogen peroxide was used to directly oxidize contaminants to more bio-degradable components and to improve the dissolved oxygen levels resulting in a 50% saving in clean-up time.[43,44] A pilot plant window containing creosote and copper, chromium and arsenic contaminated wood preservative was composted at Ilmayoki, Finland.[45] Hydrogen peroxide was employed, and the level of total hydrocarbon pollutant was reduced by 76% from 10 000 ppm to 2000 ppm in 163 days.

Treatment of an underground gasoline pipeline spill in the USA. A collection system recovered most of the gasoline from an area of impermeable soil on top of fractured limestone. Hydrogen peroxide assisted bioremediation was chosen to treat the remainder of the gasoline to improve the rate of remediation. Treatment of a gasoline station in the Netherlands was also successfully achieved with minimal disruption to the activities of the filling station.[46]

Treatment of chlorinated solvent contamination. An industrial firm in the USA, treating groundwater contaminated with trichloroethene, vinyl chloride and other solvents, was concerned with emissions from its current process. A UV/H_2O_2 system was installed which made the existing process redundant.

7 Waste Water Treatment

Peroxygens are used in a variety of applications in waste water treatment:

- Industrial effluent pre-treatment
- Odour and corrosion control
- Enhancement of biological treatment and prevention of biological upsets
- Sludge bulking control
- Tertiary effluent treatment
- Storm water overflows and emergency treatments for receiving waters

Oxidative pre-treatment of industrial effluents with hydrogen peroxide can reduce toxicity and COD, thereby improving the biodegradability of waste water. Hydrogen peroxide reacts with many commonly occurring inorganic and organic pollutants, *e.g.* chlorine, cyanide, formaldehyde, nitrite, sulfide, sulfite and thiosulfate. Such pollutants are oxidized to less noxious products, thus reducing their contribution to the oxygen demand of the effluent (Table 5.2).

Applied in conjunction with a catalyst, hydrogen peroxide will also oxidize a range of the more robust organic pollutants, such as phenols, polyaromatics, halogenated aliphatics and alkenes. Powerful oxidative hydrogen peroxide

Table 5.2 *Peroxygen treatments and products of treatment for commonly occurring pollutants*

Pollutant or condition	Peroxygen treatment	Products
Anaerobiosis	Hydrogen peroxide	Dissolved oxygen
Chlorine	Alkaline hydrogen peroxide	Chloride
COD from organic species, including dye colour, organohalogens, organosulfur, phenolics and surfactants	Catalysed hydrogen peroxide	Biodegradable products and inorganic salts
Cyanide	Alkaline hydrogen peroxide	Cyanate (which hydrolyses to ammonia and carbonate)
Filamentous bacteria	Hydrogen peroxide	Dissolved oxygen
Formaldehyde	Alkaline hydrogen peroxide	Formate and carbon dioxide
Microbial contamination	Peracetic acid	Acetic acid and dissolved oxygen
Nitrite	Acidic hydrogen peroxide	Nitrate
Odour	Pollutant dependent-acidic or alkaline hydrogen peroxide	Pollutant dependent
Sulfide	Hydrogen peroxide	Sulfate and sulfur (pH dependent)
Sulfite	Hydrogen peroxide	Sulfate
Thiosulfate	Hydrogen peroxide	Sulfate and tetrathionate (pH dependent)

systems, including iron-catalysed H_2O_2 (Fenton's system), ozone-catalysed H_2O_2 and UV-catalysed H_2O_2, are known as advanced oxidation processes (AOPs; see Section 2). Whilst all AOPs are capable of breaking down bio-recalcitrant species, the best system for a particular application will be dependent upon several factors:

- Nature of the polluted medium
- Total load of oxidizable species
- Existing and target pollutant levels
- Effluent characteristics
- Discharge constraints

Treating the pollutant when it is at its highest strength in the waste water reduces consumption of hydrogen peroxide. Arsenic from contaminated waste waters has been successfully removed using a Fenton's system (Figure 5.11).[47]

The effluent from a photographic process containing reduced sulfur species has been treated using hydrogen peroxide in the presence of an ion-exchange resin containing ammonium molybdate.[48] The resulting water can be safely discharged to drain.

Reduced sulfur compounds plague many industrial and municipal facilities, *e.g.* refineries, chemical plants, tanneries, pulp and paper mills, and waste water treatment plants. These compounds, *e.g.* hydrogen sulfide, can cause serious problems to plant and personnel through formation of noxious, odorous, poisonous and flammable atmospheres and corrosion of equipment. Treatment with hydrogen peroxide provides a cost-effective way of removing these compounds from waste streams. Hydrogen peroxide is often added as a preventative measure. As an oxygen supplier, its addition helps to maintain aerobic conditions in waste waters, and prevents the production of reduced sulfur species by anaerobic micro-organisms.

For best performance aerobic biological waste water treatment plants must be maintained at a constant dissolved oxygen (DO) level. Under certain conditions, *e.g.* adverse weather conditions or under biological oxygen demand (BOD) overloaded conditions, DO levels drop and treatment effectiveness can be seriously reduced. Continuous running under lower than optimum DO levels leads to biological upsets such as proliferation of anaerobic or filimentous

$$2Fe^{2+} + H_2O_2 + 2H^+ \longrightarrow 2Fe^{3+} + 2H_2O$$

$$HAsO_2 + H_2O_2 \longrightarrow H_3AsO_4$$

$$Fe^{3+} + H_3AsO_4 \longrightarrow FeAsO_4(s) + 3H^+$$

$$H_3AsO_4 + nFe(OH)_3(s) \longrightarrow H_3AsO_4(ads.)[Fe(OH)_3]_n(s)$$

Figure 5.11   *Removal of arsenic from waste waters using a Fenton's system.*

organisms. The production of anaerobic species leads to odorous emissions whilst filamentous bacteria contribute to sludge bulking (see below).

Hydrogen peroxide can be used as a supplemental source of oxygen to prevent or cure biological upsets under low DO conditions. It is also used as an oxygen supply for long term use to intensify or enhance biological treatment in plants facing constant overload conditions. Hydrogen peroxide applied to waste water treatment plants, lagoons or ponds assists in reducing the BOD load at the plant outfall, and can prevent growth of sewage fungi in receiving waters. Hydrogen peroxide is usually dosed into the waste water inlet at the entrance to an activated sludge plant or into the waste water transport line for a trickle bed filter plant, waste water pond or lagoon. The oxygen gas bubbles formed from the dismutation of the hydrogen peroxide are extremely small, allowing fast dissolution to DO and efficient oxygen transfer to biomass. As the active biomass is constantly taking up oxygen, an excess of hydrogen peroxide is required to maintain a measurable DO. To allow for changes in the DO uptake rate by the active biomass, the hydrogen peroxide feed rate is adjusted during dosing periods to maintain a constant DO level. Periodic monitoring of the effluent for residual hydrogen peroxide is also normally performed.

Electrolytic processes using hydrogen peroxide have been used for waste water treatment, particularly in destroying disease-carrying bacteria in drinking water, pools, spas, *etc.*[49] Hydrogen peroxide is fed from a peristaltic pump into an electrolytic cell containing copper and silver electrodes. The waste water flows through the cell and is purified.

Low DO levels, nutrient imbalance, shock loading, overloading and extremes of temperature can all contribute to proliferation of filamentous bacteria in biological treatment plants. Filamentous bacteria cause a condition called sludge bulking, where poor sludge settlement leads to the carryover of biomass into the final effluent, reducing the effectiveness of the biological treatment process and increasing the suspended solids in the outflow. Hydrogen peroxide normally applied to the activated sludge line can cure sludge bulking by weakening the sheath in which filamentous bacteria grow, causing microscopically observable fractures in the sheath which ultimately prevent the bacteria reproducing.

Peracetic acid is a particularly powerful biocide used for disinfection of raw or partially treated sewage effluents and sewage sludge. Peracetic acid can also be successfully used for the disinfection of drinking water.[50]

For problematic waste waters, secondary biological treatment is insufficient to meet the increasingly strict legislative standards on COD, colour and priority pollutants such as pesticide and herbicide residues. Whilst pollution prevention will always be the environmentally preferable option, this is not always practical, and in such cases tertiary treatment with catalysed hydrogen peroxide can present an economical and effective way to achieve compliance.

Photo-Fenton's can be used to treat a number of surface waters containing fluoresence whitening agents.[51] Tinopal CBS-X is degraded rapidly using the photo-Fenton's system which causes de-aromatization of the molecule and rapid hydroxylation prior to complete mineralization. Contaminated water can be pre-treated with ozone followed by an ozone/hydrogen peroxide treatment under pressure to remove toxic organic compounds prior to discharge.[52] Chlorinated phenols and polynuclear hydrocarbons are efficiently removed from water in a wood preservation facility using $Fe(III)/H_2O_2$.[53] The process shows a COD reduction of 88% and no chlorinated dioxins of furans are produced. Pesticides have been decomposed under irradiation of waste waters containing hydrogen peroxide.[54]

Following periods of heavy rain, storm water build-up in waste treatment plants can present serious problems. Sewage-contaminated storm water rapidly becomes oxygen depleted, emitting noxious odours and presenting a biological and toxicological threat to receiving waters. Even relatively clean storm water presents a BOD overload threat to receiving waters. To prevent anaerobicity developing in storm or receiving waters, hydrogen peroxide can be applied as an oxygen source delivered directly into the storm water overflow. In cases where significant contamination with sewage has occurred, peracetic acid can be applied to disinfect storm waters prior to discharge. Hydrogen peroxide can also be applied to receiving waters to remediate water courses following accidental adverse releases or spillage. Hydrogen peroxide can be used to treat spillages of toxic or potentially polluting chemicals by directly oxidizing the polluting substance. For example, chlorine, cyanide, sulfide and thiosulfate spills can all be effectively handled with hydrogen peroxide. Hydrogen peroxide is also applied in cases of natural substances such as molasses, animal slurry and silage effluent to prevent development of anerobicity due to natural breakdown of the substance by biological activity in the water course.

Rapid treatment with hydrogen peroxide can therefore reduce fish kills following storm water discharge or accidental spillage of polluting substances into water courses. Whilst storm water overflows are regular events occurring at known and accessible outfalls, accidental spillage of polluting substances may occur from more inaccessible outfalls in open water, or may not even be detected until they have passed through a length of water course. In such cases, a solid peroxygen such as sodium percarbonate can provide advantages in ease of transport and handling. Sodium percarbonate can be added in its powder form from a boat, convenient bridge or jetty.

8   Conclusion

The application of hydrogen peroxide for environmental clean-up has a pivotal rôle to play within the chemical industry. Indeed, such treatments have breathed new life into industries which have been classed as damaging to the environment, thus extending their lifetime. Whilst it will always be desirable to aim for zero effluent processes, it is only realistic to have in the armoury the ability to treat waste products 'end of pipe'. Such 'end of pipe' applications for peroxygens will enable more economic routes to industrially important products to be developed.

References

1 A.P. Jackman and R.L. Powell, in 'Hazardous Waste Treatment Technologies', Noyes Publications, Park Ridge, NY, 1991.
2 J.R. Bolton and S.R. Carter, in 'Surface and Aquatic Environmental Photochemistry', Eds. R. Helz, R. Zepp and D. Crosby, CRC Press, Boca Raton, FL, 1994, pp. 467–490.
3 O. Legrini, E. Oliveros and A.M. Braun, *Chem. Rev.*, 1993, **93**, 671.
4 C.N. Kurucz, T.D. Waite, W.J. Cooper and M.J. Nickelson, *Adv. Nucl. Sci. Technol.*, 1991, **22**, 1.
5 A.D. Donaldson, R.P. Apa, T.L. Eddy and J.E. Flinn, *Am. Chem. Soc., Mech. Eng.*, 1991, **161**, 41.
6 J.W. Tester, H.R. Holgate, F.J. Armellini, P.A. Webley, W.R. Killilea, G.T. Hong and H.E. Barner, in 'Emergency Technologies in Hazardous Waste Management III', *A. C. S. Symposium Series*, 1993, **518**, pp. 35–76.
7 H.M. Cheung, A. Bhatnager and G. Jansen, *Environ. Sci. Technol.*, 1991, **25**, 1510.
8 J.R. Bolton, K.G. Bircher, W. Tumas and C.A. Tolman, *J. Adv. Oxid. Technol.*, 1996, **1**, 13.
9 G.F. Stolarlk and J.J. King, in 'The New Advanced Oxidation Process Plant of the Los Angeles Department of Water and Power', Presented at the Advanced Oxidation Workshop International Water Supply Association A.W.W.A., Coral Gables, FL, April 1992.
10 E.M. Aeita, K.M. Regan, J.S. Lang, L. McReynolds, J.W. Kang and W.H. Glaze, in 'Advanced Oxidation Processes for Treating Groundwater Contaminated with TCE and PCE: Pilot Scale Evaluation', J.A.W.W.A., May 1998.
11 H.J.H. Fenton, *J. Chem. Soc.*, 1884, **65**, 889.
12 C.H. Walling, *Acc. Chem. Res.*, 1975, **65**, 125.
13 D.L. Sedlak and A.W. Andern, *Environ. Sci. Technol.*, 1991, **25**, 777.
14 E. Lipzynska-Kochany, *Water Poll. Res. J. Canada*, 1992, **27**, 97.
15 G. Rupert, R. Bauer and G.J. Heisler, *Photochem. Photobiol. A. Chem.*, 1993, **73**, 75.
16 R.J. Watts, M.D. Dudell and R.M. Monsen, *Water Environ. Res.*, 1993, **65**, 839.
17 B.W. Tyre, R.J. Watts and G.C. Miller, *J. Environ. Qual.*, 1991, **20**, 831.
18 B.C. Faust and J. Hoigne, *J. Atm. Environ.*, 1990, **24**, 79.
19 S.W. Leung, R.J. Watts and G.C. Miller, *J. Environ. Qual.*, 1992, **21**, 377.
20 J.J. Pignatello, *Environ. Sci. Technol.*, 1993, **27**, 304
21 Y. Sun and J.J. Pignatello, *Agric. Food. Chem.*, 1992, **40**, 322.
22 S.R. Carter, P.H. Brown, J.A.Buckley and R.D.S. Stevens, US 848242.
23 S. Lunck, P. Sedlak and T. Verpek-Siska, *React. Kinet. Catal. Lett.*, 1984, **26**, 39.
24 V.C. Balzani and V. Carassitii, 'Photochemistry of Co-ordination Compounds', Academic Press, New York, 1970, pp. 172–174.
25 L.H. Hall and J. Lamber, *J. Am. Chem. Soc.*, 1968, **90**, 20.
26 A Sararzadeh-Amiri, US 5266214.
27 R. Gierzatowicz, L. Pawlowski and E. Smulkowska, *Eff. Water Treat. J.*, 1986, 26.
28 Ugilor, UK 1318164.
29 C. Henry and J.C. Boeglin, Tribune du Cebedeau, 1971, 331.
30 R. Gierzotowicz, L. Pawlinski and E. Smulkowski, *Eff. Water Treat J.*, 1985, 426.
31 E.A Shilov, *Zh. Fiz. Khim.*, 1953, **27**, 1103.
32 M.J. Buck, J.A. Clucas, C.F. McDonaghi and S.E. Woods, Proc. 1st Intl. Symp., Chem. Oxid. Technol. for the Nineties, Nashville, 1991, p. 78.
33 M.R. Hoffman, *Environ. Sci. Technol.*, 1977, **11**, 61
34 J.A.L. Fraser and A.F.E Simms, IWPC Ann. Conf., 1982, Eastbourne, UK, p. 1.
35 J.P. Zumbrunn and E.J. Lafarte, *Eau Industriei*, 1981, **53**, 49.
36 S.R. Lindstrom, *Pollut. Ing.*, 1975, 40.
37 A.F.E. Simms, *Prog. Wat. Tech.*, 1980, **12**, 609.
38 Solvay Interox, EP 166557.

39 Goodyear Tyre and Rubber, US 4211680.
40 Solvay Interox, EP 221689B.
41 B. Hicks and J. Caplan, *Poll. Eng.*, 1993, January, p. 30.
42 S.B. Wilson and R.A. Brown, *GWMR*, Winter 1989, p. 173.
43 P.K. Aggerwal, 'Methods to select chemicals for in-situ biodegradation of fuel hydrocarbons', Air Force Engineering and Services Centre, July 1990.
44 J.J.M. Staps, 'International evaluation of in-situ biorestoration of contaminated soil', National Institute of Public Health and Environmental Protection, January 1990.
45 R.J. Valo, 'In-situ and on-site Biorem.', Pap. Int. In-situ On-site Biorem. 1944.
46 DHV report to Interox on the Velson Site, 1992.
47 L.A. Tescira, M.M. Kohler and A.G. Monteiro, *Extr. Process Treat. Minimization Wastes*, Proc. 2nd Int. Symp., 1996, 191, Ed. V. Romachandran.
48 Eastman Kodak Co., EP 0787689.
49 Austech, WO 97/19896.
50 M. Profaiser, A. Massone, C. Nurizzo and A.F. Bandera, Mededelingen Facultkit Land bouwkundige En Toegepaste, *Biologische Wetenschappen*, 1997, **62**, 1785.
51 C.O. Pulgarin and P.A. Peringer, *J. Adv. Oxid. Technol.*, 1996, **1**, 85.
52 Legerlede-Holding-IO, NL 1001474.
53 Domtar, CA 2187568.
54 E. Schule, *Stuttgarter Berichte zur Siedlungswasserwirtschaft*, 1997, **142**, 77.

CHAPTER 6

Miscellaneous Uses for Hydrogen Peroxide Technology

1 Introduction

This chapter will focus on three important industries where peroxygens have had a significant impact, particularly with respect to environmental issues: chemical purification, pulp and paper bleaching and hydrometallurgy and metal finishing.

2 Chemical Purification

The range of chemicals which require some form of purification is extremely diverse. The need for chemical purification is very much a function of what the manufacturer perceives is necessary for their product, and normally reflects the wishes of the customer, product grade, application, *etc*. Typical improvements made include the removal of colour, and the chemical removal of minor by-products, since the chemical manufacturer generally requires a product which has good colour, odour and purity.

Adsorptive materials have been used for many years for the purification of chemicals. These materials include carbons, clays, and synthetic alumino-silicates. The disposal or regeneration of spent adsorbent is usually expensive, and generally viewed as a costly inconvenience.

Hydrogen peroxide is by far the most widely used peroxygen reagent for chemical purification. Several methods exist for adding hydrogen peroxide to the chemical, either directly or in the presence of a solvent. The purification process can be carried out under acid, neutral or alkaline conditions. Hydrogen peroxide can also be used as part of other purification processes, *e.g.* in conjunction with decolorizing charcoal, or as a secondary oxidizing agent to other chemicals. Chemical purification by peroxygens can be carried out when the chemical cannot easily be purified by other methods, such as distillation or recrystallization, and the chemical itself is inert to peroxygens, *i.e.* the main reactions that occur should be with the impurities.

The various types of compounds which have been purified with peroxygens and which will be discussed here are petroleum products, miscellaneous organic chemicals, surfactants, natural oils, waxes and gums, natural sugars and starches, synthetic polymers, inorganic acids and salts, clays, talc and minerals.

Hydrogen peroxide can also be used to purify starting materials as well as products and this has the added benefit of reducing the amount of effluent associated with the product.

Chemical purification processes of petroleum products have been known for many years. The resulting products need to be colourless with no bad odours and chemically pure. The main contaminants are either those derived from chemicals present in the crude oil or chemicals added during processing.

The isolated products derived from crude oil can contain contaminants which allow, after a period of time, some discoloration, usually phenols, alicyclic unsaturated or heterocyclic nitrogen compounds. This process can be prevented by treatment with hydrogen peroxide (0.01–1.0% m/m of petroleum product)[1] used in conjunction with alkali metals, alloys and salts,[2] as the colouring components in the oil undergo selective oxidation. Clarification of petroleum hydrocarbon distillates after atmospheric pressure distillation can also be achieved using hydrogen peroxide.[3] The colour of phenolic products, which deteriorates over time, can be maintained for longer periods by treatment with a hydrogen peroxide/alkali metal hydroxide system.[4] The reduction in the odour of petroleum products has been achieved by chemical purification; the major contaminants are reduced sulfur compounds which can be treated with peroxygens to give oxidized compounds.[5]

Various hydrocarbon 'sweetening' processes have been devised based on hydrogen peroxide with various additives, *e.g.* alkali metal hydroxides,[6,7] alkali nitrites,[8] surfactants,[9] acetic acid,[10] and cationic exchange resins.[11] In addition, peroxysulfates have been used for the purification of sulfolane.[12] The deodorization of phenol obtained from coal tar can be achieved using hydrogen peroxide, although other reagents such as alkaline earth peroxides, percarbonates, and perborates[13] have been used. By contrast, acidified hydrogen peroxide is used to remove phenols, indoles, quinolines, linolenes and nitriles from naphthalenes.[14,15] Hydrogen peroxide is also used for the desulfurization of coal and coke when used in domestic or industrial applications.[16,17] Solvay Interox has developed a method to deodorize organic mixtures, employing hydrogen peroxide catalysed by a transition metal, *e.g.* tungsten optionally supported on an inert material.[18] The reduced sulfur compounds are oxidized to sulfones or sulfonic acids, amines to hydroxylamines or amine oxides. The oxidized products may be removed by water washing or fractional distillation.

Hydrogen peroxide is used in conjunction with an alkyl phosphate or acetone (caution), in order to improve phase contact,[19–21] to chemically purify crude heavy oils. Microbiological contamination of hydrocarbon oils, particularly jet fuels, can be eliminated by treatment with hydrogen peroxide, particularly in the presence of $FeCl_3$ or $CuCl_2$.[22]

The chemical purification of a range of organic chemicals using peroxides has been known since the 1950s. Formerly, these applications have been confined to bulk organic chemicals, however, more applications have now been found, particularly in the treatment of intermediate products or reaction streams during chemical processing and in cleaning up re-cycle streams.

There are many organic chemicals that have been purified using peroxygens, and Tables 6.1 to 6.6 are a summary to illustrate the range of substrates treated, and their associated application.

Table 6.1 *Purification of carboxylic acids and derivatives*

Material	Reagent	Comments
Acetic acid[23]	Peracetic acid	Recycle stream in the preparation of ε-caprolactone
Dilute acetic acid[24]	Hydrogen peroxide or peracetic acid and a phosphoric acid catalyst	Glycerine production plant gives a widely usable acetic acid
Acetic acid/acetic anhydride mixture	Peracetic acid[25] Hydrogen peroxide/ Pd systems[26]	Carbonylation reaction, liquors containing iodine effluent
Acetic anhydride	Hydrogen peroxide, peracetic acid,[27] sodium perborate[28]	Purification for use in the pharmaceutical industry
Fatty acids	Organic peroxides (methyl ethyl ketone peroxide)	Improved colour stability to heat and oxidation
Ellagic acid[30]	Acetyl peroxide, aqueous alkaline hydrogen peroxide	Reacidify to generate the acid
Monochloroacetic acid[31]	Hydrogen peroxide	—
Acrylic acid esters[32]	Hydrogen peroxide	—

Table 6.2 *Purification of polycarboxylates and derivatives*

Material	Reagent	Comments
Phthalic acid esters	Alkaline hydrogen peroxide[33]	Used in addition to other purification methods
Phthalic acid esters	Peracetic acid,[34] UV radiation with peracetic acid[35]	Bleaching
Pyromelitic anhydride[36]	Hydrogen peroxide	Use aromatic hydrocarbon solution and a lower aliphatic ketone **(CAUTION)**
Maleic anhydride[37]	Hydrogen peroxide/sulfuric acid	Colour improvement
Maleic anhydride[38]	Hydrogen peroxide/ formaldehyde	Removal of resinous materials
Citric acid[39]	Hydrogen peroxide	Purification from fermentation, suitable for food industry applications
Sodium citrate liquors[40]	Hydrogen peroxide	Recovery from waste liquors
Folic acid[41]	Alkaline hydrogen peroxide	—
Aliphatic carboxylic acids[42]	Hydrogen peroxide	Dodecanoic acid[43]
Terephthalic acid[44]	Hydrogen peroxide/sodium hypochlorite	Polymer production

Table 6.3 *Purification of alcohols and phenols*

Material	Reagent	Comments
Isopropanol[45]	Hydrogen peroxide plus exchange resin	Produced by catalytic hydration of propene and contains odorous non-ionic impurities
2,2,2-Trifluoroethanol[46]	Hydrogen peroxide	Removal of sulfur-containing impurities
Ethylene glycol[47]	Hydrogen peroxide	Polyethylene terephthalate plant-recycle stream
Polyethylene glycol[48]	Hydrogen peroxide	Followed by an activated charcoal treatment
Glycerin[49]	Hydrogen peroxide/low boiling solvent with V, Mo, W, Ru or Cr catalyst	Glycerin is distilled from the liquors
Phenol[50]	Hydrogen peroxide	Removal of sulfur contaminants from benzene sulfonic acid hydroxylation
Phenol[51]	Alkaline hydrogen peroxide	From cumene hydroperoxide process
Ethanol[52]	Hydrogen peroxide	Aldehyde removal

Table 6.4 *Purification of other oxygen-containing organic compounds*

Material	Reagent	Comments
Acetone[53] **(CAUTION)**	Alkaline hydrogen peroxide	Large volume, purify before distillation
Tetrahydrofuran[54]	Alkaline hydrogen peroxide	Removal of methacrolein impurity
Polyvinyl pyrrolidone[55]	Alkaline hydrogen peroxide	Aldehyde removal
Aspartame[56]	Acidic hydrogen peroxide (formic acid/hydrogen chloride)	Major sugar replacement product
Oxo aldehydes[57]	Hydrogen peroxide and Co(II)	Reducing iron carbonyls-Co(III) is the purifying reagent regenerated with hydrogen peroxide

Table 6.5 *Purification of sulfur- and nitrogen-containing organic compounds*

Material	Reagent	Comments
2-Mercaptobenzothiazole[58]	Alkaline hydrogen peroxide	Starting material for a range of rubber chemicals
Sulfolene[59]	Peroxymonosulfate	Takes out sulfur dioxide and speeds up subsequent catalytic hydrogenation
4,4'-Hydroxydiphenyl sulfone[60]	Alkaline solution of an inorganic peroxide	Metal blocking agent added to improve purification
Benzonitrile[61]	Hydrogen peroxide	Removes hydrogen cyanide
Acetonitrile[62]	Hydrogen peroxide plus a base	Purified, suitable for processing solvent
Amine products;[63] amines, oxides, betaines, quaternary ammonium compounds, *etc.*	Hydrogen peroxide	Decoloration of pink coloured products
Amine oxides[64]	Hydrogen peroxide	Prior discoloration with carbon dioxide and light
Formamide[65]	Acidic hydrogen peroxide	Removes nitriles and sulfur compounds
Lactams[66]	Hydrogen peroxide/ orthophosphoric acid	Followed by cation and an anion-exchange resin
Alkyl amides[67]	Hydrogen peroxide	Colour improvement in fibres

Table 6.6 *Purification of miscellaneous organic chemicals*

Material	Reagent	Comments
Tertiary phosphites from olefins[68]	Hydrogen peroxide	Oxidation to phosphine oxide
Arylsulfonyl halides[69]	Hydrogen peroxide	Dark product purified to give light coloured one
Benzoyl chloride[70]	Hydrogen peroxide	—
Commercial carbon blacks[71]	Urea/hydrogen peroxide	For printing inks, plastics, paints, *etc.*
Activated carbon	Hydrogen peroxide	Regeneration

Surfactant product bleaching is probably the largest area for peroxygen chemical purification. The surfactant products require a chemical bleaching step in order to give a material of acceptable odour and colour to the customer. The main sources of odour are from the substrates and/or the processing required to generate the product, *e.g.* sulfonation of aromatics used in the manufacture of many anionic surfactants and hydrotropes. Sulfonated surfactants are generally made by sulfonation of a substrate using oleum or sulfur trioxide. The substrate is either derived from natural sources or chemically produced. The products are, in general, unacceptably coloured or odorous owing to the nature of the starting material and impurities produced from the vigorous sulfonation reaction (charring, polycondensation). There are several types of sulfonated surfactant, and the chemical bleaching methods for each are reviewed.

α-Sulfo fatty acid esters have the following general formula:

$$\text{R} \longrightarrow \underset{\underset{\displaystyle \text{H}_3\text{C}}{\big|}}{\text{CH}} \longrightarrow \text{SO}_3\text{H} \qquad \text{R} = \text{Up to C}_{20}$$

These compounds are made via the sulfonation of modified natural oils, *e.g.* methyl ester of beef tallow fatty acid,[73,74] ethyl ester of hardened palm oil acid,[75] methyl ester of palm oil fatty acid[76] and other oils including coconut oil and palm oil products.[77,78] The main chemical bleaching agent is hydrogen peroxide.[79–81] Other additional reagents used are low relative molecular mass alcohols[82] or, in the case of dialkyl sulfosuccinates, alcoholic hydrogen peroxide and peracetic acid.[83]

Hydrogen peroxide bleaching has been incorporated into the process for producing pastes of the α-sulfo fatty acid ester salt.[84–86] During processing, the neutralization and bleaching steps are done simultaneously. No loss of the α-sulfo fatty acid ester salt is observed. In general, chemical bleaching of these compounds can be achieved using up to 2% *m/m* hydrogen peroxide to oil at temperatures up to 80 °C.

Alkyl aryl sulfonates have the following general formula:

$$\text{R} = \text{C}_{12}\text{ average}$$

They are prepared by sulfonation of an alkyl aryl compound under vigorous conditions. This reaction leads to coloured products. Peroxygens are used to chemically bleach these compounds to give lightly coloured materials.[87–91] To obtain better quality products, a result of superior chemical bleaching, hydrogen peroxide has been used in conjunction with other reagents, *e.g.* sodium hypochlorite,[92,93] the sodium salt of ethylene diamine tetraacetic acid,[94] and with alcohols.[95] Several other peroxygen reagents have also been used for bleaching, *e.g.* sodium peroxide,[96] persulfates, percarbonates and perborates.[97]

Olefin sulfonates are made via sulfonation of an olefin, producing highly coloured products. Their general formula is:

$$R-\underset{\underset{SO_3H}{|}}{CH}-CH_3$$

$$R = \text{Up to } C_{20}$$

The reduction in colour to give a more acceptable product is addressed by additional processing using hydrogen peroxide.[98] However, colourless products have been obtained using hydrogen peroxide in conjunction with UV-VIS irradiation,[99] phosphate salts,[100] or alkaline conditions.[101]

Alkyl glycosides are produced by the esterification between a glycoside and an alcohol. The product is treated with an acid salt, activated carbon and a complexing agent, and finally bleached using hydrogen peroxide.[102–104] The level of hydrogen peroxide is 0.5–5% *m/m* of surfactant at 80–140 °C for 0.5–4 hours. Hydrogen peroxide bleaching under alkaline conditions[105,106] gives pale products with good storage stability with respect to colour[107] and microbial attack.[108] Chemical bleaching can be improved by treatment of the alkyl glycoside with hydrogen peroxide followed by sulfur dioxide at levels of 500 to 2500 ppm. The source of the sulfur dioxide can be sulfite salts.[109]

Fatty acid esters or amides derived from rape seed oil[110] and coconut oil[111] can also be bleached using hydrogen peroxide. Synthetic carboxylic acid esters can also be decolorized using hydrogen peroxide,[112,113] although improvements can be made by incorporation of UV radiation,[114] using alkaline conditions,[115] or using peracetic acid.[116] Hydrogen peroxide has also been used to decolorize betaine amphoteric[117] and quaternary ammonium surfactants.[118]

The oils, fats and waxes are extracted from vegetable and animal sources. To produce acceptable, soluble and usable products with respect to colour, odour and impurities, they require chemical purification. The coloured and odorous materials result from polyunsaturates and aerial oxidation products. Hydrogen peroxide has been used successfully for many years. Tables 6.7 and 6.8 summarizes the current methods used to chemically bleach oils, fats and waxes.

Other methods employed to refine natural oils involve the use of ion-exchange resins[135] and bleached earths.[136] Hydrogen peroxide is, however, employed to regenerate these materials for further use.

Sugar is a high volume product that is isolated from natural sources and used in the food industry. Colour is generated as a result of caramelized sugar and polycondensation of sugar with proteins. Since the final product has to be colourless, a bleaching/deodorization process is incorporated into the sugar

Table 6.7 *Purification of oils*

Oil	Reagent	Comments
Soya bean	Catalytic hydrogen peroxide Potassium permanganate Sodium chlorate	Methods devised in the 1940s and 1950s
Rice[122] Tung[123] Cotton seed[124,125] Allium vegetable[126]	Hydrogen peroxide	Using up to 2% m/m at temperatures of 130 °C
Peppermint[127]	Acidified hydrogen peroxide	Clear mellow odour
Palm[128]	Hydrogen peroxide	Chlorine and boron residue
Coconut	Peracids	
Rice bran	Peracids	
Bay leaf	Hypochlorites	
Sal	Dithionites	
Tallow	Borohydrides	
Tall[129]	Iron catalysed hydrogen peroxide at 70 °C	Best system found; large transmittance increase
Crude sulfated turpentine[130]	Iron catalysed hydrogen peroxide at 70 °C	Sulfur level falls dramatically
Pine[131]	Hydrogen peroxide plus tungsten catalyst	Increased efficiency
Rice[132] Olive husk[133] Dehulled oats[134]	Hydrogen peroxide plus acetone **(CAUTION)** Hydrogen peroxide plus alcohols	Solvent increases contact between oil and hydrogen peroxide

Table 6.8 *Purification of fats and waxes*

Fat or wax	Reagent	Comments
Animal fat	Peroxygens[137,138]	—
Wool fat	Hydrogen peroxide in the presence of phosphoric acid[139–141]	Product for food or therapeutic use[142]; fats can be bleached before being extracted
Lanolin purification	Hydrogen peroxide and steam at 110 °C[143] Alkaline hydrogen peroxide[144,145]	Decolorizing wool wax
Crude Japan wax	Acidified hydrogen peroxide[146,147]; oxalic acid[148]	—
Raw wax from haze	Hydrogen peroxide, ozone, and a phase-transfer agent[149]	Similar wax obtained to Japan wax
Yellow beeswax	Neutral or alkaline hydrogen peroxide at 70 °C[150,151]	Gives white beeswax
Rice wax	Hydrogen peroxide, and methyl ethyl ketone[152,153] **(CAUTION)**	Ketones can form hazardous hydroperoxides which are explosive
Carnauba wax	Hydrogen peroxide[154]	Initial separation of impurities using a cyclone then bleaching with hydrogen peroxide

extraction stage, normally at the aqueous syrup or molasses stage. Hydrogen peroxide is used to decolorize and purify sugar juices and syrup at alkaline pH.[155–158] Active carbon[159] or zinc chloride can be used as a post treatment.[160] Quaternary ammonium exchange resins are also used to chemically purify sugar solutions. Regeneration of these resins is via treatment with hydrogen peroxide,[161] which oxidizes the extracted coloured sugar contaminants. The use of hydrogen peroxide in conjunction with other reagents has resulted in better discoloration performance. Deep red-brown sugar solution obtained by the alkaline treatment of dextrose is neutralized and decolorized using hydrogen peroxide with ascorbic acid.[162] Other additives reported are active carbon in conjunction with ion-exchange resins,[163] zeolites,[164] surfactants, polyamides, phosphates, ureas, *etc.*[165] Hydrogen peroxide has also been used to decolorize sugar solutions from other sources, *e.g.* sugar beet,[166–168] polydextroses[169] and glycosides.[170]

Hydrogen peroxide has been used to bleach soya beans,[171] lechitan and phosphatides.[172] The alkali metal peroxides are alternative reagents.[173] It is also used to decolorize fatty acid sarcosides,[174] whilst xanthines require sodium perborate or percarbonate[175] and starches require some irradiation.[176]

The chemical purification of polymers is carried out to decrease colour and odour, and to increase the long-term stability. Hydrogen peroxide is used to obtain colourless, transparent polyfluorinated olefins,[177] polyvinyl alcohols[178] and to improve the colour of polymaleic acid[179] and polybutadienes.[180] In addition, peracetic acid has been used with thermoplastic polymers to reduce colour, *e.g.* bisphenol-A polyacrylate.[181] Hydrogen peroxide in conjunction with UV radiation improves the decoloration of α-hydroxy acrylic acid.[182] Acrylonitrile polymers are stabilized to heat[183] and colour[184] by treatment with hydrogen peroxide during processing. Hydrogen peroxide is used to chemically purify polymers that contain trace metal ions[185] or sulfur residues[186] to give more acceptable products. Treatments involving hydrogen peroxide improve the colour stability of polysulfide-based adhesives and sealants.[187]

Anionic and cationic exchange resins have found a wide variety of applications in processing industries for chemical purification. Hydrogen peroxide is used to regenerate these resins.[188–190] Synthetic resins used in optical lenses must be transparent to light. Decolorization is achieved using sodium perborate or percarbonate.[191]

Inorganic salts and acids are purified either to produce products meeting customer specifications or to 'clean-up' reaction streams on a plant. Salt solutions are treated with hydrogen peroxide to produce a decolorized product on crystallization. This has been used in the production of fluorosilicic acid.[192] zirconium oxide,[193] sodium tripolyphosphate,[194] boric acid,[195] calcium phosphate,[196] *etc.* Decolorized products can also be produced with sodium sulfate, a by-product of methionine processing.[197] On the plant, metals can be removed from sludges. Arsenic and fluorine have been removed from ammonium phosphate using hydrogen peroxide and inorganic salts such as aluminium sulfate and ferric sulfate as flocculants.[198] Much work has been done on two acids in particular, phosphoric acid and sulfuric acid; purification is normally carried out either as part of the manufacturing process or via the re-cycle stream.

As with ammonium phosphate above, hydrogen peroxide is used in conjunction with a flocculant to purify phosphoric acid.[199–201] An ultra-pure grade, suitable for semi-conductor use, can be prepared with a combination of hydrogen peroxide and hydrogen sulfide.[202] Organic compounds can also be removed using hydrogen peroxide and an organic solvent,[203–206] to remove the oxidized contaminant.[207]

The decolorization and purification of both concentrated and dilute sulfuric acid has been known for many years.[208–210] Spent acids from nitration processes are purified using hydrogen peroxide in the presence of an inert solvent to extract organic by-products.[211]

Hydrogen peroxide is used to purify clays, talcs and minerals to afford a white, bright product or to remove organic and inorganic contaminants. Kaolin used for coatings, in the paper industry and as a catalyst support has been chemically purified in this way. Aqueous kaolin slurries are bleached with hydrogen peroxide, at elevated temperatures under alkaline or acidic conditions.[212,213] Organic contaminants can also be removed using hydrogen peroxide alone[214] or in combination with ozone.[215] Further improvements can be made by post-treatment with cation or anion exchange resins.[216] As alternatives to hydrogen peroxide, sodium perborate or percarbonate can be employed as alkaline treatments.[217,218] Peroxide is also being used to bleach other minerals, such as silicates,[219] natural chalk,[220] holmite type clays[221] and rutile.[222]

In conclusion, purification finds applications in a wide range of natural products, organic and inorganic chemicals. In the future, customers of chemical producers will demand more chemically pure products to satisfy the demands of their customers. As the environmental lobby gathers momentum, re-cycle of chemical reagents, re-cycle streams, *etc.*, will become more widespread resulting in significant efforts being made to chemically purify these items. Unwanted contaminants will be extracted, chemically purified and passed onto customers for use in other industries, not just disposed to waste. Consequently, chemical purification technology involving hydrogen peroxide has the potential for major expansion in the years ahead.

3 Pulp and Paper

The production of paper is the largest and fastest growing single application of hydrogen peroxide. This growth has fundamentally been driven by environmental issues, and a large quantity of new and improved technology has now been devised in response to this demand.

There are broadly three types of operation carried out to produce paper and board products: mechanical pulping, chemical pulping and re-cycled paper pulping. For mechanical pulping, the fibre source, commonly wood chips, is sheared at a temperature and pressure above the glass transition temperature of the lignin (Figure 6.1), resulting in the separation of the cellulose fibres (Figure 6.2).[223] A substantial fraction of the lignin and the non-crystalline carbo-hydrates are left in the pulp, thus driving yields of pulp on dry fibre to over 95%. These pulps are often referred to as high-yield pulps.

Chemical pulps are produced by an initial 'cooking' process involving chemical digestion of the lignin fragments with sulfur nucleophiles, followed by a series of further steps to remove lignin and increase pulp brightness.[224] The removal of some of the lignin and non-crystalline carbohydrates lowers the yield

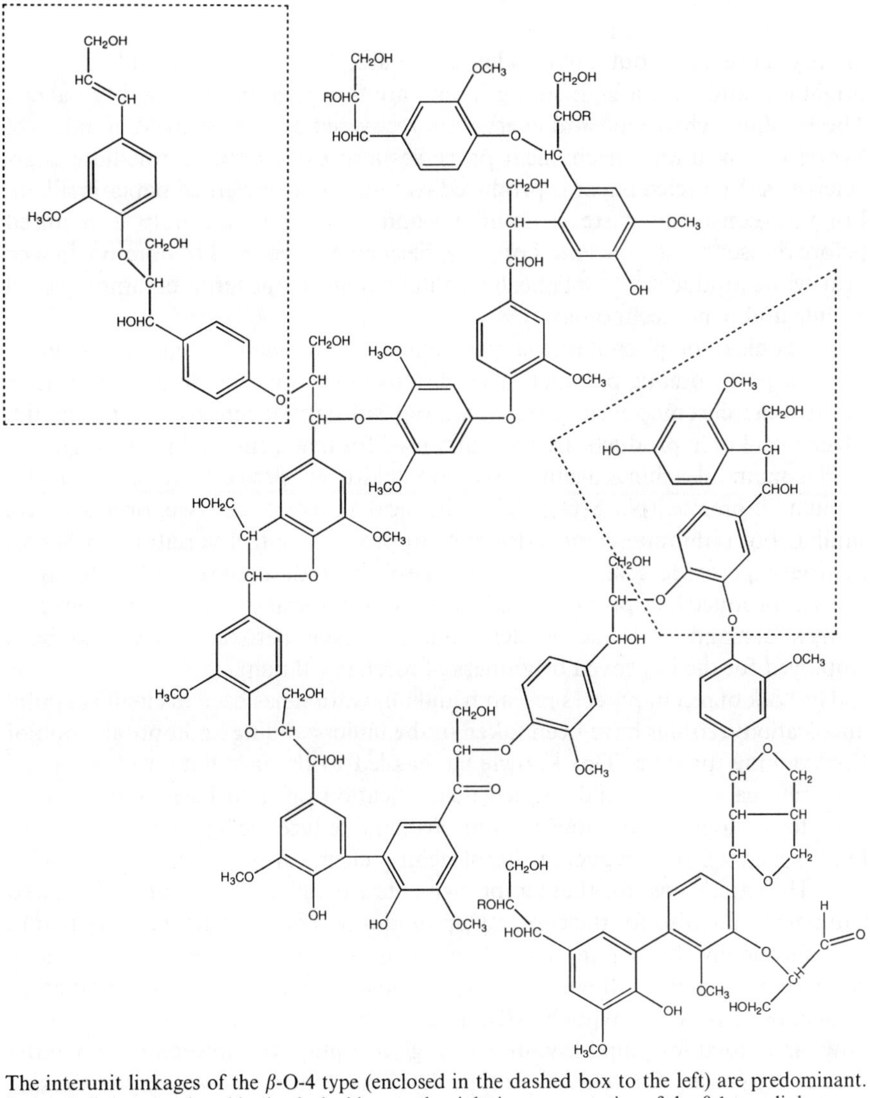

The interunit linkages of the β-O-4 type (enclosed in the dashed box to the left) are predominant. Interunit linkage enclosed in the dashed box to the right is representative of the β-1 type linkage.

Figure 6.1 *Representative structure of coniferous lignin.*

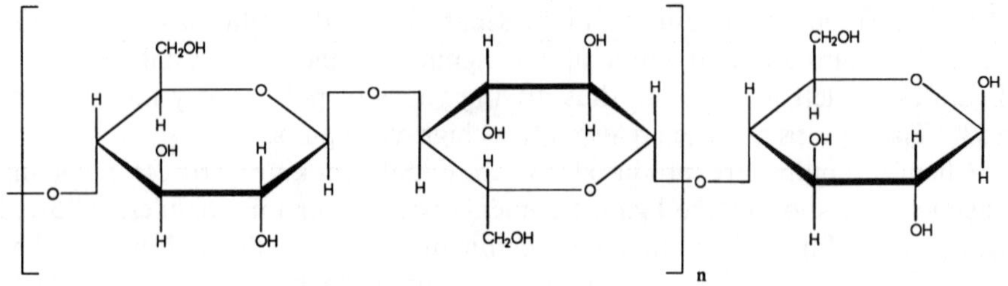

Figure 6.2 *Representative structure of cellulose.*

on dry fibre to about 50%. The pulps are, however, bleachable to high brightness and have a high strength, and are thus used in high quality papers. The 'cooking' chemicals and energy are recovered and recycled. A number of hybrid chemical and mechanical processes also exist with intermediate characteristics. Re-cycled paper is produced by re-pulping unwanted paper in alkali. For post-consumer waste, a de-inking and cleaning of the pulp is required before re-use.[225] For all classes of pulp, bleaching is required to obtain white or near-white products.[226,227] Table 6.9 contains some of the terms commonly used in pulp and paper technology.

Mechanical pulp contains approximately 20% *m/m* lignin, a coloured, random polyaromatic polymer. The objective of the bleaching operation is to destroy the chromophoric groups without substantial removal of the lignin. Mechanical pulp products are currently used for newsprint and packaging.

The main bleaching agents used are hydrogen peroxide (oxidative) and sodium dithionite ($Na_2S_2O_4$). The chemical costs of the two products are similar, but dithionite is more tolerant to process control variations, whereas hydrogen peroxide gives higher brightness. In addition, oxidative bleaching gives a product less prone to yellowing and, naturally, an effluent lower in oxygen demand.[226] Peracetic acid and hydrogen peroxide have also been employed for the improved brightness of mechanical pulp.[228]

The bulk of recent process research and innovation has been in chemical pulp, and dramatic strides have been taken in the understanding and optimization of the 'cooking' process. This knowledge has led to the installation of extended 'cooking' capabilities, and oxygen delignification prior to bleaching. Both of these technologies have a low running cost, and reduce the lignin going in to the final bleaching, thus reducing the bleaching chemical requirements and effluents. The capital cost for these improvements are high not only for the required equipment, but also for the capacity up-grades necessary in the recovery plant.

Traditionally, the lignin content of chemical pulps has been reduced with chlorine treatment, followed by an alkaline extraction, and a number of bleaching steps with hypochlorite, and chlorine dioxide.[226] Hypochlorite is now rarely used for pulp bleaching, as it gives a pulp with lower strength and a greater propensity for yellowing after bleaching. Since the discovery of dioxins in bleach plant effluents in 1986,[229] there has been constant regulatory and market pressure to reduce the emission of organo-chlorine compounds from

Table 6.9 *Glossary of terms used in the pulp and paper industry*

Basic terms
Stock: the pulp stream within the manufacturing process
Consistency: the % dry matter in the stock

Mechanical and high yield pulp
SGW: stone ground wood
TMP: thermomechanical pulp
CMP: chemomechanical pulp
CTMP: chemothermomechanical pulp
APMP: alkaline peroxide mechanical pulp

Chemical or Kraft pulp
C: chlorine stage
O: oxygen (alkaline) stage
E: extraction stage
D: chlorine dioxide stage
P: hydrogen peroxide stage
Z: ozone stage
Q: metal sequestration stage
E_0: extraction plus oxygen
E_p: extraction with hydrogen peroxide
$\overset{\circ}{P}$*: hydrogen peroxide stage at high temperature
$\overset{\circ}{A}$: acidic oxidizing stage (peracetic acid or Caro's acid)
ECF: elemental chlorine free process
TCF: totally chlorine free process
TEF: totally effluent free process
C, O, E, D, P, Z: any subscripted numbers after these letters refer to % of chlorine equivalent used

---

Recycled pulp
Generally, the recycle is similar to mechanical pulping operations but product can be higher quality. De-inking and adhesive ('stickies') removal are peculiar to recycled pulp.

---

bleach plants. This can be achieved by reducing the lignin content of the pulp going into the bleach plant (extended 'cooking', oxygen delignification) and by a reduction or elimination of the use of elemental chlorine. The latter can be entirely replaced by chlorine dioxide, if there is sufficient generation capacity on the mill site, a process referred to as elemental chlorine free (ECF) bleaching. The addition of oxygen to the first extraction stage, and of hydrogen peroxide to the first and second extraction stages not only reduces the chlorine dioxide required for complete elimination of the chlorine, but also improves the pulp cleanliness and bleachability.

Totally chlorine free (TCF) bleaching has also been developed in recent years because of regulatory and market pressure in German-speaking and Scandinavian countries. TCF bleaching started with sulfite pulps, as the pulps are brighter and easier to bleach than Kraft pulps (they are however somewhat weaker). All German chemical pulp mills are sulfite mills, and all bleach with O P sequences.[230] For Kraft mills, the lower initial brightness and the more residual resistant lignins make it more difficult to bleach the pulp. Scandinavian pulp mills started to produce TCF pulps with a Q P P P P sequence in their existing bleach plants. The first pulps had lower brightness, and required very large quantities of hydrogen peroxide. With experience and the use of more optimized bleaching conditions, obtained brightness can be almost the same as those produced by chlorine bleaching.[231]

The traditional conditions for alkaline hydrogen peroxide bleaching have called for modest temperatures in the region of 70–80 °C, but temperatures as high as 120 °C can give superior brightening without detriment to the pulp properties.[232]

Alkaline hydrogen peroxide is not very effective at eliminating the last traces of lignin from the pulp. Acidic, electrophilic oxidizing agents, such as ozone and peracids, are more effective. Ozone has been installed in a number of mills in Scandinavia in sequences such as O Z E$_o$ P O Z E D. The capital investment for the ozone reactor and generation equipment is high; gas/pulp mixing is critical to the operation, and the gas is both toxic and corrosive, and thus residuals have to be destroyed. Despite these difficulties, ozone can be used industrially for pulp delignification.[233] A lower capital alternative is to use peracids, the most common being Caro's acid[234] and peracetic acid.[235] For efficient and selective delignification, the peracid stages are operated within the pH range of 2–4.5, with stabilization to ensure no hydroxyl radicals are generated from any residual hydrogen peroxide.[232] A number of mills world-wide have used peracids within their existing mill configuration in the sequences based on the Q P A P concept.[236]

A number of other technologies have been proposed for TCF bleaching. Dimethyldioxirane has also been found to be a very selective oxidant in pulp bleaching,[237] and technology for *in situ* generation of this reagent has been disclosed.[238] Polyoxometallates have been suggested as oxidation catalysts suitable for pulp delignification[239] and biodelignification continues to be a subject of academic interest. The relative merits of TCF and ECF bleaching have been vigorously debated in recent years. These developments are now increasingly seen as a step towards 'totally effluent free' (TEF) processes, which aim to prevent all emissions from bleach plants. It is currently believed that the prevention of mill closures could be achieved with TCF and ECF bleaching technologies; in practice, the former is likely to be cheaper and easier to implement.[230]

A number of experimental chemical wood bleaching technologies have recently been disclosed: hydrogen peroxide and tetraacetyl ethylene diamine,[240] anthracene derivatives with hydrogen peroxide[241] and *in situ* produced performic acid[242] are but a few.

Not all re-cycled papers are de-inked and bleached, but those which are often need hydrogen peroxide to reach the target brightness and cleanliness of the pulps. For wood-containing papers, hydrogen peroxide is required in the pulping operation to prevent yellowing of the pulp. The pulp can also be bleached with hydrogen peroxide or dithionite after the de-inking and cleaning processes. As for virgin fibre, the bleaching effect of dithionite is limited.

4 Hydrometallurgy and Metal Finishing

The chemistry of hydrogen peroxide and Caro's acid can be used in many applications involving metals: metal extraction and separation from ores or waste, hydrometallurgy, and surface treatment of metals and alloys. Compared to alternative treatments, the direct chemical costs are sometimes higher, but the difference is often outweighed by advantages in simplicity of operation (cost savings on equipment), and in lower overall effluent production.

Hydrogen peroxide can act as an oxidant or reductant towards elements in various oxidation states. In some cases (*e.g.* iron, copper, selenium, tellurium), no net reductive behaviour is exhibited at any pH. Where either net reduction or oxidation is observed (*e.g.* vanadium, chromium, manganese, cerium), reductive behaviour is usually shown at acid pH and oxidative behaviour at alkaline pH. This can be useful in precipitating or separating metal from solution, since higher oxidation states are often less soluble or form anionic species (*i.e.* opposite charge to other components). Where it is required to oxidize these metals at acidic pH, Caro's acid ($H_2O_2 + H_2SO_4 \leftrightarrow H_2SO_5 + H_2O$) is a useful alternative to hydrogen peroxide, from which it can be generated on site.

Two types of indirect oxidation using hydrogen peroxide are used. The first involves re-oxidation of an intermediate, or 'relay' metal, which in turn acts as an oxidant in the recovery process; examples include iron(III) oxidation of uranium(IV) and of thiourea in gold production. The second involves oxidation to assist acid dissolution of metals, chiefly in finishing of stainless steel, brass or copper. Several metals form peroxo complexes in their highest oxidation states when treated with hydrogen peroxide (*e.g.* vanadium, molybdenum, tungsten) and these are much more water soluble than the corresponding oxides. This, therefore, provides a more efficient means of extracting these metals selectively into solution in, for example, scrap re-cycling (*e.g.* tungsten carbide), and also as a potential method of purifying crude metal oxides (zirconium oxide).

Gold extraction is mainly practised by leaching finely ground ore with alkaline, aerated cyanide solution for up to 24 hours (Figure 6.3). The reaction may involve intermediate formation of hydrogen peroxide, but the mechanism has not yet been fully established. Peroxygens such as hydrogen peroxide[243] and calcium peroxide[244] can be added to improve the recovery. In some cases, aeration is difficult owing to the viscosity of the ore slurry, and this may require use of a peroxygen as the sole oxidant.[245] Some ores are more refractory, and require a more vigorous treatment, for which oxygen under pressure (expensive equipment) or gaseous chlorine (corrosive, difficult to handle) are conventional solutions. Caro's acid, generated on site, is a useful alternative, as it attacks sulfur and other oxidizable materials surrounding the gold particles, making the latter more accessible.[246] Although Caro's acid is itself corrosive, it is

$$4Au + 8CN^- + O_2 + 2H_2O \longrightarrow 4Au(CN)_2^- + 4OH^-$$

Figure 6.3 *Extraction of gold using aerated cyanide solution.*

$$2Cr^{3\oplus} + 3H_2O_2 + 10OH^{\ominus} \rightleftharpoons 2CrO_4^{2\ominus} + 8H_2O$$

Figure 6.4   *Oxidation of chromium with alkaline hydrogen peroxide.*

compatible with several stainless steel grades, unlike chlorine which requires more expensive metals such as titanium or tantalum for plant construction.

Cyanide disposal from gold mining is commonly via a tailings pond in which the cyanide is gradually broken down by air and sunlight. Treatment with hydrogen peroxide or Caro's acid can be used either to moderate the level of cyanide entering the lagoon or to treat any residues in the overflow, which may arise from insufficient residence time (pond too small, rainfall, *etc.*).

Redox chemistry with hydrogen peroxide is pH dependent; for example, cerium(IV) is reduced to cerium(III) in acid, whereas cerium(III) is oxidized to cerium(IV) in alkali. The reductive step may be used to solubilize cerium in extraction from ores.

The two main oxidation states of chromium usually display opposite charges in their compounds. Chromium(VI) is particularly toxic and all operations are subject to stringent legislation in most countries with significant process industries. Chromium(III) is oxidized by hydrogen peroxide to chromium(VI) in neutral or alkaline conditions, but reduced from chromium (IV) to (III) in acid (Figure 6.4).

Oxidation in acidic solution can be achieved by Caro's acid, for example at $85\,^{\circ}C$.[247] This chemistry is employed in the extraction of chromium from superalloy scrap.[248] It can also be used to detoxify, recover or re-cycle effluents from chromic acid oxidation or pickling liquors. Detoxification and recovery of chromic acid solutions is by the addition of hydrogen peroxide which reduces the chromium(III) salts, followed by alkali to precipitate and separate chromium hydroxide.[249] Re-cycle can be by direct re-oxidation with Caro's acid, or by first raising the pH to alkaline levels, oxidation with hydrogen peroxide and re-acidification with subsequent salt by-product production.

In the nuclear industry, actinide elements are extracted with a solution of tri-n-butyl phosphate in kerosene. In order to completely remove these metals before disposal of the organic solvent, the tri-n-butyl phosphate must be decomposed. It has been found that this can be done with hydrogen peroxide and a chromate catalyst at near-neutral pH under reflux, in a two-phase system.[250] This application links environmental protection and metal recovery.

Extraction of copper from ores is enhanced (both rate and yield) by the addition of hydrogen peroxide or Caro's acid (Figure 6.5).[251]

$$CuS + H_2O_2 + 2H^{\oplus} \longrightarrow Cu^{2\oplus} + S + 2H_2O$$

$$CuS + H_2SO_5 + 2H^{\oplus} \longrightarrow CuSO_4 + S + H_2O$$

Figure 6.5   *Extraction of copper using hydrogen peroxide or Caro's acid.*

At higher pH, the sulfur is oxidized to sulfate; this is also true for the extraction of copper from chalcopyrite ($CuFeS_2$) in the presence of iron and is effective for other mixed sulfidic ores.[252] Copper metal can be dissolved by non-oxidizing acids if hydrogen peroxide is added.[253] This is used in the secondary recovery of copper from anode slimes arising from copper refining; copper is dissolved, leaving a residue rich in selenium, tellurium and precious metals.[254] Recently, a ferric leaching process, applicable to secondary copper sulfides, has been developed,[255] which is economically attractive due to low hydrogen peroxide consumption.

Cobalt can be separated from solutions containing nickel, zinc and manganese by selective oxidation to Co(III) with Caro's acid, leading to precipitation as the hydroxide, with efficiencies $> 99\%$ and leaving only ppm levels of cobalt in solution (Figure 6.6).[256] The hydroxide is readily filtered, the process comparing well with product prepared using chlorine as the oxidant. The preferred pH range is 3–4.8, alkali being added to maintain this level. The cobalt(III) hydroxide produced can be redissolved as a cobalt(II) salt under acidic conditions by using the corresponding acid with hydrogen peroxide as a reducing agent (Figure 6.7).[249] The process is stoichiometric at ambient temperature and hence simpler and cleaner than alternatives using methanol in formaldehyde.

Cobalt can also be separated from nickel in alkaline conditions by first adding ammonia and hydrogen peroxide to form cobalt(III) pentamine complex, and then crystallizing the bulk of the nickel as the sparingly soluble nickel ammonium sulfate (Figure 6.8). The separation of nickel from cobalt can be achieved by precipitation via oxidation to nickel(III) hydroxide with Caro's acid at near-neutral pH (Figure 6.9).[257]

$$2Co + H_2SO_5 + 6OH^{\ominus} \longrightarrow 2Co(OH)_3 + SO_4^{2\ominus} + H_2O$$

Figure 6.6 *Selective oxidation of cobalt with alkaline Caro's acid.*

$$2Co(OH)_3 + H_2O_2 + 2H_2SO_4 \longrightarrow 2CoSO_4 + O_2 + 6H_2O$$

Figure 6.7 *Dissolution of cobalt(III) hydroxide in the presence of acidic hydrogen peroxide.*

$$2Co^{2\oplus} + 10NH_3 + H_2O_2 + 2H^{\oplus} \longrightarrow 2[Co(NH_3)_5H_2O]^{3\oplus}$$

Figure 6.8 *Separation of cobalt from nickel solutions using hydrogen peroxide in the presence of ammonia.*

$$2Ni^{2\oplus} + H_2SO_5 + 6OH^{\ominus} \longrightarrow 2Ni(OH)_3 + SO_4^{2\ominus} + H_2O$$

Figure 6.9 *Separation of nickel from cobalt using Caro's acid under near-neutral pH.*

$$Mn^{2\oplus} + H_2SO_5 + H_2O \longrightarrow MnO_2 + H_2SO_4 + 2H^{\oplus}$$

Figure 6.10   *Separation of manganese in the presence of Caro's acid.*

The pH is controlled by addition of base. If sodium carbonate is used, a basic carbonate is precipitated. Caro's acid can also be used to extract nickel preferentially from mixed sulfide ores such as pentlandite $(Ni.Fe)_9S_9$.[258]

Extraction of manganese, like other redox metals, with hydrogen peroxide is pH dependent. Hydrogen peroxide acts as a reductant in acid but oxidant in alkali, whereas Caro's acid shows oxidizing behaviour throughout the pH range (Figure 6.10).

By control of pH, sequential recovery of cobalt and manganese from mixed solutions can be achieved.[256] Manganese dioxide, as the mineral pyrolusite or from separations as above, is a major source of manganese compounds. It can be converted readily to manganese(II) salts using the respective acid and hydrogen peroxide as a reducing agent (Figure 6.11).[258]

The cost of iron compounds is insufficient to support the stoichiometric use of peroxygens to recover them. However, hydrogen peroxide has a role in removing iron compounds from mixtures with more valuable metals, and also in the re-oxidation of iron used as a relay oxidant in recovery of other metals. It is also used as an oxidant in steel pickling. For example, solutions containing 50 ppm each of iron, nickel and copper can be purified by addition of hydrogen peroxide at 90 °C; filterable iron oxide hydrates are precipitated. Clearly, the decomposition of hydrogen peroxide is strangely favoured under such conditions and carefully controlled addition is required to obtain near-stoichiometric oxidation.

Zinc exhibits no solution redox chemistry under normal conditions, and peroxygens can only be used to remove impurities from zinc solutions prior to electrolytic formulation of metallic zinc. For example, ferrous iron can be clearly oxidized and precipitated from zinc solutions using hydrogen peroxide at 95 °C and pH 1.5 without addition of foreign ions. Manganese(II) can also be oxidized and precipitated from zinc solutions as high quality manganese dioxide using Caro's acid.

Arsenic is a frequent contaminant in metal ores (especially sulfides) and therefore appears in hydrometallurgical processes applied to these ores. Because of its toxicity, it must be efficiently removed from aqueous process streams before discharge. Traditionally this has been done with lime, but a large excess is required and the product (calcium arsenite and arsenate) can re-release some arsenic under the influence of atmospheric carbon dioxide. A more reliable process is co-precipitation with iron salts under near-neutral conditions. This requires both element ores to be in their higher oxidation states and hydrogen

$$MnO_2 + H_2O_2 + H_2SO_4 \longrightarrow MnSO_4 + 2H_2O + O_2$$

Figure 6.11   *Conversion of manganese(IV) oxide to manganese(II) sulfate in the presence of acidic hydrogen peroxide.*

peroxide can achieve this efficiently in dilute solution. Ferric ion can be added or generated *in situ* after arsenic oxidation (Figure 6.12).

The solubility product of ferric arsenate is so low that residual levels of 50 ppb arsenic in solution can be achieved.[259] Hydrogen peroxide is also used to produce arsenic acid from arsenious oxide,[260] where the cleanliness of the reaction offers advantages over other oxidants (Figure 6.13).

Selenium and tellurium can be dissolved and separated using hydrogen peroxide.[261] In copper-refining anode slimes, Se and Te are dissolved at 95–100% recovery using hydrogen peroxide at natural pH forming tetravalent acids (Figure 6.14).[261] The same chemistry is used to refurbish photocopier drums containing selenium; under alkaline conditions, Se(IV) is oxidized to Se(VI) by hydrogen peroxide.

Because of the extreme toxicity of cyanide, a number of alternative leaching processes have been devised for gold ore. The only one of those to be commercialized involves thiourea, which uses ferric sulfate as an oxidant (Figure 6.15). The iron(III) oxidizes thiourea to formamidine disulfide, which in turn oxidizes the gold to its thiourea complex. A number of oxidants can be used to regenerate the iron(III), of which hydrogen peroxide is among the cleanest and most convenient.[262] The use of oxygen or ozone requires gas/slurry mixing equipment, whilst the use of permanganate or hypochlorite results in the introduction of foreign ions.

Leaching uranium ores requires an acid and an oxidant in order to convert the insoluble uranium(IV) oxide to a soluble uranium(VI) salt. Iron(III) is the usual oxidant (Figure 6.16).

$$Fe^{II}, As^{III} \xrightarrow{\ H_2O_2\ } Fe^{III}, As^{V}$$

$$Fe^{III} + AsO_4^{3\ominus} \longrightarrow FeAsO_4$$

Figure 6.12 *Removal of arsenic salts from mixed ores using hydrogen peroxide.*

$$As_2O_3 + 2H_2O_2 + H_2O \longrightarrow 2H_3AsO_4$$

Figure 6.13 *Preparation of arsenic acid using hydrogen peroxide.*

$$Se + 2H_2O_2 \longrightarrow H_2SeO_3 + H_2O$$

Figure 6.14 *Recovery of selenium from anodic slimes.*

$$Au + 2(NH_2)_2CS + Fe^{3\oplus} \longrightarrow [Au((NH_2)_2CS)_2] + Fe^{2\oplus}$$

Figure 6.15 *Extraction of gold using a hydrogen peroxide/thiourea system.*

$$UO_2 + 2Fe^{3+} \longrightarrow UO_2^{2+} + 2Fe^{2+}$$

Figure 6.16   *Extraction of uranium via the oxidation of uranium(IV) to uranium(VI).*

The ferrous ion is traditionally re-oxidized with manganese dioxide or sodium chlorate, but Caro's acid has also been introduced for this application.[263] Under these conditions, Caro's acid is more efficient than hydrogen peroxide.

Tungsten can be removed from mixed metal wastes by solubilization with acidified hydrogen peroxide (Figure 6.17). In the case of tungsten carbide, the reaction takes place at 30–35 °C, often in the presence of a promoter.[264] Tungsten can be subsequently precipitated as hydrated tungstic oxide (by heating to decompose the peroxygen complex), or as calcium tungstate by lime treatment. In the case of tungsten, copper and nickel scrap, acidic hydrogen peroxide preferentially dissolves the other metals, allowing tungsten powder to be recovered directly.[265]

The chemistry of molybdenum is similar to that of tungsten and it can be recovered by dissolution of metallic or oxide scrap in acidic hydrogen peroxide.[266] High purity molybdic oxide can be precipitated by heating solutions to decompose the peroxide complexes. Peroxo complexes of tungsten and molybdenum can be separated by mixing with tri-octylphosphine oxide in kerosene, which extracts molybdenum selectively.[267] Molybdenum occurring in uranium ores is brought into solution during acid extraction of the uranium, but is partly in the (V) oxidation state. Caro's acid treatment completes oxidation to molybdenum(VI) to facilitate recovery.[268]

Recovery of vanadium with peroxygens involves both oxidation and complexation. In solution, conversion of lower oxidation states into vanadium(V) allows separation by solvent extraction (Figure 6.18).[269] This chemistry can be used for vanadium by-products in uranium extractions. With hydrogen peroxide, vanadium(IV) is not oxidized in acidic solution, but rather in alkaline conditions, e.g. 60 °C at pH 9 (Figure 6.19).[270] Use of excess hydrogen peroxide readily forms peroxo complexes and this is of value in selective dissolution of vanadium from secondary sources.

$$W + 4H_2O_2 + 2H_2SO_4 \longrightarrow [WO(O_2)(SO_4)_2]^{2-} + 2H^+ + 5H_2O$$

Figure 6.17   *Removal of tungsten from mixed metal wastes using acidic hydrogen peroxide.*

$$2VO^{2+} + H_2SO_5 \longrightarrow 2VO_2^+ + H_2SO_4$$

Figure 6.18   *Oxidation of vanadium to vanadium(V) using Caro's acid.*

$$V_2O_4 + H_2O_2 + 2OH^- \longrightarrow 2H_2VO_4^-$$

Figure 6.19   *Extraction of vanadium by-products from uranium ores using alkaline hydrogen peroxide.*

Zirconium has no significant redox chemistry in solution. Formation of soluble peroxy complexes in dilute aqueous acid can be used to purify zirconium compounds by extraction in organic media. Subsequent decomposition of the complex yields ultra-pure zirconia ZrO_2, which is of interest as a high performance ceramic material.[271]

5 Conclusion

Hydrogen peroxide has diverse uses throughout a range of applications; three important areas have been briefly discussed in this chapter. The use of hydrogen peroxide in pulp and paper bleaching has increased significantly over the last decade due to environmental legislation. The versatility of hydrogen peroxide as a reducing agent has also been highlighted for the extraction of metals from their ores. The purification of industrially important chemicals has been known for many years, and is likely to remain a strength of the reagent into the future.

References

1 Sun Oil Co., US 2767204.
2 J.R. Ermead, *Can. J. Technol.*, November 1995, 1.
3 Y. Motoyama, JP 10426.
4 Nippon Steel Chemical, JP 62/0164.
5 P.S. Tam, J.R. Kittrell and J.W. Eldridge, *Ind. Eng. Chem. Res.*, 1990, **29**, 324.
6 Shell Co., US 2744954.
7 Alst, JP 11260977.
8 Universal Oil, US 2790752.
9 V. Kelly, US 364683.
10 S. Kanakzi, JP 397957.
11 V.P. Tutbalina, *Khim. Technol. Pererat. Nefti. Gaza.*, 1985, 14.
12 Phillips Petroleum, US 4558138.
13 Timbrae, AU 147351.
14 E. Coal, *Chem. Res.*, SU 1047897.
15 Nippon Steel Co. Ltd., JP 7982147.
16 J.T. Riley, US 4701183.
17 Juan Cierva, BE 778671.
18 Solvay Interox, unpublished results.
19 H. Fushimi, JA 3028477.
20 Quyoe Chemical, US 362067.
21 P.L. Deobene, *J. Am. Oil Chem. Soc.*, 1985, **13**, 34.
22 Biolex Co., AU 568889.
23 Daicel Chem., JA 61/56151.
24 Daicel Chem., JP 02/134342.
25 Hoechst AG., EP 0217191.
26 Eastman, WO 92/03403.
27 Courtaulds, BR 749098.
28 M.K. Rastoyi, *Res. Ind.*, 1988, **33**, 25.
29 Kao Corp., JP 02123192.
30 Kikkoman Co., JP 04/046180.
31 Daicel Chem., JA 60/139641.
32 Y. Kanzaki, JA 74/11207.

33 Huls AG., UR 285637.
34 Interox, BR 8103842.
35 Ashland, US3 668092.
36 Japan Cat. Chem., JA 4800437.
37 Standard Oil, US 260546.
38 Mitsubishi Gas Chem., EP 0168897.
39 Jiri Pendl, CS 246781.
40 Jiri Pendl, CS 244220.
41 Upjohn, US 2626947.
42 I.B. Blanshtein, SU 326173.
43 Dupont, US 4149013.
44 Toray Industries, JP 8140640.
45 Shell Development, US 2628986.
46 Misubishi Metals, JA 62/056446.
47 NPEC, JA 14725.
48 R.P. Petrovnina, *Mater. Nauchno-Tekh. Konf. Kazan. Khim.-Tekhnol. Inst. Kazan. Zavoda. Org. Sint.*, 1973, Ed. E.V. Kuznetsov, Publ. Kazan. Khim.-Tekhnol. Inst., Kazan, USSR, p. 157.
49 N.I. Stepanenko, *Khim. Prom.-st.*, 1980, 735.
50 Stamicarbon, BR 729377.
51 Hercules, CA 592836.
52 S.E. Kharin, *Ferment. Spirit. Prom.*, 1972, **5**, 7.
53 Sumitomo, DE 3305954.
54 DuPont, US 4413136.
55 General Aniline, US 2914516.
56 Farmitalia, BE 900317.
57 Hoechst AG., DE 1028556.
58 R. Oakamoto, BR 1321181.
59 Phillips Petroleum, US 4558138.
60 Tokai Denka, JA 82/77666.
61 Japan Cat. Chem., JA 70/21500.
62 G.K. Analyt, BE 1002441.
63 Ethyl Co., US 4347381.
64 Ethyl Co., US 5068430.
65 P. Hampl, CZ 1152899.
66 Toyo, JA 77/111585.
67 DuPont, US 3316298.
68 BASF, US 2485859.
69 Stauffer, US 4696774.
70 Sicedizon, BR 845072.
71 A. Sosnowski, PL 126422.
72 V.M. Ropot, SU 1404457.
73 Lion Corp., JA 59/16870.
74 Lion Corp., DE 3334517.
75 Lion Corp., DE 3345892.
76 Henkel, DE 3620158.
77 Henkel, GB 1001284.
78 S. Ahmed and H. Kifli, AOCS 2nd World Conf., 1986.
79 Henkel, GB 1001283.
80 Henkel, DE 1468032.
81 Henkel, DE 255076.
82 Lion Corp., JA 59/25369.
83 Shell Chem., NE 84725.
84 Henkel, KGAA, DE 4017466.
85 Henkel, KGAA, DE 4017467.

86 Henkel, KGAA, DE 4016468.
87 Monsanto, US 2874061.
88 Continental Oil, CA 593392.
89 Henkel, DE 617770.
90 Lion Corp., JA 1958161.
91 Hoechst AG., EP 0131913.
92 Colgate Palmolive, US 2804466.
93 Henkel, DE 3319591.
94 Tula. Chem. Ind. Res., SU 1432053.
95 Lion Fat, DE 902255.
96 Degussa, FR 1196533.
97 Rudolf, DE 902255.
98 Proctor & Gamble, US 876002.
99 Proctor & Gamble, US 3585116.
100 Lion Fat, DE 2622057.
101 Henkel, DE 2647523.
102 Huls AG., EP 306650.
103 Huls AG., EP 306651.
104 Huls AG., EP 306652.
105 Henkel KGAA, EP 362671.
106 Henkel KGAA, DE 4019175.
107 Jao Corp., JP 02/264789.
108 Henkel KGAA, DE 4035722.
109 A.E. Stanley, US 4557729.
110 Huels Chem., DE 362331.
111 Hydriewerke, DT 1068690.
112 Mathieson, US 2861108.
113 Hoechst AG., EP 65267.
114 Ashland Oil, US 3668092.
115 Ashland Oil, US 3701767.
116 J. Smidrkal, CS 223721.
117 Nippon Oils & Fats, JP 62/260894.
118 Stepan, US 4370272.
119 Unilever, BP 577879.
120 DuPont, US 2795595.
121 Interox Internal Report 1961.
122 S. Nojima, JA 50/3633.
123 J. Greenfield, *Am. Tung. Oil Topics*, 1959, **14**, 5.
124 Swift, US 2915538.
125 P.L.N. Rao, *Indian Oil Seeds J.*, 1962, **6**, 130.
126 Y. Araki, JA 67/27302.
127 Nabiscon Brands, EP 019354.
128 Unilever, EP 28488.
129 Interox Internal Report 1986.
130 Interox Internal Report 1984.
131 Interox Internal Report 1992.
132 R. Koyama, JA 52/2584.
134 DuPont, US 4053492.
135 Rohm & Haas, US 2771480.
136 D.A. Morgan, World Conf. on Processing Palm, Palm Kernel and Coconut Oils, 1984, Malaysia.
137 Kali-Chemie, GB 742233.
138 J. Maslowska, *Gospod Miesna*, 1981, **33**, 19.
139 K. Heidler, CZ 138902.
140 AB Separator, DE 1467495.

141 Croda, GB 706422.
142 Armour, US 3338932.
143 Toyo Spinning, JA 70/23617.
144 J. Kromer, CZ 91063.
145 Interox Internal Report 1985.
146 T. Shima, JA 51/41.
147 Y. Yoshiwara, JA 53/2231.
148 T. Wachi, JA 53/4933.
149 S. Tachibana, *Mokuzai Gakkaishi*, 1984, **30**, 185.
150 Interox Internal Report, 1984.
151 M. Pauilicek, CS 241594.
152 Noda Wax, JA 76/30204.
153 Oriza Yuka Co. Ltd., JP 7113261.
154 Cerias Johnson, BR 9006044.
155 Montecatini, FR 1093406.
156 Fuyi Sugar, JA 64/5427.
157 Coca Cola Co., US 4288551.
158 Holly Sugar Co., US 4196017.
159 Montecatini, Italy 08605.
160 T. Veki, JA 57/8673.
161 Rohm & Haas, CA 567125.
162 S. Katoka, JA 66/16099.
163 M. Irita, JA 74/54554.
164 Kiev Food Ind. Tech. Inst., SU 1208074.
165 R. Riffer, *Proc. Tech. Sess. Cane Sugar Refin. Res.*, 1981, 84.
166 D. Khadzhikinov, *Khranitelnoprom Nauka*, 1986, **2**, 56.
167 Evidenzbuero Oesterreichischer Zuckerfabriken GmbH, AT 324248.
168 Atochem, EP 419384.
169 Pfizer, US 4622233.
170 Corn Products, GB 1139110.
171 Yakuhin, JA 7986647.
172 Nisshin Oil Mills, JP 8020661.
173 Glidden, US 2839546.
174 FMC Co., JP 66/17971.
175 Grace, DE 2026623.
176 L. Sant Lebe and G. Berger, *Nuch Sci. Abstr.*, 1973, **26**, 43236.
177 Pechiney, FR 1140964.
178 Veb Carl Zeiss, DD 261792.
179 Petrotex, US 3634503.
180 VBE Ind., JP 7950088.
181 Union Carbide, EP 0117539.
182 Nippon Peroxide, JP 02/305823.
183 Celanese, US 3656883.
184 Toray, JA 4723492.
185 S. Loeschav, DD 148883.
186 Celanese, US 3833550.
187 Yokohama Rubber, JA 63/251462.
188 Rohm & Haas, US 3791866.
189 Ionics, US 3787339.
190 Mitsubishi, JA 82/75150.
191 Suwa Seiko, JA 77/129690.
192 Fisons, BP 1309086.
193 Kokai, JA 61/270217.
194 FMC, US 459 0051.
195 Zhanyiaba Salt, Chinese Patent, 87/103625.

196 Agency of Ind. Sci. Tech., US 4742457.
197 Kuhlmann, FR 2183550.
198 S. Scholle, CZ 231123.
199 Exxon Res. and Eng., EP 79190.
200 Farmland Ind., US 4808391.
201 W.R. Grace, DE 3045890.
202 FMC, US 4808391.
203 Mobil Oil, US 4457899.
204 Hoechst AG., EP 30372.
205 Hoechst AG., US 4279878.
206 Hoechst AG., US 4279878.
207 Mitsui Taatzu Chem., JA 57/166303.
208 BASF, BE 780103.
209 Mitsubishi, JP 35296.
210 Mitsui Min., JP 91985.
211 Enichem Syn., EP 460745.
212 Georgia Kaolin, FR 1549125.
213 Engelhard, US 3589922.
214 Huber, US 3661515.
215 FMC, US 4935391.
216 Gerogia Kaolin, US 4451440.
217 VEB Vereingtui Papier, DE 231367.
218 Georgia Kaolin, BP 1228759.
219 Engelhard, BP 1228759.
220 Peroxid Chemie, BP 2103197.
221 Toyota Cent., JA 03/008711.
222 UOP Inc., US 42255643.
223 Pulp and Paper Manufacture, Vol. 2: 'Mechanical Pulping', TAPPI and CPPA.
224 Pulp and Paper Manufacture, Vol. 5: 'Alkaline Pulping', TAPPI and CPPA.
225 Pulp and Paper Manufacture, Vol. 3: 'Secondary Fibres and Non-wood Pulping', TAPPI and CPPA.
226 'The Bleaching of Pulp', 3rd Edition, R.P. Singh, TAPPI Press, 1979.
227 'Encyclopedia of Chemical Technology', Kirk Othmer, 4th Edition, Vol. 4, Wiley Interscience, 1992, p. 293.
228 Z. Li, Y. Ni and A. Van Heinningen, *Pulp Paper Canada*, 1997, **98**, 34.
229 K.P. Kringstad, L. Johnsson, M.C. Kolar and F. de Sousa, *TAPPI Journal*, June 1989.
230 'Why TCF', Jaoko Poyry, Consulting, October 1993, Helsinki and Stockholm.
231 G. Gellerstedt and I. Petterson, *J. Wood Chem. Tech.*, 1982, **2**, 231.
232 P. Tibbling and B. Dillner, 25th EUCEPA Conference, Vienna, 1993.
233 P.M. Gottlieb, S.R. Miller and T. Macas, TAPPI Pulping Conf., Atlanta, Georgia, 1993.
234 J.L. Minor and E.L. Springer, TAPPI Conf. Proc. Wood and Pulping Chem., 1989, 133.
235 N.A. Troughton, F. Desprez and J. Devenyns, Int. Non-chlorine Bleach Conf., Amelia Island, Florida, 1994.
236 R.A. Seccombe, N.A. Troughton, R.T. Hill, H. Martens and A. Haakana, Int. Non-chlorine Bleach Conf., Amelia Island, Florida, 1994.
237 C.L. Lee, R. Hodgikyan, A. Skothes, G. Sacchiadis and J.T. Wearing, Int. Pulp Bleach Conf., Vancouver, Canada, 1994.
238 Solvay Interox, WO 94/18286.
239 C.L. Hill, Int. Pat. Applic., US 93/08068.
240 Warwick Int. Group, WO 97/25402.
241 Mita A., JP 09158074.
242 Rousu-Consult-Oy-Eza, WO 97/26403.

243 J. Laroesch, H. Knome, F. Mertz, S. Gos and H.J. Morais, in 'Perth Gold 88', Rondal International Ltd., Golden, Colarado, pp. 202–205.

244 S.P. Ball, A.J. Monhemius and P.J. Wyborn, in 'Precious Metals 89', ed. M.C. Jha and S.D. Hill, TMS of AIME 1988, pp. 149–163.

245 J. Loroesch, H. Knome and A.J. Griffiths, SME Annual Meeting, Phoenix, Arizona, 1988.

246 Ontario Research Foundation, US 4537628.

247 Solvay Interox, unpublished results.

248 D.B. Mobbs, P. Groppi and D.M. Anderson, Second Congress, Vobalt Metallurgy and Uses, Venice, 1985, pp. 417–419.

249 D.B. Mobbs and S.B. Magid, in 'Recycle and Second Recovery of Metals', ed. R.S. Taylor, H.Y. Sohn and N. Jarrett, TMS of AIME 1985, Fort Lauderdale, pp. 291–311.

250 Interox Chemicals, US 4950425.

251 Interox Chemicals, South African Patent 82/6573.

252 A.R. Burkin, G.D. Manning and A.J. Monhemius, Proc. Symp. Hydromet., 1981, Paper B5.

253 E. Salkowsky, *Chem. Seit.*, 1916, **40**, 448.

254 Solvay Interox, unpublished results.

255 S.E. Yuri, *Minerals*, 1996, **51**, 47.

256 D. Mobbs and D.M. Mounsey, *Trans. IMM Sec. C*, 1981, **90**, C103.

257 Solvay Interox, unpublished results.

258 M. Martinon, *Bull. Soc. Chim. Fr.*, 1985, **43**, 355.

259 N. Papassioni and M. Stefanakis, in 'Arsenic Metallurgy Fundamentals and Applications', ed. R.G. Redoly, TMS Annual Meeting, 1988, p. 321.

260 Interox Chemicals, EP 213214.

261 Interox Chemicals, EP 127357.

262 D. Deshenes and E. Ghali, *Hydrometallurgy*, 1988, **20**, 179.

263 G.C. Lucas, E.J. Fulton, F.E. Vautier, D.J. Waters and R.J. Ring, *Proc. Australas Inst. Min. Metals*, 1983, **287**, 27.

264 D.B. Mobbs and S.B. Magid, in 'Recycle and Secondary Recovery of Metals', ed. R.S. Taylor, H.Y. Sohn and N. Jarrett, TMS of AIME 1985, Fort Lauderdale, pp. 291–311.

265 D.M. Anderson and P.J. Wyborn, in 'Recycling of Metaliferous Materials', IMM, Birmingham, England, 1990, pp. 1–10.

266 E. Ozensov, PhD Thesis, University of London, 1990.

267 Interox Chemicals, US 4275039.

268 A.R. Burkin and A.J. Monhemius, 'Acid Leaching of Uranium Ores Using Hydrogen Peroxide and Caro's Acid', Joint CIM Annual Conference and 8th Hydrometallurgy Meeting, Montreal, 1978.

269 Interox Chemicals, EP 49064.

270 A.J. Mattus, S.B. Magid, M.F. Lopez and D.B. Mobbs, *CIM Bulletin*, 1985, **76**, 102.

271 Solvay Interox, unpublished results.

Subject Index